# Mammalian Dispersal Patterns

The Effects of Social Structure on Population Genetics

Edited by B. Diane Chepko-Sade
and Zuleyma Tang Halpin

The University of Chicago Press • Chicago and London

B. Diane Chepko-Sade is a researcher with the
Department of Anthropology, Northwestern University.
Zuleyma Tang Halpin is associate professor of biology
at the University of Missouri—St. Louis.

The University of Chicago Press, Chicago 60637
The University of Chicago Press, Ltd., London

© 1987 by The University of Chicago
All rights reserved. Published 1987
Printed in the United States of America

96 95 94 93 92 91 90 89 88 87    5 4 3 2 1

Library of Congress Cataloging-in-Publication Data
Mammalian dispersal patterns.

Based on a symposium, "Patterns of dispersal
among mammals and their effects on the genetic
structure of populations," that was presented at
the annual meeting of the American Society of
Zoologists held in Denver, Colorado. December 27–30.
1984—P.
Includes bibliographies and index.
1. Mammal populations—Congresses.   2. Mammals—
Genetics—Congresses.   3. Mammals—Behavior—Congresses.
4. Population genetics—Congresses.   5. Social behavior
in animals—Congresses.   I. Chepko-Sade, B. Diane.
II. Halpin, Zuleyma Tang   III. American Society of
Zoologists.
QL708.6.M36   1987        599'.056        87-6026

ISBN 0-226-10266-1
ISBN 0-226-10268-8 (pbk.)

To Sewall Wright, whose view of the processes involved in the evolution of populations has been an inspiration to the editors in the planning and organizing of this book.

# Contents

# Preface

Population geneticists have developed sophisticated mathematical models for evolutionary processes, based on the pioneering work of Sewall Wright, J. B. S. Haldane, and R. A. Fisher in the early part of this century. Although these models are essentially identical with respect to many of the mathematical equations involved, they disagree on the relative importance of the four major forces of evolution: mutation, selection, migration, and genetic drift. Fisher ([1930] 1958) and Haldane (1932) emphasized mutation and selection, believing that populations were generally too large and panmictic in their breeding patterns for migration or drift to be of great significance. Sewall Wright, while recognizing the importance of natural selection, also gave migration and drift important roles in evolution in his shifting balance theory (1932, 1982, and others). The controversy continues today (Lewontin 1974; Kimura 1968; Wade 1982; Templeton 1981).

Based on evidence from the fossil record and rates of karyotypic diversity, Bush et al. (1977) have suggested that mammals with certain types of social organization show much faster rates of evolution than mammals with other types of social organization. Specifically, aspects of social organization that decrease the effective population size of groups (e.g. territoriality, philopatry, and social structures involving skewed sex ratios and/or large variances among individuals in reproductive success) increase the rate of evolution by increasing the importance of genetic drift. Bush et al.'s study indicates that Wright's shifting balance theory of evolution is of great importance among many mammalian species—particularly those showing a high degree of social structure that constrains breeding and migration in certain ways. Still, more information is needed on how natural populations of many species are actually structured (if at all), and whether, and to what extent, social organization affects the genetic structure of the population.

Geneticists have addressed these issues primarily with the use of mathematical models based on studies conducted in the laboratory. Studies of natural populations, on the other hand, have generally been limited to measurements of gene frequencies in a given population at a given time. Few of these studies have succeeded in measuring actual rates of migration, effective population sizes, individual fitnesses, and the dynamic shifts in gene frequencies that result in evolution. Such measurements would provide the empirical figures necessary to enter into population genetical equations in order to measure the relative importance of the various forces of evolution.

In the last twenty years, however, a number of animal behaviorists and wildlife biologists have carried out long-term longitudinal studies on populations of identified individuals. The object of these studies has not always been to examine the population genetics of the species under study, but many of these studies provide or could provide information of crucial interest to the population geneticist. For the first time, lifetime reproductive rates, migration rates between groups per generation, and effective group sizes can be measured for a number of diverse species, based on demographic information collected by field researchers.

Physical anthropologists have long been interested in the demography and the dynamics of population genetics in the human and primate groups that they study. Along with human population geneticists and demographers, they have developed mathematical techniques for dealing with such data. However, humans tend to be highly variable in parameters such as migration rates, lifetime reproductive rates, and effective group sizes, depending upon what culture is studied. It would be of interest to see how humans as a group compare with other mammals in these parameters, and in the potential rate of evolution based on calculations made from these parameters.

The purpose of this book is to bring together the work of animal behaviorists, wildlife biologists, and anthropologists who are doing long-term longitudinal studies of social organization and dispersal patterns with that of population geneticists who are interested in the evolutionary causes and consequences of different patterns of social organization. We hope the book will make anthropologists, wildlife biologists, behaviorists, and population geneticists more aware of the types of data and analytical methods that are relevant to questions they have in common. This greater awareness and appreciation may in turn lead to more collaborative efforts between those who study populations in depth on the one hand, and geneticists and theoreticians on the other, all of whom would benefit from the kind of data collected in long-term longitudinal studies.

This book is divided into five parts. The introduction, by William M. Shields, discusses the evolutionary importance of dispersal patterns and mating systems in mammals. This part also reviews the current literature on mammalian dispersal and defines the terms used in the dispersal literature and in the rest of the book. Differing uses of the term *inbreeding* have generated some controversy. *Inbreeding* commonly refers to incestuous matings between very closely related individuals: mother-son matings or brother-sister matings, for example. However, in the population genetical literature, there is a very specific definition of the word *inbreeding*. According to Crow and Kimura (1970, 61), "inbreeding occurs when mates are more closely related than they would be if they had been chosen at random from the population." There are specific short-term as well as evolutionary consequences of inbreeding, depending on the levels involved and the history of its occurrence and maintenance in the population. Shields discusses these in the first chapter, and then in greater detail in the final chapter (see also Templeton, chap. 17, and Wade and Breden, chap. 18, this vol.).

Part 2, "Empirical Studies of Dispersal in Populations with Known Individuals," is by far the longest. It contains detailed descriptions of social structure, mating patterns, and dispersal patterns for a number of mammalian species. Each study in this section is based on a long-term longitudinal study of a population or a subpopulation of individually recognizable individuals. Migration rates and mating patterns are based on empirical observations. Data of this nature take many years of painstaking work to collect, but are the necessary foundation of everything else we would like to be able to say about the evolutionary effects of dispersal or mating patterns. If these chapters seem to deal too much with the details of what each animal did and not enough with the evolutionary overview of what it all means, it may be because field biologists are frequently not versed in the theories of population genetics and have not considered their findings in that theoretical framework. In other cases the researchers are still in the process of summarizing all of those life-historic events, collected over a number of annual seasons, that make up the demography of a population and that are needed to calculate finally such refined parameters as effective population size and migration rate per generation. Nevertheless, only from the work of the indefatigable field-worker can all of the necessary demographic details be obtained for the building of life tables or the calculating of migration rates, fertility rates, age at first reproduction, levels of inbreeding, and effective population sizes within living populations of real animals.

Part 3, "Dispersal Patterns and Genetic Structure," deals with populations in which gross patterns of population structure and migration are known, but individual life histories and genealogical information are not. Genetic information, in the form of blood polymorphisms, is known for populations and for samples of populations, and evolutionary inferences can be made from changes in gene frequencies over time.

The fourth part, "Demography, Dispersal Patterns, and Genetic Structure," contains three studies in which both individual life-historical information and genetic structure of populations are known. Not surprisingly, two of these chapters are about human groups, since humans have the useful habit of remembering (and often even recording) life-historical information, and of sharing it with the researcher. The collection of comparable data on any other species requires constant detailed observations of an entire population over a number of generations of individuals. Analysis of data in this section is necessarily more complex than that in previous sections because of the complexity and detail of the data to be integrated. Here demographic and population genetic methods of analysis are pushed to their limits and new methods are being invented.

Part 5, "Mathematical Models of Population Structure," examines some of the empirical data now available for characterizing the structure of populations and builds models based on these populations in an attempt to comment on the evolutionary significance of some of the different patterns seen.

In the concluding chapter, Chepko-Sade and Shields et al. discuss the value of one population parameter, $N_e$, in characterizing and comparing populations and predicting the relative importance of Wright's shifting balance theory in the mode and tempo of evolution in a given population. An attempt is made to calculate $N_e$ for each of the populations described in Part 1, based on the data presented in those chapters and on additional raw data graciously supplied by the authors. Finally, a set of guidelines is presented for researchers who are planning field studies, indicating what sorts of data should be collected in the future to further our understanding of population structure of different species and different populations of the same species.

We are only just beginning to make the crucial link between data and theory, although both are now becoming increasingly available. The problem seems to center on the fact that few of the researchers who collect long-term population data are well versed in the detailed and mathematical theories of population genetics. The sociobiological approach has done much to raise the consciousness of scientists in all fields to the need for more interdisciplinary research, particularly research informed by sound evolutionary theory. However, this research must not be superficial. It is not enough to ask whether (or how) animals behave to increase their individual fitness (this conclusion often appears to precede the research). Rather we should ask how populations evolve, and what effects different mating structures and migration patterns have on the mode and tempo of evolution of all traits.

# REFERENCES

Bush, G. L., S. M. Case, A. C. Wilson, and J. L. Patton. 1977. Rapid speciation and chromosomal evolution in mammals. *Proceedings of the National Academy of Sciences* 74(9): 3942–46.

Crow, J., and M. Kimura. 1970. *An introduction to population genetics theory.* Minneapolis, Minn.: Burgess Publishing.

Fisher, R. A. [1930] 1958. *The genetical theory of natural selection.* Rev. ed. New York: Dover.

Haldane, J. B. S. 1932. *The causes of evolution.* London: Longmans, Green.

Kimura, M. 1968. Genetic variability maintained in a finite population due to mutational production of neutral and nearly neutral isoalleles. *Genetic Research* 11: 247–69.

Lewontin, R. C. 1974. *The genetic basis of evolutionary change.* New York: Columbia University Press.

Templeton, A. 1981. Mechanisms of speciation: A population genetic approach. *Annual Review of Ecology and Systematics* 12:23–48.

Wade, M. J. 1982. Group selection: Migration and the differentiation of small populations. *Evolution* 36(5): 949–61.

Wright, S. 1932. The roles of mutation, inbreeding, crossbreeding, and selection in evolution. *Proceedings of the Sixth International Genetics Congress* 1:356–66.

―――. 1982. Character change, speciation, and the higher taxa. *Evolution* 36(3): 427–43.

# Acknowledgments

The present volume is based on a symposium, "Patterns of dispersal among mammals and their effects on the genetic structure of populations," that was presented at the annual meeting of the American Society of Zoologists held in Denver, Colorado, December 27–30, 1984. The symposium was sponsored by the Division of Animal Behavior and the Division of Ecology. We are grateful to the American Society of Zoologists for their sponsorship of the symposium, and we are especially grateful to Mary Wiley for her help in arranging for an extended session in which to present the symposium and for arranging travel reimbursement for the junior members of the symposium. Discussions among the participants were most valuable in rewriting the papers as chapters of this book. Chapters 1–18 were presented as papers at that symposium, and the concluding chapter was written afterwards. Two papers presented at the symposium were withdrawn from this publication.

During the years that have been spent planning and preparing this volume, a number of people have provided support and assistance of various kinds. Susan Abrams gave us pointers on how to organize the symposium so that it could be made into a useful book. David Mech and Devra Kleiman helped during the formative stages of this project by locating scientists who had done (or were in the process of doing) long-term studies of mammals that would be likely to produce the kinds of quantitative data we were interested in presenting. Bill Shields and Jim Cheverud contributed their enthusiasm and knowledge of the literature.

Northwestern University and the University of Missouri–St. Louis shared the costs of telephone calls and of photocopying and mailing the many revisions of manuscripts that were sent back and forth to contributors. Donald Stone Sade wrote a computer program, which facilitated preparation of the index.

Halpin is grateful to Connie Quinlan for assistance in proofreading manuscripts and to Arlene Zarembka for her encouragement and support. Chepko-Sade is grateful to her youngest son, Omen, for postponing his birthdate until the week after the manuscript was delivered to the University of Chicago Press, and to his babysitter, Deanne Beaugrand-Place, for such loving care that he never felt neglected during the revision and publication phases of the book.

# I. INTRODUCTION

# 1. Dispersal and Mating Systems: Investigating Their Causal Connections

*William M. Shields*

The genetic structure of any population is expected or known to have profound consequences for the social behavior (e.g., Hamilton 1964; Wade and Breden, chap. 18, this vol.), the population dynamics (e.g., Lidicker 1975; Lidicker and Patton, chap. 10, and Gaines and Johnson, chap. 11, this vol.), and the tempo and mode of evolution characterizing a species (for reviews, see Mayr 1963; Wright 1978). This construct property of "genetic structure," in turn, is determined primarily by the dispersal, demography, and mating system characterizing a species (for reviews, see Crow and Kimura 1970; Wright 1978; Wilson 1980; Shields 1982). Since these are intrinsic properties of the organisms in question, their evolutionary (ultimate) and immediate (proximate) causes and consequences, and how they influence each other, are currently the topic of much research in behavioral (e.g., Greenwood 1980; Waser and Jones 1983; Murray 1984), ecological (e.g., Baker 1978; Swingland and Greenwood 1983), genetical (e.g., Endler 1977; Wright 1978; Wilson 1980; Shields 1982), and evolutionary arenas (e.g., Mayr 1963; Dobzhansky 1970; Gould and Eldredge 1977). In addition, the interrelated problems of dispersal and mating systems and inbreeding and outbreeding have received a great deal of attention in the context of park and reserve design and the conservation and captive management of rare and endangered species (e.g., Soule and Wilcox 1980; Frankel and Soule 1981; Schonewald-Cox et al. 1983).

This increase in basic and applied research on the same topics in such disparate subdisciplines has resulted in confusion as well as new synthetic perspectives. Since the researchers in the different fields approach dispersal, mating systems, population size, and population structure with different interests and goals, they also define and use many of the same words and concepts in different ways. For example, Waser and Jones (1983) were interested in interactions among parents and their adult progeny in the context of the evolution of different kinds of social systems. Since their interest was in parental retention of adult progeny in or adjacent to the parent's home range, they defined *philopatry* as nondispersal of young.

In contrast, those who focused on the genetic rather than the behavioral consequences of limited dispersal defined philopatry as nondispersal or short-range dispersal whether parents and their adult progeny had opportunities to interact or not (e.g., Mayr 1963; Greenwood 1980; Shields 1982, 1983). While the second definition can be applied more broadly than the first (e.g., to plants and sedentary invertebrates as well as to social mammals), its meaning would have been too broad to reflect the immediate interests of Waser and Jones (see also Jones, chap. 8, Waser, chap. 16, and Smith, chap. 9, this vol.).

Since definitions made clearly and used consistently are always valid, I would not criticize such differences in use. Rather I wish to point out that in this field it may be necessary to provide definitions rather than allow each reader to interpret a word or concept in his or her own often unique way. To forestall this possibility, and the confusion that could result, I define the key concepts used throughout this work before I use them to explore relationships among dispersal, mating systems, and population structure and their proximate and ultimate causes and consequences. I also provide "synonyms" that other researchers use for the same, similar, or related concepts.

## DEFINITIONS

*Dispersal.* The movement of an organism or propagule from its site or group of origin to its first or subsequent breeding site or group. The zygote-to-zygote distance dispersed is the absolute distance between natal and subsequent breeding sites. *Natal* dispersal is the movement of a propagule between birthplace or natal group and first breeding site or group. *Breeding* dispersal is the movement between consecutive breeding sites or groups of adult breeders (Greenwood, Harvey, and Perrins 1979; Shields 1979, 1982). Dispersal (or migration) rate ($m$) is the proportion of individuals moving between demes or breeding groups per generation (Wright 1977, 1978). (Here *migration* is intended to imply gene dispersal and is often used as a synonym of dispersal in genetic models, e.g., see Smouse and Wood, chap. 14, Sade et al., chap. 15, and O'Brien, chap. 13, this vol.).

*Effective Dispersal.* The number of home ranges an organism moves before settling to breed. Effective dispersal may be calculated by dividing the mean or median dispersal distance by the mean home range diameter (or the mean distance separating two adults in sessile organisms) for solitary species. For group-living (gregarious) species, it might be the mean or median number of groups separating natal and breeding areas or the number of groups exchanging breeders (Shields 1982, 1983).

*Successful Dispersal.* Dispersal is successful only if the propagule obtains an opportunity to breed and raises young that breed successfully, that is, if genes migrate (Howard 1960; Lidicker 1975; Endler 1979).

*Philopatry.* Limited dispersal such that the average propagule moves no more than 10 home ranges away from its site of origin (synonyms: vicinism, ortstreue, site tenacity, site fidelity, results in "viscous" populations). A species is philopatric only if both the natal and breeding dispersal of both sexes are limited in magnitude (Mayr 1963; Greenwood 1980; Shields 1982).

*Vagrancy.* Wide dispersal such that the average propagule moves more than say 30 home ranges from its site of origin. Since dispersal magnitudes occur as a continuum, the distinction between philopatry and vagrancy is always somewhat arbitrary (Shields 1982).

*Dispersion.* The interindividual spacing characterizing the members of a particular species or population during the breeding season.

*Deme.* The random-mating population. It is the breeding group within which parents are drawn nearly at random, that is, all breeding males and females have equal probabilities of mating (Mayr 1963; Dobzhansky 1970).

*Census Population Size.* The number of individuals or breeders censused in a delimited area or breeding group (synonyms: $N$, $N_i$) (Crow and Kimura 1970; Wright 1978).

*Effective Population Size.* The size of an ideal deme that would produce the same level of inbreeding and opportunity for drift as a real population with a specific census size and other characteristics. There are two estimates: an "inbreeding" and a "variance" effective size, which differ primarily in growing or declining populations (synonym: $N_e$) (Crow and Kimura 1970; Wright 1978).

*Neighborhood.* In a continuously distributed group of organisms, the population of a region from which the parents of individuals born near the center may be treated as if drawn at random (Wright 1946, 1978).

*Neighborhood Area.* In a continuously distributed species, the size of an area that would include 86.5% of the parents of individuals born near the center (see Wright 1978, 302–3).

*Neighborhood Size.* The number of individuals breeding in a neighborhood area (synonyms: effective population size, deme size, $N_e$) (Kimura and Ohta 1971; Wright 1978).

*Inbreeding.* The mating of relatives that share greater common ancestry than if they had been drawn at random from an entire species. *Extreme* inbreeding is the mating of members of one nuclear family, including selfing, full-sib, half-sib, and parent-offspring matings. Inbreeding is generated in two ways: by nonrandom mating of close pedigree kin in any deme or by random mating in small demes ($N_e < 1,000$), with the degree of inbreeding increasing with decreasing deme size (Allen 1965; Wright 1978; Shields 1982).

*Outbreeding.* Random mating in large demes ($N_e > 10,000$) or nonrandom mating among individuals with fewer common ancestors than would be expected by chance in a species (Allen 1965; Shields 1982).

*Genetic Drift.* The change in allele or genotype frequencies resulting from sampling error in finite populations (synonyms: drift, and in some cases, founder effect) (Wright 1977, 1978).

*Mating System.* The sex ratio of simultaneous breeders. Usually can be characterized by the sex differences in variance in reproductive success (Trivers 1972; Emlen and Oring 1977; Murray 1984).

*Monogamy.* There are an equal number of breeding males and females and some evidence that each yearly set of progeny (litter) usually has one father and one mother. Variances in male and female reproductive success are equal or nearly so.

*Polygyny.* More females than males breed, with the result that variance in reproductive success is greater in males than in females. The greater the difference in variance, the greater the degree of polygyny.

*Polyandry.* More males than females breed, so variance in female reproductive success is likely greater than for males.

*Polygyny-Polyandry.* Each male and female can breed with more than one member of the opposite sex, while some may fail to breed. The variance in reproductive success can be greater for either sex or may be equal.

*Variance in Reproductive Success.* The second moment of the distribution of family sizes (the number of progeny produced by each individual during its lifetime) of progeny that successfully reproduce in a deme for either the males, females, or both (synonym: $V_k$).

Using these definitions, I discuss some of the proposed explanations for associations among different kinds of mating systems, demographic conditions, and the patterns of dispersal observed in different mammals. I also explore the types of data and research designs most likely to allow testing of such hypotheses.

## DEMOGRAPHY, MATING SYSTEMS, AND DISPERSAL

Most discussions of the adaptive value of dispersal or nondispersal have focused on genetic effects or somatic effects and less often on both (for reviews, see Baker 1978; Shields 1982; Swingland and Greenwood 1983). A conclusion that one of the advantages of dispersal may be that it increases the level of outbreeding or, as it is more often stated, decreases the probability or level of inbreeding, recurs regularly (e.g., Howard 1960; Lidicker

1962, 1975; Mayr 1963; Bengtsson 1978; Greenwood 1980; Pusey and Packer 1987; Ralls, Harvey, and Lyles 1986). An apparently opposite conclusion, that one of the advantages of limiting dispersal (philopatry) might be to maintain an adaptive level of inbreeding or to avoid maladaptive levels of outbreeding, is also traditional (e.g., Fisher [1930] 1958; Mayr 1963; Dobzhansky 1970; Shields 1982). The only unequivocal conclusion one can draw is that the degree and pattern of dispersal have genetic consequences, and these must be considered in exploring its adaptive value.

The somatic value of dispersal has usually been seen as escape from locally crowded conditions that result in intense competition for important resources (e.g., food, space, nests, or mates). The net result is that a disperser should face more favorable conditions in its new locale than if it had remained in its natal area (e.g., Howard 1960; Lidicker 1962, 1975; Murray 1967; Gauthreaux 1978; Gaines and Johnson, chap. 11, this vol.). More recently, Hamilton and May (1977) have shown that global competition for space may result in dispersal being favored, even if individual propagules have little chance of successfully establishing in a new but less crowded area. The advantage in their model accrues to the individual that "forces" propagules to disperse rather than to the propagules themselves (e.g., a parent forcibly dispersing seeds; see also Horn 1983). The somatic advantages of nondispersal, if any, are expected to stem from reduced risk and energy expenditure. Dispersing propagules are often thought to be at greater risk to predation and certainly use more energy during their travels and afterward in their novel ranges than sedentary individuals (e.g., Bengtsson 1978; Baker 1978). An obvious conclusion is that both dispersal and philopatry entail somatic costs and benefits that need to be considered in adaptive explanations of either phenomenon.

In the past decade, realization of the importance of dispersal and genetic structure to the evolutionary process has resulted in a host of theoretical, comparative, and single-species studies aimed at elucidating the proximate and ultimate causes of different patterns of dispersal. Although the topic has been the same, the conclusions reached have often been confusing or contradictory. I review some of the prominent examples before attempting my own interpretation of where we are and where we should be going.

## THEORETICAL AND COMPARATIVE STUDIES

In a series of seminal papers, Greenwood and colleagues (e.g., Greenwood 1980, 1983; Greenwood, Harvey, and Perrina 1979; Greenwood and Harvey 1982) were prominent in focusing greater attention on the commonly observed sex bias in dispersal distances (in solitary species) or dispersal probabilities (in gregarious species) in most philopatric birds and mammals. In the majority of mammal species investigated, males had a higher probability of leaving their natal groups or of dispersing greater distances than did their sisters. In birds, however, the reverse was more often documented (table 1.1).

Greenwood (1980) proposed two alternative, but not mutually exclusive, hypotheses to explain the observed patterns of vertebrate dispersal. He suggested that the traditional somatic advantages of philopatry were sufficient to explain localized dispersal under most conditions. Familiarity with an area and the reduced risks and energetic costs associated with philopatry would normally favor site tenacity in all but deteriorating environments. He also supposed that there were significant genetic costs to "too close" inbreeding that would favor some dispersal—explicitly, a sexual bias in dispersal—that could prevent extreme inbreeding while allowing most of the somatic benefits of philopatry. As he noted, however, the prediction of a sexual bias to reduce extreme inbreeding did not offer suggestions about which sex would disperse further or more frequently, unless there were some way of determining which sex was more likely to bear the burden of inbreeding (e.g., see Smith 1979).

To explain which sex moved, he suggested that sexual differences in the degree or method of competing for critical resources could influence the sexes differently and result in the costs and benefits of dispersal being different for the sexes. Explicitly, he noted that most mammals are polygynous and that males often compete directly for females that are more or less clumped in space and time. This polygyny could favor male-biased dispersal in two ways. Since subordinate males are likely to be excluded from mating in their natal areas (as well as elsewhere), they might disperse more widely than females in order to increase their probability of mating or of gaining more mates. In many mammals, female kin are often organized into matrilineal social groups. In such species the social benefits of being philopatric and remaining near close kin may be very strong for females (e.g., Sherman 1977; 1981; Wrangham 1980). Under such conditions males might disperse further than females to avoid maladaptive levels of inbreeding. Greenwood concluded that male-biased dispersal in mammals might be favored for both somatic and genetic reasons and that the somatic benefits might be related to *mate-defense* mating systems.

Table 1.1 Number of Species and Families of Birds and Mammals Showing Predominantly Male, Female, or No Sex Bias in Natal Dispersal

| | Propagules | | | | | |
| --- | --- | --- | --- | --- | --- | --- |
| | Mammals | | | Birds | | |
| | Male | Female | Both | Male | Female | Both |
| Species | 45 | 5 | 15 | 3 | 21 | 6 |
| Families | 23 | 4 | 7 | 1 | 11 | 5 |

*Source:* Greenwood 1980.

In the mostly monogamous birds (and in many polygynous species too), males usually compete for territories that attract mates, rather than for females directly. Under such conditions familiarity with an area might be more important to a male than to a female. If somatic benefits force males into greater philopatry, then females might be forced to disperse more than males in order to avoid maladaptive levels of inbreeding. Alternatively, since females benefit most by gaining the best or most resources, they may gain more by dispersing as they seek access to the territories or males with greater resources (e.g., Shields 1984). In either case *resource-defense* mating systems should be associated with female-biased dispersal. Greenwood (1980, 1983) concluded that both the genetic and somatic benefits of sex-biased dispersal could help explain the phenomenon in both birds and mammals, despite the sex-role reversal in the two taxa.

Greenwood and his colleagues stimulated renewed interest in functional analyses of philopatry, dispersal, and the sexual biases in dispersal systems. At the same time, I was reviewing many of the earlier hypotheses about why organisms might disperse characteristic distances (Shields 1979, 1982, 1983). I also mentioned the somatic costs and benefits of different dispersal strategies, but focused more on genetic consequences. I echoed Greenwood and many others in noting that wide dispersal is likely to carry many potential somatic costs, and philopatry many potential somatic benefits. Rather than emphasizing the genetics costs associated with an absence of dispersal (inbreeding depression), however, I suggested that too much dispersal might carry genetic costs in the form of a potential for outbreeding depression.

If dispersal carried propagules into areas with different selective conditions or with local genomes that were reproductively incompatible, then outbreeding depression would result and this genetic cost could have been a selective factor in the evolution or maintenance of philopatry. I also was seduced by the simple elegance of the extreme inbreeding ("incest") avoidance hypothesis and noted that philopatry combined with a sexual bias in dispersal distances could provide optimal mate relatedness with optimal genetic consequences (Shields 1982, 151; see also Baker and Marler 1980; Bateson 1983).

In an elegant mathematical analysis, Bengtsson (1978) also explored the relationship between the genetic and somatic consequences of different dispersal strategies. He explicitly assumed that the genetic costs of inbreeding associated with philopatry would sooner or later be balanced by the increased somatic costs of greater dispersal. Reasonable numerical estimates of inbreeding depression and the increased mortality associated with increased dispersal resulted in an intermediate and optimal dispersal probability or distance for each set of conditions. As with all optimality models, however, the model produces the same predictions if one assumes that the actual cost of philopatry is somatic (e.g, increased competition rather than inbreeding depression) and the actual benefit genetic (e.g., reduced outbreeding depression rather than mortality). In the sense that the causes of the costs and benefits driving the model are assumed rather than known, the model, while elegant, is content free.

Dobson (1982) reanalyzed much of the mammal data and concluded that sex-biased dispersal was a response to different levels of competition between the sexes for mating opportunities. His analysis highlighted the strong association between male-biased dispersal and polygyny and an absence of sex-biased dispersal in monogamous species of mammals (table 1.2). He did note, however, that he felt there was still strong evidence that avoiding extreme inbreeding was a major factor favoring dispersal in at least some species, including his own California ground squirrel (*Spermophilus beecheyi*, Dobson 1979). His model failed to address the predominantly female-biased dispersal observed in monogamous birds and the predominantly female dispersal in resource-defending polygynous mammals such as the white-lined bat (*Saccopteryx bilineata*, Bradbury and Vehrencamp 1976, 1977).

In a comprehensive comparative analysis of *philopatry* (which they defined as absolute nondispersal, i.e., progeny sharing home ranges with their parents) in solitary mammals, Waser and Jones (1983) concluded that their data were consistent with Greenwood's general notion that dispersal is a strategy to avoid the genetic costs of extreme inbreeding. They also concluded that the data were consistent with the notion that it was somatic costs and benefits that determined the pattern of dispersal observed in many species. They also added new and needed emphasis on the important role that demography could play in determining dispersal patterns regardless of which of the several functional explanations were correct. Explicitly, they suggested that longer-lived species (e.g., the sciurid rodents) would, if they were too sedentary, find themselves surrounded by close relatives as both potential mates *and* competitors. The potential for incest, competition, or both could favor greater dispersal by the more competitive sex. Shorter-lived species (e.g., such smaller rodents as the murines and microtines), in contrast, would be less likely to accumulate kin because most close kin would have died in any case. Members of such species might be expected to show greater philopatry or reduced sexual biases in dispersal

TABLE 1.2 Number of Mammal Species with Particular Mating Systems and Types of Sexual Bias in Dispersal

| MATING SYSTEM | PRIMARY PROPAGULES | | |
|---|---|---|---|
| | Male | Female | Both Sexes |
| Monogamous | 0 | 1 | 11 |
| Polygynous or promiscuous | 46 | 2 | 9 |

*Source:* Dobson 1982.

since the risks of inbreeding or competing with close kin would be lower (also see Waser 1985; Waser, chap. 16, and Jones, chap. 9, this vol.).

Finally, Moore and Ali (1984; see also Moore 1984) reviewed much of the underlying theory and data on mammals, and especially the primates. They found little to recommend inbreeding avoidance as an explanation of the patterns observed in all but one of the mammal species they reviewed (the exception being the chimpanzee, *Pan troglodytes* (Pusey 1980). They reached this conclusion despite their examination of some of the classic "avoiders," such as the olive baboon (*Papio anubis,* Packer 1975, 1979). As support for their conclusion, they cited many species that appeared to inbreed regularly, the inefficiency of biased dispersal in avoiding extreme inbreeding (i.e., the population may be less inbred but the probability of an individual avoiding extreme inbreeding is reduced marginally at best for most of the dispersal distributions actually observed in nature), and the absence of more efficient behavioral mechanisms for avoiding extreme inbreeding in species with sex-biased dispersal. Their main conclusion was that much of the dispersal data could be explained most parsimoniously in the context of Greenwood's (1980, 1983) competition hypotheses, without recourse to the more complicated inbreeding avoidance. In essence they offered a new rule analogous to Williams's (1966) canon about levels of selection. One should exhaust the somatic possibilities to explain dispersal patterns before invoking any genetic consequences as causal factors. They did not deny the relevance of genetic consequences, but they did suggest that such explanations should be invoked only when absolutely necessary (Moore and Ali 1984; Moore 1984).

## SINGLE-SPECIES STUDIES

The earliest and best-accepted examples of sex-biased dispersal as inbreeding avoidance mechanisms have been drawn from single-species field studies of a variety of primates. Since primate studies have a history of being used to explore the biological bases of human behavior, it is perhaps understandable that primatologists might seek and find rudimentary "incest" taboos in nonhuman groups. Itani (1972) explicitly discussed how incest is "avoided" in the Japanese macaque (*Macaca fuscata*) because males invariably leave their natal troops while females invariably stay. Similar observations have been made for many other primate species (e.g., Sade 1968) and have often been interpreted as functional responses to the specter of inbreeding depression (e.g., Packer 1975; 1979; Pusey 1980; Cheney and Seyfarth 1983; Pusey and Packer 1987). Indeed, so enshrined is this dogma that a recent research news article in *Science* (Lewin 1983) reporting on Cheney and Seyfarth's detailed work on the nonrandom dispersal of the vervet monkey (*Cercopithecus aethiops*) is entitled "Brotherly Alliances Help Avoid Inbreeding," despite Cheney and Seyfarth's conclusion that the primary function of the sibling alliances was to increase the competitive ability of dispersers entering new troops.

Cheney and Seyfarth documented the typical primate dispersal pattern of males leaving and females remaining in their natal troops. Their larger data base also enabled them to document the fact that male dispersal was nonrandom. Each troop appeared to have an exchange relationship with one other troop, such that males transferred reciprocally between such "sibling" groups much more often than would be expected by chance. The pattern was remarkably similar to the kinds of lineage exchange practiced by many human hunter-gatherers (for review, see van den Berghe 1980).

As they noted, the effect of the reciprocal exchange would be to reduce the effective population size much below what would characterize the same set of troops exchanging breeders at random over larger geographic and genetic distances. They concluded that the increased inbreeding that resulted from the nonrandom dispersal might not be costly enough to outweigh the social benefits of dispersing into troops with kin. The somatic benefits would include more easy entrance to the troop, especially if dispersers traveled in groups, and more amicable relations between exchanging troops when they do meet. They also noted that the level of inbreeding induced by the pattern they observed might reflect an inbreeding optimum along the lines of those proposed by Bateson (1983) and Shields (1982).

In one of the classic studies on the topic, Packer (1975, 1979) reported that with one exception, male olive baboons (*Papio anubis*) leave their natal troops, apparently to avoid inbreeding depression. As evidence, Packer showed that the single male that bred in his natal troop and his female kin suffered reduced reproductive success relative to troops with immigrant breeders. Moore and Ali (1984) criticized this interpretation, noting that the troop in question was suffering reduced reproductive success before the male in question became a breeder, so the difference could not be attributed solely to inbreeding depression (but see Packer 1985).

While Packer's data are consistent with an inbreeding avoidance interpretation in his study area, they lose the weight of generality for this species in light of more recent work in Kenya (Bercovitch, pers. comm.). During a long-term study of the olive baboon in Gilgil, a number of males (> 30% of those with known origins) born in study troops bred in their natal troop. In addition, many males remained in single troops long enough for their putative daughters to have matured (> 10 years). Rather than avoiding such kin as mates, the females appeared not to discriminate between natal males and presumably less related immigrants. Bercovitch believes that such data are inconsistent with a strict inbreeding avoidance paradigm and that some sort of somatic or social benefits must accrue to dispersers in his study area. The question of relations among inbreeding depression, inbreeding avoidance, competition, and dispersal, then, remains open for the olive baboon.

On the basis of his data on the black-tailed prairie dog (*Cynomys ludovicianus*, Hoogland (1982) offered an ingenious variation on the theme of avoiding inbreeding; he suggested that males do disperse more than females in order to gain access to mating opportunities. He proposed that the reason they leave their natal area is not that the number of potential mates are fewer, or that their ability to compete with males is less, but that local females recognize them as close kin and would reject them as potential mates because of the inbreeding depression that would result. He also provided convincing data that females do avoid breeding with their close kin (fathers and full sibs). Rather than fight useless battles to win access to females that will not breed with them, such males disperse to areas with less related females that will accept them as mates (for a similar argument for Belding's ground squirrel, *Spermophilis beldingi*, and an even more extensive data base, see Holekamp 1984).

More recently, Cockburn, Scott, and Scotts (1985) analyzed dispersal in the marsupial genus *Antechinus*. Because females were always philopatric, while juvenile males always dispersed, bred, and then died, they assumed that mothers, who unlike males could survive to breed more than once, forced their sons to disperse and accepted "unrelated" or nonprogeny juvenile males into their territories. They also provided convincing evidence that the usual competitive or ecological hypotheses for explaining such a sex bias were unlikely in their species. Their data, then, like Hoogland's (1982) on the prairie dog, and Pusey's (1980) on the chimpanzee, would be difficult to explain without recourse to some sort of adaptive avoidance of extreme inbreeding argument.

On the other side of the coin, however, are species known to show little dispersal despite the high levels of inbreeding that could or does result. For example, Smith and Ivins (1983) examined dispersal in the pika (*Ochotona princeps*). They found that successful pikas almost always settled within 50 m of their birthplace. Some individuals tried to move greater distances but were met with extreme aggression by local residents, despite the fact that the net result was very high probabilities (and documented instances; Chepko-Sade and Shields et al., chap. 19, this vol.) of extreme inbreeding. Smith and Ivins concluded that a history of such inbreeding would likely reduce the genetic costs associated with strong philopatry (for reviews, see Shields 1982; Templeton, chap. 17, this vol.) and would allow pikas the somatic benefits of reduced movement in an otherwise hostile social environment (also see Smith, chap. 9, this vol.).

Similar dispersal syndromes and explanations might apply to other mammal species with high levels of pedigree-documented inbreeding (i.e., studies with > 1% of matings involving extreme inbreeding among nuclear family kin, or where a larger proportion of the matings occur between more distant pedigree kin such as cousins; for mice, see Howard 1949; carnivores, including mongooses, Rood, chap. 6, this vol.; some primates, gibbons, Tilson 1981; gorillas, Fossey 1983; and even including some human groups, e.g., Darlington 1960; Sangvhi 1966; Rao and Inbaraj 1977; for a general review of birds and mammals, see Ralls, Harvey, and Lyles 1986). That inbreeding may be more common than is often supposed is made even more likely by the many mammals suspected, on the basis of indirect genetic or dispersal evidence, of inbreeding fairly intensely (e.g., mice, Selander and Yang 1969; other rodents, Jones 1984, and chap. 8, this vol.; Bowen and Koford, chap. 12, and Lidicker and Patton, chap. 10, this vol.; deer, Smith 1979; Nelson and Mech 1984, and chap. 2, this vol.; bear, Rogers, chap 5, this vol.; wolves, Mech 1970, and chap. 4, this vol.; and may primates, vervets, Cheney and Seyfarth 1983; bonnet macaques, Moore and Ali 1984; reviewed in Pusey and Packer 1987; and general review in Shields 1982).

## SYNTHESIS

The general consensus appears to be that there are many potential costs and benefits associated with dispersal (tables 1.3 and 1.4). Most of the costs are mirror images of one another, with the costs of extreme dispersal (e.g., outbreeding depression) being alleviated by philopatry and yet potentially being replaced by analogous or opposite costs (e.g., inbreeding depression). Such cost-benefit functions generate optimality models that are quite seductive. Which costs and benefits an investigator focuses on and which optimality models result, however, appear to depend as much on personal preference as on hard data. The result is a plethora of plausible models and hypotheses with perhaps too little empirical grounding (certainly including Shields 1982). The models are available and potentially useful, however, since they do provide testable predictions that allow some discrimination among them. It is certainly time that more long-term studies (such as those in this volume) be done that can actually measure, or at least estimate, some of the actual costs and benefits and so begin the winnowing process.

## CAUSES OF DISPERSAL OR SEX BIASES IN DISPERSAL
Ultimate

Although it may require a huge effort to measure explicit costs and benefits, it may be possible to begin in specific cases with less precision. For example, one bone of contention among many workers in the field is the relative importance of somatic and genetic contributions to the positive or negative value of different patterns of dispersal. The genetic costs of either inbreeding or outbreeding depression are expected to surface primarily in the reproductive rather than the survivorship component of an individual's fitness (tables 1.3 and 1.4). An individual that fails to disperse or disperses too far is not expected to suffer any individual consequences (unless it has dispersed into habitat for which it is genetically unsuited, generating a

TABLE 1.3 Potential Costs and Benefits Likely to Be Associated with Philopatry

POTENTIAL COSTS
  I. Genetic
     A. Inbreeding Depression (reviewed in Shields 1982)
        1. Unmasking deleterious recessives
        2. Reducing adaptive variability in progeny
  II. Somatic
     A. Direct Fitness Effects
        1. Reduced access to critical resources because of shortages due to crowding or the intrinsic subordination of young in the local social system. Could reduce survivorship, fecundity, or both, relative to that of dispersers (e.g., Murray 1967).
     B. Indirect Fitness Effects
        1. Increased competition with close kin over access to critical resources. Could reduce the survivorship of fecundity of the sedentary individual's kin relative to that of the kin of those that choose to or are forced to disperse (e.g., Hamilton and May 1977).

POTENTIAL BENEFITS
  I. Genetic
     A. Optimal Inbreeding (reviewed in Shields 1982)
        1. Maintain locally adapted gene complexes when particular genes are adapted to peculiar local conditions. Propagules may themselves suffer reduced survivorship or their "hybrid" young may not be adapted to the conditions associated with either parental type.
        2. Maintain intrinsically coadapted gene complexes when the phenotypic response of a gene is conditioned by its interactions with other genes.
  II. Somatic
     A. Direct Fitness Effects
        1. Increased survivorship and fecundity as a result of reduced risk and energy use (reviewed in Baker 1978; Shields 1982).
           a. Familiarity with local area allows more efficient use of refuges from predators and weather.
           b. Familiarity increases competitive efficiency at locating and controlling food and nest resources.
           c. Reduced movement reduces time at risk to predators.
           d. Reduced movement reduces energy requirements allowing greater investment in maintenance and reproduction.
           e. Familiar social environment allows for reduced investment in competing for resources (i.e., interactions with known individuals are usually less intense than with conspecific strangers; see Shields 1984).
           f. Promotes maintenance of adaptive local traditions that can be transmitted culturally (Bateson 1983).
     B. Indirect Fitness Effects
        1. Increased survivorship and fecundity of kin of sedentary individuals who benefit from the assistance or cooperation of those kin in competitive social interactions.
           a. Maintaining kin associations allows a larger group of kin to control a larger territory resulting in increased per capita reproductive success (e.g., Woolfenden and Fitzpatrick 1978).
           b. Competitive interactions among kin are ameliorated so that more energy can be allocated to maintenance and reproduction by sedentary groups of kin compared to the kin of dispersers who are more likely to be surrounded by more competitive or even disruptive nonkin (e.g., Greenwood, Harvey, and Perrins 1979).

TABLE 1.4  Potential Costs and Benefits of Dispersal, Especially Overwide Dispersal (vagrancy)

POTENTIAL COSTS
  I.  Genetic
    A.  Outbreeding Depression (reviewed in Shields 1982).
      1.  Disrupting genetically coadapted complexes of alleles.
      2.  Producing hybrid young with allele combinations that are not well adapted to either of their parents' environments.
    B.  Migration Load (reviewed in Crow and Kimura 1970).
      1.  Entering a novel environment with alleles that are less suited to survival and reproduction in that environment than the alleles carried by natives.
  II.  Somatic
    A.  Direct Fitness Effects
      1.  Decreased survivorship and fecundity because of the greater risks and energy used during or after dispersal (for reviews, see Baker 1978; Shields 1982; Bateson 1983; Waser and Jones 1983; Moore and Ali 1984).
        a.  Lack of familiarity with area reduces efficiency of finding, using, and defending resources from unrelated competitors.
        b.  Increased locomotion increases time at risk to predation and requires energy that could have been invested more directly in maintenance and reproduction.
        c.  Early experience may have equipped propagule with learned or culturally transmitted traditions that are less adaptive in its new environment.
        d.  Lack of early experience with pathogens may have left propagule more susceptible to local diseases.
    B.  Indirect Fitness Effects
      1.  Decreased survivorship and fecundity for the relatives of widely dispersing individuals usually because those left behind are at a competitive disadvantage to those whose kin are also philopatric.
POTENTIAL BENEFITS
  I.  Genetic
    A.  Outbreeding Enhancement
      1.  Avoid inbreeding depression and enhance the variability of progeny.
  II.  Somatic
    A.  Direct Fitness Effects
      1.  Avoid locally crowded conditions so that the propagule has greater access to important resources (e.g., mates, food, nest sites, or nests) than if it had remained in its natal area.
    B.  Indirect Fitness Effects
      1.  Avoid competing with sedentary kin and increase their survivorship and fecundity relative to the kin of individuals that do not disperse under the same conditions (i.e., dispersal as a form of phenotypic altruism).

migration load; tables 1.3 and 1.4, and see Crow and Kimura 1970). Rather, an individual is expected to produce less fertile or vigorous progeny, owing to segregational or recombinational load. In contrast, the strict somatic arguments suggest that the reductions in fitness associated with maladaptive dispersal will be visited upon the individual propagule itself. It is expected to survive less well, but if it survives it should show the same levels of reproduction and its progeny should be as vigorous as is true of its more or less sedentary competitors.

If we can measure whether dispersal or philopatry has a greater effect on the survivorship or reproductive component of fitness in any particular organism, we can tentatively assign a greater weight to either the somatic or genetic consequences in that particular case (table 1.5). Such analyses of the reproductive success associated with differences in dispersal and/or levels of inbreeding and outbreeding, and whether they occur by affecting survivorship or fecundity, are absolutely necessary to test ultimate hypotheses about adaptive value. Despite their acknowledged importance, however, few analyses of natural systems have been done, and those that have, have produced conflicting results (e.g., Packer 1979; Moore and Ali 1984; Greenwood, Harvey, and Perrins 1978; Van Noordwijk and Scharloo 1981; Mech, chap. 4, Rood, chap. 6, and Rogers, chap. 5, all in this vol.).

Proximate

As we wait for additional measurements of the ultimate costs and benefits associated with dispersal, we must be wary of accepting easy or obvious answers based primarily on proximate considerations before all of the data are examined. Many (e.g., Baker 1978; Lidicker 1975; Shields 1982; Waser and Jones 1983; Moore and Ali 1984) have cautioned that the answer for one species might be a red herring for others. The notion that all dispersal decisions could be made for universal reasons or have evolved as a result of a single causal process has little chance of being true or of leading to fruitful research or knowledge. If the question is the value of dispersal behavior, the answer is likely to be plural. Given such plurality as an assumption, however, we may still begin evaluating the relative contributions of different factors in determining dispersal and mating behavior by using simple contrasts when critical factors vary either naturally or in experimental situations.

To illustrate such a program let us assume that we are interested in both the proximate and ultimate causes and consequences of dispersal and

TABLE 1.5 Contrasts Useful in Discriminating between Alternative Ultimate Hypotheses to Explain the Function of Dispersal

| | | SOMATIC COSTS | |
| --- | --- | --- | --- |
| | | Present | Absent |
| GENETIC COSTS | Present | Survivorship = − <br> Fecundity = − | Survivorship = 0 <br> Fecundity = − |
| | Absent | Survivorship = − <br> Fecundity = 0 | Survivorship = 0 <br> Fecundity = 0 |

Notes: A minus sign (−) indicates a negative effect, and a zero (0) no effect, on the particular fitness component.

breeding suppression in a communal or cooperatively breeding organism. Early study might have indicated that individuals born into a particular group often disperse from their natal groups before breeding or refrain from breeding in their natal groups if they stay, despite a physiological capacity to do so. Such observations might lead one to suspect that the dispersal and suppression of breeding might be adaptations to avoid close inbreeding. This conclusion would be supported if the presence of opposite-sex parents as potential mates appeared to proximately stimulate dispersal and suppress breeding. One might feel even more sanguine if others remain in their natal troops and breed if their opposite-sex kin die, even if their parent or other same-sex kin remain in the group. Such a data set could and often has led to conclusions that dispersal and mating behavior are controlled proximately by the presence and absence of kin as potential mates, owing to the adaptive value of avoiding (extreme) inbreeding (inbreeding avoidance or IA hypothesis; for examples of such arguments, see Packer 1975, 1979; Greenwood 1980; Greenwood and Harvey 1982; Pusey 1980; Koenig and Pitelka 1979; Shields 1982, 1983; Dobson 1982; Hoogland 1982; Holekamp 1984; Pusey and Packer 1987; Berger, chap. 3, Bowen and Koford, chap. 12, Halpin, chap. 7, Mech, chap. 4, Rogers, chap. 5, and Rood, chap. 6, all in this vol.).

In contrast, someone else looking at the same data set might note that individuals that disperse or suppress breeding are often subordinate to other group members and may merely be seeking an opportunity to breed in the face of overwhelming competition. They may not be avoiding incest, but rather may be avoiding worse somatic consequences at the hands of dominants in their natal groups should they remain and/or try to breed. They might also be leaving in order to reduce the level of competition between themselves and the kin (same or opposite sex) they leave behind (reproductive competition or RC hypotheses, Packer 1979; Greenwood 1980; Moore and Ali 1984; Shields 1983; Koenig, Mumme, and Pitelka 1984; Hannon et al. 1984; Pusey and Packer 1987; Berger, chap. 3, Halpin, chap. 7, Jones, chap. 8, Mech, chap. 4, Rogers, chap. 5, Rood, chap, 6, Smith, chap. 9, and Waser, chap. 16, all in this vol.).

Given the right conditions, these two hypotheses can generate alternative predictions even though they themselves are not mutually exclusive explanations (tables 1.6 and 1.7). The right conditions here are simply the presence of one of the supposed triggers (opposite-sex kin or dominant competitors) in the *absence* of the second (tables 1.6 and 1.7). It is not enough to show that data are consistent with one interpretation or the other since the hypotheses may make identical predictions (cells A and D). Those cells are important, however, in showing that the behavior in question is not random with respect to at least one of the proposed controlling factors. To determine the relative importance of each, however, it is necessary to examine the hypotheses' contrasting predictions (cells B and C).

TABLE 1.6 Contrasts Useful in Discriminating between Alternative Proximate Hypotheses to Explain Dispersal in a Communally Breeding Organism

|  |  | DOMINANT COMPETITORS | |
|  |  | Present<br>Cell A | Absent<br>Cell B |
|---|---|---|---|
| RELATED MATES | Present | IA: Disperse<br>RC: Disperse<br><br>Disperse: 6<br>Remain: 0 | IA: Disperse<br>RC: Remain<br><br>Disperse: 2<br>Remain: 6 |
|  | Absent | IA: Remain<br>RC: Disperse<br><br>Disperse: 6<br>Remain: 14 | IA: Remain<br>RC: Remain<br><br>Disperse: 0<br>Remain: 6 |
|  |  | Cell C | Cell D |

*Notes:* IA = inbreeding avoidance; RC = reproductive competition. The predictions of the two hypotheses are presented in the top half of each cell, while the actual number of times animals dispersed or remained in their natal groups for each set of conditions is found in the lower half of the cell. The data are for acorn woodpeckers and include natural and experimental cases.
*Sources:* Koenig, Mumme, and Pitelka 1984; Hannon et al. 1985.

In this case, the data are less than straightforward. The relative paucity of dispersal and absence of breeding suppression in groups with close kin of the opposite sex (in the absence of dominant same-sex competitors) are unequivocal evidence *against* an IA hypothesis (cell B, tables 1.6 and 1.7). The occurrence of significant amounts of nondispersal and shared breeding in the absence of opposite-sex kin (but in the presence of dominant competitors, cell C, tables 1.6 and 1.7) is tentatively consistent with IA and unequivocal in denying the simplest RC hypothesis as well. When these contrasts are examined, it appears that something else needs to be considered.

Since the data are not hypothetical, but rather actual data from the acorn woodpecker (Koenig, Mumme, and Pitelka 1984; Hannon et al. 1985), it may be interesting to note that all but two of the individuals that remained and all but one that bred in the absence of opposite-sex kin, but in the presence of presumably dominant same-sex competitors (cell C), did so only when the dominant individual was close kin (usually a parent) of the same sex. Nonetheless, the data are equivocal with respect to the relative importance of IA and RC in shaping dispersal and breeding patterns. Since 10 of 49 of the groups in question (20%) were known or suspected to have closely inbred (usually within nuclear families), I believe on balance, the data are more convincing of the importance of competition than of avoiding inbreeding in this species.

TABLE 1.7 Contrasts Useful in Discriminating between Alternative Proximate Hypotheses to Explain Suppression of Breeding in a Communally Breeding Organism

|  |  | DOMINANT COMPETITORS | |
|---|---|---|---|
|  |  | Present<br>Cell A | Absent<br>Cell B |
| RELATED MATES | Present | IA: Suppress<br>RC: Suppress<br><br>Breed: 0<br>Suppressed: 6 | IA: Suppress<br>RC: Breed<br><br>Breed: 10<br>Suppressed: 2 |
|  | Absent | IA: Breed<br>RC: Suppress<br><br>Breed: 9<br>Suppressed: 14 | IA: Breed<br>RC: Breed<br><br>Breed: 6<br>Suppressed: 0 |
|  |  | Cell C | Cell D |

*Notes:* See table 1.6.

*Sources:* See table 1.6.

## CONCLUSIONS

Obviously, the sort of comprehensive data, careful and critical analyses, and long-term effort exemplified by these elegant studies on the acorn woodpecker, as well as by all of the long-term empirical studies of mammals reported in this volume, will be necessary to begin, much less complete, exploration of alternative explanations of the causes of dispersal. The ideal here is to continue seeking situations, occurring naturally or induced experimentally, where hypothesized controlling factors, such as the presence of potential mates that are close kin or related or unrelated competitors of either sex, vary or are varied independently. Contrast tables can be generated that permit evaluation among these real or imagined alternatives as the sole, primary, or jointly necessary or sufficient causes of the different dispersal patterns and mating systems we observe in nature. In this way, we can finally go beyond our theoretical biases and let the organisms tell us directly what might be going one.

## ACKNOWLEDGMENTS

I would like to thank our editors and organizers, Z. T. Halpin and B. D. Chepko-Sade, for the opportunity and assistance they provided during the development and completion of this work. I also am grateful to all of the symposium participants, especially P. Smouse and P. Waser, for their direct comments on the manuscript and the stimulating general discussions that resulted in significant revisions. I must also thank the nonparticipants who provided manuscripts, ideas, and critical comments in abundance, including F. Bercovitch, J. Crook, S. Hannon, K. Holekamp, R. Mumme, F. Pitelka,

L. Wolf, and E. Waltz. Finally, I am especially indebted to W. Koenig, who provided free access to unpublished data and spent an inordinate amount of time tutoring me about acorn woodpeckers, despite knowing that we disagreed, in part, in our interpretations of those data. His openness and collegiality are both admirable and refreshing.

## REFERENCES

Allen, G. 1965. Random and non-random inbreeding. *Eugenics Quarterly.* 12:181–98.

Baker, M. C., and P. Marler. 1980. Behavioral adaptations that constrain the gene pool. In *Evolution of social behavior: Hypotheses and tests,* ed. H. Markl, 59–60. Deerfield Beach, Fla.: Verlag-Chemie.

Baker, R. R. 1978. *The evolutionary ecology of animal movement.* London: Hodder and Staughton.

Bateson, P. 1983. Optimal outbreeding. In *Mate choice,* ed. P. Bateson, 257–77. Cambridge: Cambridge University Press.

Bengtsson, B. O. 1978. Avoiding inbreeding: At what cost? *Journal of Theoretical Biology* 73:439–44.

Bradbury, J. W., and S. L. Vehrencamp. 1976. Social organisation and foraging in Emballonurid bats, pt. 1: Field studies. *Behavioral Ecology and Sociobiology* 1:337–81.

———. 1977. Social organisation and foraging in Emballonurid bats, pt. 2: Mating systems. *Behavioral Ecology and Sociobiology* 2:1–17.

Chency, D. L., and R. M. Seyfarth. 1983. Nonrandom dispersal in free-ranging vervet monkeys: Social and genetic consequences. *American Naturalist* 122:392–412.

Cockburn, A., M. P. Scott, and D. J. Scotts. 1985. Inbreeding avoidance and male-biased natal dispersal in *Antechinus* spp. (Marsupialia:Dasyuriadae). *Animal Behavior* 33:908–15.

Crow, J. F., and M. Kimura. 1970. *An introduction to population genetics theory.* New York: Harper and Row.

Darlington, C. D. 1960. Cousin marriage and the evolution of the breeding system in man. *Heredity* 14:297–332.

Dobson, F. S. 1979. An experimental study of dispersal in the California ground squirrel. *Ecology* 60:1103–9.

———. 1982. Competition for mates and predominant juvenile male dispersal in mammals. *Animal Behavior* 30:1183–92.

Dobzhansky, T. 1970. *Genetics of the evolutionary process.* New York: Columbia University.

Emlen, S. T., and L. W. Oring. 1977. Ecology, sexual selection, and the evolution of mating systems. *Science* 197:215–23.

Endler, J. A. 1977. *Geographic variation, speciation, and clines.* Princeton: Princeton University Press.

———. 1979. Gene flow and life history patterns. *Genetics* 93:263–84.

Fisher, R. A. [1930] 1958. *The genetical theory of natural selection.* Rev. ed. New York: Dover.

Fossey, D. 1983. *Gorillas in the mist.* Boston: Houghton-Mifflin.

Frankel, O. H., and M. E. Soule. 1981. *Conservation and evolution.* Cambridge: Cambridge University Press.

Gauthreaux, S. A., Jr. 1978. The ecological significance of behavioral dominance. In *Perspectives in ethology,* vol. 3, ed. P. P. G. Bateson and P. H. Klopfer, 17–54. London: Plenum Press.

Gould, S. J., and N. Eldredge, 1977. Punctuated equilibria: The tempo and mode of evolution reconsidered. *Paleobiology* 3:115–51.

Greenwood, P. J. 1980. Mating systems, philopatry, and dispersal in birds and mammals. *Animal Behavior* 28:1140–62.

———. 1983. Mating systems and the evolutionary consequences of dispersal. In *The ecology of animal movement,* ed. I. R. Swingland and P. J. Greenwood, 116–31. Oxford: Clarendon Press.

Greenwood, P. J., and P. H. Harvey. 1982. The natal and breeding dispersal of birds. *Annual Review of Ecology and Systematics* 13:1–21.

Greenwood, P. J., P. H. Harvey, and C. M. Perrins. 1978. Inbreeding and dispersal in the great tit. *Nature* 271:52–54.

———. 1979. The role of dispersal in the great tit (*Parus major*): The causes, consequences and heritability of natal dispersal. *Journal of Animal Ecology* 48:123–42.

Hamilton, W. D. 1964. The genetical evolution of social behavior, pts. 1 and 2. *Journal of Theoretical Biology* 7:1–52.

Hamilton, W. D., and R. M. May. 1977. Dispersal in stable habitats. *Nature* 269:578–81.

Hannon, S. J., R. L. Mumme, W. L. Koenig, and F. A. Pitelka. 1985. Replacement of breeders and within-group conflict in the cooperatively breeding acorn woodpecker. *Behavioral Ecology and Sociobiology* 17:303–12.

Holekamp, K. B. 1984. Natal dispersal in Belding's ground squirrels (*Spermophilus beldingi*). *Behavioral Ecology and Sociobiology* 16:21–30.

Hoogland, J. L. 1982. Prairie dogs avoid extreme inbreeding. *Science* 215:1639–41.

Horn, H. S. 1983. Some theories about dispersal. In *The ecology of animal movement,* ed. I. R. Swingland and P. J. Greenwood, 54–62. Oxford: Clarendon Press.

Howard, W. E. 1949. Dispersal, amount of inbreeding and longevity in a local population of prairie deer mice on the George reserve, southern Michigan. *University of Michigan Laboratory of Vertebrate Biology Contributions* 43:1–50.

———. 1960. Innate and environmental dispersal of vertebrates. *American Midland Naturalist* 63:152–61.

Itani, J. 1972. A preliminary essay on the relationship between social organization and incest avoidance in non-human primates. In *Primate socialization,* ed. F. Poirier, 165–71. New York: Random House.

Jones. W. T. 1984. Natal philopatry in bannertailed kangaroo rats. *Behavioral Ecology and Sociobiology* 15:151–55.

Kimura, M., and T. Ohta. 1971. *Theoretical aspects of population genetics.* Princeton: Princeton University Press.

Koenig, W. D., R. L. Mumme, and F. A. Pitelka. 1984. The breeding system of the acorn woodpecker in central coastal California. *Zeitschrift für Tierpsychologie* 65:289–308.

Koenig, W. D., and F. A. Pitelka. 1979. Relatedness and inbreeding avoidance: Counter plays in a communally nesting acorn woodpecker. *Science* 206:1103–5.

Lewin, R. 1983. Brotherly alliances help avoid inbreeding. *Science* 222:148–49.

Lidicker, W. Z., Jr. 1962. Emigration as a possible mechanism permitting the regulation of population density below carrying capacity. *American Naturalist* 96:29–33.

———. 1975. The role of dispersal in the demography of small mammals. In *Small mammals: Their productivity and population dynamics,* ed. F. B. Golley, K. Petrusewicz, and L. Ryszkowski, 103–28. Cambridge: Cambridge University Press.

Mayr, E. 1963. *Animal species and evolution.* Cambridge: Harvard University Press.

Mech, L. D. 1970. *The wolf: The ecology and behavior of an endangered species.* Garden City, N.Y.: Natural History Press.

Moore, J. 1984. Female transfer in primates. *International Journal of Primatology* 5:537–89.

Moore, J., and R. Ali. 1984. Are dispersal and inbreeding avoidance related? *Animal Behavior* 32:94–112.

Murray, B. G., Jr. 1967. Dispersal in vertebrates. *Ecology* 48:975–78.

———. 1984. A demographic theory on the evolution of mating systems as exemplified by birds. *Evolutionary Biology* 18:71–140.

Nelson, M. E., and L. D. Mech. 1984. Home-range formation and dispersal of deer in northeastern Minnesota. *Journal of Mammalogy* 65:567–75.

Packer, C. 1975. Male transfer in olive baboons. *Nature* 255:219–20.

———. 1979. Inter-troop transfer and inbreeding avoidance in *Papio anubis. Animal Behavior* 27:1–36.

———. 1985 Dispersal and inbreeding avoidance. *Animal Behavior* 33:676–78.

Pusey, A. E. 1980. Inbreeding avoidance in chimpanzees. *Animal Behavior* 28:543–52.

Pusey, A. E., and C. Packer. 1987. Dispersal and philopatry. In *Primate societies,* ed. D. L. Cheney, R. M. Seyfarth, B. B. Smuts, T. Struhsaker, and R. W. Wrangham. Chicago: University of Chicago Press.

Ralls, K., P. H. Harvey, and A. M. Lyles. 1986. Inbreeding in natural populations of birds and mammals. In *Conservation biology: The science of diversity,* ed. M. Soule. Sunderland, Mass.: Sinaver.

Rao, P. S., and S. G. Inbaraj. 1977. Inbreeding effects on human reproduction in Tamil Nadu of South India. *Annals of Human Genetics* 41:87–97.

Sade, D. S. 1968. Inhibition of son-mother mating among free-ranging rhesus monkeys. *Science and Psychoanalysis* 12:18–27.

Sanghvi, L. D. 1966. Inbreeding in India. *Eugenics Quarterly* 13:291–301.

Schonewald-Cox, C. M., S. M. Chambers, B. MacBryde, and L. Thomas, eds. 1983. *Genetics and conservation.* Menlo Park: Calif.: Benjamin/Cummings.

Selander, R. K., and S. Y. Yang. 1969. Protein polymorphism and genic heterozygosity in a wild population of the house mouse *Mus musculus. Genetics* 63:653–67.

Sherman, P. W. 1977. Nepotism and the evolution of alarm calls. *Science* 197:1246–53.

———. 1981. Reproductive competition and infanticide in Belding's ground squirrels and other animals. In *Natural selection and social behavior,* eds. R. D. Alexander and D. Tinkle, 311–31. New York: Chiron Press.

Shields, W. M. 1979. Philopatry, inbreeding, and the adaptive advantages of sex. Ph.D. diss., Ohio State University, Columbus, Ohio.

———. 1982 *Philopatry, inbreeding, and the evolution of sex*. Albany: State University of New York Press.

———. 1983. Optimal inbreeding and the evolution of philopatry. In *The ecology of animal movement*, ed. I. R. Swingland and P. J. Greenwood, 132–59. Oxford: Clarendon Press.

———. 1984. Factors affecting nest and site fidelity in Adirondack barn swallows (*Hirundo rustica*). *Auk* 101:780–89.

Smith, A. T., and B. L. Ivins. 1983. Colonization in a pika population: Dispersal vs philopatry. *Behavioral Ecology and Sociobiology* 13:37–47.

Smith, R. H. 1979. On selection for inbreeding in polygynous animals. *Heredity* 43:204–11.

Soule, M. E., and B. A. Wilcox, eds. 1980. *Conservation biology: An evolutionary-ecological perspective*. Sunderland, Mass.: Sinauer.

Swingland, I. R., and P. J. Greenwood, eds. 1983. *The ecology of animal movement*. Oxford: Clarendon Press.

Tilson, R. L. 1981. Family formation strategies of Kloss's gibbons. *Folia Primatologica* 35:259–87.

Trivers, R. L. 1972. Parental investment and sexual selection. In *Sexual selection and the descent of man, 1871–1971*, ed. B. Campbell, 136–79. Chicago: Aldine.

van den Berghe, P. L. 1980. Incest and exogamy: A sociobiological reconsideration. *Ethology and Sociobiology* 1:151–62.

Van Noordwijk, A. J., and W. Scharloo. 1981. Inbreeding in an island population of the great tit. *Evolution* 35:674–88.

Waser, P. M. 1985. Does competition drive dispersal? *Ecology* 66:1170–75.

Waser, P. M., and W. T. Jones. 1983. Natal philopatry among solitary mammals. *Quarterly Review of Biology* 58:355–90.

Williams, G. C. 1966. *Adaptation and natural selection*. Princeton: Princeton University Press.

Wilson, D. S. 1980. *The natural selection of populations and communities*. Menlo Park, Calif.: Benjamin/Cummings.

Woolfenden, G. E., and J. W. Fitzpatrick. 1978. The inheritance of territory in group-breeding birds. *Bioscience* 28:104–8.

Wrangham. R. W. 1980. An ecological model of female-bonded primate groups. *Behaviour* 75:262–99.

Wright, S. 1946. Isolation by distance under diverse systems of mating. *Genetics* 31:39–59.

———. 1977. *Evolution and the genetics of populations, vol. 3: Experimental results and evolutionary deductions*. Chicago: University of Chicago Press.

———. 1978. *Evolution and the genetics of populations, vol. 4: Variability within and among natural populations*. Chicago: University of Chicago Press.

# II. EMPIRICAL STUDIES OF DISPERSAL IN POPULATIONS WITH KNOWN INDIVIDUALS

# 2. Demes within a Northeastern Minnesota Deer Population

*Michael E. Nelson and L. David Mech*

Vertebrate populations in general are genetically subdivided (Smith, Garten, and Ramsey 1975), but large mobile vertebrates with greater dispersal capacity might be expected to have less heterogeneous populations than smaller vertebrates (Wright 1978). However, physical and biochemical evidence suggests that white-tailed deer (*Odocoileus virginianus*) populations are genetically subdivided into demes across short distances (Rees 1969; Harris, Huisman, and Hayes 1973; Manlove et al. 1976; Ramsey et al. 1979; Chesser et al. 1982). Cothran et al. (1983) hypothesized that deer demes result from the matriarchal society of white-tailed deer, which facilitates father-daughter incest.

The relationship between genetics and deer social organization, however, has just begun to be defined, and many questions remain unanswered. Ramsey et al. (1979) suggested that genetic analyses of multiple, geographically dispersed samples could be used for detection of genetic subdivision. Nevertheless, temporal and spatial characteristics of deer demes have not been described and the effects of movement traditions, home range tenacity, juvenile dispersal, and breeding movements on deme structure and discreteness have not been examined. This chapter analyzes deer movements and social behavior as a basis for identifying deer demes in part of northeastern Minnesota.

## STUDY AREA

The study was conducted in the east-central Superior National Forest of northeastern Minnesota, 48° N, 92° W, from 1974 through 1984. This area is roughly 50 km square (2,500 km²) and is near the northeastern limit of white-tailed deer distribution. The climate is cool temperate, with snow cover averaging over 1 m during 5 months of winter starting in mid-November. The region is relatively flat, with mixed coniferous-deciduous

forests (Nelson and Mech 1981). The deer population in this area declined
from 1968 through 1977 by 80% to 90% as a result of severe winters, a
high wolf (*Canis lupus*) population, and maturing habitat (Mech and Karns
1977). Deer densities varied from 0.2 to 0.4 deer/km² during this study
(Floyd, Mech, and Nelson 1979; Nelson and Mech, 1986). Several areas
that became devoid of wintering deer after 1974 (Mech and Karns 1977)
remained so through 1984.

   Most deer in the study area concentrate during winter in four areas we
refer to as *deeryards* (fig. 2.1). The Garden Lake deeryard is the largest,
encompassing 33 km² and holding no more than 800 deer. The Isabella
yard is the second largest, covering 27 km² with 400 deer present in March

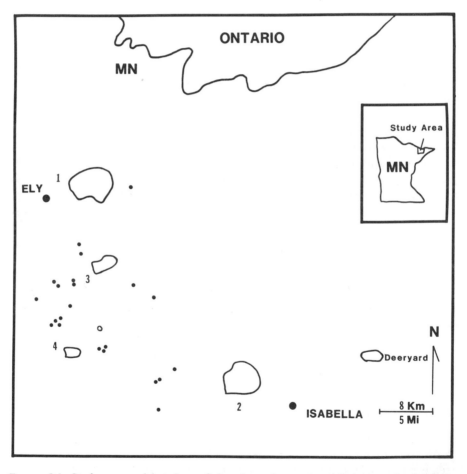

FIGURE 2.1. Study area and locations of the winter deeryards studied: 1 = Garden Lake
yard; 2 = Isabella yard; 3 = Kawishiwi Campground yard; and 4 = Snort Lake yard. Dots
represent individual deer wintering outside the yards, and the blank region northeast of the
yards indicates that no deer used that region during winter. Ely and Isabella are towns.

1984 (Nelson and Mech, 1986). The Kawishiwi Campground and Snort Lake yards are smaller, measuring 7 km² and 6 km² respectively and probably have no more than 40 to 50 deer each. Study area regions outside the major deeryards are devoid of wintering deer, except for the west and southwestern portions of the area where a few singles and small groups of deer are observed outside of yards (fig. 2.1).

In spring, most deer migrate a mean of 17 km (range = 4–40 km) to individual summer ranges that average 319 ha for adult males and 83 ha for adult females (Hoskinson and Mech 1976; Nelson and Mech 1981). An average migration takes 1.8 days for deer traveling 20 km or less and 7 days for deer traveling farther (Nelson and Mech 1981). Migrations are typically very directional and site-specific with no extensive wandering.

Fawns are born on summer ranges in June and accompany their mothers during their first year (Hoskinson and Mech 1976; Nelson and Mech 1981, 1984). Seventy percent of the 1- and 2-year-old males disperse an average of 7 km (range = 4.0–9.6 km) from their birth ranges to new ranges, but females continue to occupy summer ranges adjacent to, or overlapping, their mothers' ranges (Nelson and Mech 1984). During the breeding season in October and November, some adult males travel up to 8 km in search of receptive does near their summer ranges (Nelson and Mech 1981). With the onset of winter weather in November, deer return to their winter ranges (Hoskinson and Mech 1976; Nelson and Mech 1981). Migrations and home ranges are traditional, apparently perpetuated by cohesive family units that influence offspring movements. Thus, deer movements can be classified as spring and fall migration (rapid, site-oriented movement), summer range, winter range, rutting (males only), and dispersal (1- and 2-year-old males).

Wolf predation and bucks-only hunting are the major sources of mortality for deer (Hoskinson and Mech 1976; Nelson and Mech 1981). On an annual basis, wolf predation accounts for 81% and 31% of the mortality for adult females and males respectively (Nelson and Mech, 1986). Bucks-only hunting accounts for an additional 53% of the mortality for males. Adult annual survival is 47% for males and 79% for females.

## METHODS

Deer were captured on winter ranges from November 1974 through April 1984, primarily by rocket-net and clover traps. Most were immobilized (Seal, Erickson, and Mayo 1970), sexed, aged via incisor sectioning (Gilbert 1966), blood sampled (Seal, Nelson, and Hoskinson 1978), and radio tracked from the ground and air (Hoskinson and Mech 1976; Nelson and Mech 1981, 1984). Deer up to 1 year old will be called fawns, those 1 to 2 years old, yearlings, and those older than 2 years, adults.

The Garden Lake, Isabella, and Kawishiwi Campground deeryards had similar trapping effort and success throughout the study (table 2.1). Snort Lake deer, however, have only been followed since 1982.

Table 2.1 Sex, Age, and Capture Location of Radioed Deer in Northeastern Minnesota, 1974–1984

| | Sex | | |
| --- | --- | --- | --- |
| Deeryards | Male | Female | Total |
| Garden Lake | | | |
| Adult | 12 | 27 | 39 |
| Yearling | 6 | 4 | 10 |
| Kawishiwi Camp | | | |
| Adult | 2 | 15 | 17 |
| Yearling | 8 | 6 | 14 |
| Isabella | | | |
| Adult | 10 | 16 | 26 |
| Yearling | 8 | 10 | 18 |
| Snort Lake | | | |
| Adult | 1 | 8 | 9 |
| Yearling | 2 | 4 | 6 |
| Total | | | |
| Adult | 25 | 66 | 91 |
| Yearling | 24 | 24 | 48 |
| All Deer | 49 | 90 | 139 |

Radioed deer were located one to three times a week throughout the year with essentially 100% success. Home ranges were identified by the minimum-area method, that is, the area described by connecting only the outermost locations that make a convex polygon (Mohr 1947). Summer range polygons of adult does vary in size from 67 ha to 114 ha (Nelson and Mech 1981).

Radioed deer were rarely observed during summer, but were seen frequently during late fall and winter radio tracking. Fawn-doe relatedness was inferred from social cohesiveness, combined capture, and subsequent movement together (Hawkins and Klimstra 1970; Nelson and Mech 1981, 1984). Because female summer ranges are traditional (Nelson and Mech 1981), birth locations for our fawns had to have been the summer ranges of their does, and subsequent yearling movement away from them was considered dispersal.

Some fawns were captured alone, in which case their mothers' identities and home ranges were unknown. However, the mothers' home ranges were determined based on the movements of their fawns since fawns are highly associated with their does up to 12 months of age (Hawkins and Klimstra 1970; Nelson and Mech 1981, 1984). Several of these radioed fawns were observed with unmarked adult deer which presumably were their mothers or other close kin. Accordingly, when fawns of unknown maternity migrated to summer ranges, their migration patterns and home ranges were considered to mirror those of their does. Subsequent yearling movements and home range use away from those locations were then considered dispersal.

Some fawns did not migrate and were considered to be the offspring of nonmigratory does.

Calculations of average migration distances excluded data from radioed fawns of radioed does since fawns traveled with their does. The area of summer range distribution for deer from each yard was measured from the polygon formed by connecting the outermost summer ranges of individual deer.

## RESULTS

The locations of 161 deer summer ranges were determined from 139 individual deer, including 25 adult males, 66 adult females, 24 yearling males, and 24 yearling females (table 2.1). These included 22 doe ranges determined from the movements of 12-month-old fawns of unknown maternity. Fourteen of the fawns were from Isabella, 4 from Snort Lake, 3 from Kawishiwi Campground, and 1 from Garden Lake. Twelve family groups were studied, including 12 adult females, 12 yearling males, and 7 yearling females. Home range formation and dispersal for 11 of the groups were examined by Nelson and Mech (1984).

A total of 8,800 locations of deer were obtained during 1,760 hours of aerial radio tracking, supplemented by an additional 870 locations from ground tracking. Deer were tracked an average of 1.2 years each for a total of 172 deer-years of tracking. Eight adult does were followed for periods of 4.3 to 8.0 years.

### Temporal and Spatial Distribution of Deer

Spring migration typically occurred in early April but varied from early March to late April (Nelson and Mech 1981). Seventeen percent ($N = 23$) of the deer failed to migrate and remained on their winter ranges throughout the year. Seventeen of those deer were from Kawishiwi Campground and Snort Lake yards.

Migrations from Garden Lake, Kawishiwi Campground, and Snort Lake yards displayed a strong northeast to east directionality (fig. 2.2, table 2.2). Migration directions from the Isabella yard were more variable, with a general north element to them while predominating to the northwest. The summer ranges of Garden Lake and Isabella deer were distributed over similar-sized areas, larger than those of the other two yards (table 2.2). Garden Lake migrations occurred within a much smaller area than those from Isabella, but they averaged twice the distance of Isabella migrations. The Kawishiwi Campground deer had the narrowest migration pattern and the smallest summer range distribution. Snort Lake deer had the shortest migrations yet with more variable directionality.

At least 95% ($N = 132$) of the deer used summer ranges clearly within the same regions as did other members of their winter yards (fig. 2.3). Only 5% percent of the deer had summer ranges that could be considered "in-

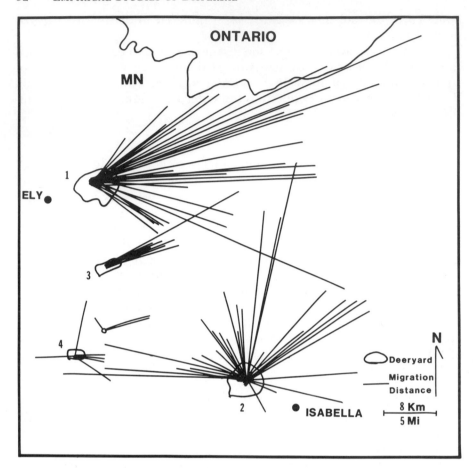

FIGURE 2.2. Distance and direction of spring migration from each of the winter yards. Each line represents data from individual deer.

TABLE 2.2 Migration Distance, Bearing, and Deme Size of White-tailed Deer from Four Deeryards in Northeastern Minnesota, 1974–1984

| YARD | NO. MIGRATING DEER[a] | MIGRATION DISTANCE (km) x̄ (SE) | MIGRATION BEARING x̄ (SE) | DEME SIZE (km²) |
|---|---|---|---|---|
| Garden Lake | 42 | 25.0 (1.8) | 77   (4) | 753–1,157[b] |
| Isabella | 41 | 12.0 (1.2) | 352 (10) | 805 |
| Kawishiwi Camp | 14 | 10.0 (1.1) | 68   (1) | 21–140[c] |
| Snort Lake | 11 | 6.0 (0.5) | 53 (20) | 125 |

a. Excludes radioed fawns of radioed does.

b. The exclusion of one outlying buck range reduced the area to 753 km².

c. The exclusion of dispersing yearling ranges reduced the area to 21 km².

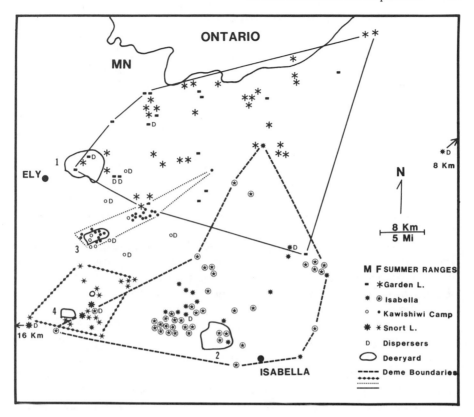

FIGURE 2.3. Summer ranges of deer from each of the winter yards. Symbols represent summer ranges of individual deer; lines are arbitrary borders indicating the authors' concept of the deme boundaries.

side" areas used by deer from other yards, depending on how borders of each yard's summer area are defined. One Garden Lake buck spent the summer 16 km within an area generally inhabited by Isabella deer. One Isabella buck summered 4.8 km inside the area used by Garden Lake deer, and another Isabella buck and two does summered 3.2 km to 8.0 km inside the Snort Lake deer summering area. A Kawishiwi Campground buck and doe summered 8.0 km and 12.8 km inside the Garden Lake deer group. The Kawishiwi buck and one of the Isabella does both had dispersed to the neighboring areas as yearlings.

Forty-two adult deer were radio tracked through two or more annual cycles, during which each continued to use the same summer and winter home ranges. Of that group, 10 adult does continued to use the same ranges during 3.1 to 8.0 years of radio tracking.

Dispersal and Home Range Formation

The movements of 22 yearling males were examined for distance, direction, and timing of dispersal. The birth ranges for 12 of them were known based on associations with radioed mothers. The birth ranges of the remaining 10 deer were assumed; 4 based on typical migrations to specific locations and 6 from the absence of migration, with continued summer use of winter ranges. By 1.5 years old, 13 (59%) of the yearlings had dispersed an average distance of 12.0 km (SE = 2.7) to new ranges. Seven dispersed during June through September and 6 in October and November. Two of the dispersals were two to nine times the distance of the others and, when excluded, lowered the mean dispersal distance to 8.2 km (SE = 0.9). Ten of the yearlings were followed to 2.5 years old, by which time 7 (70%) had dispersed to new ranges. Of those 7 deer, 6 dispersed from known birth ranges.

Of the 13 dispersing yearling males, 8 settled in areas used by deer from their same winter range. In fact, 5 deer dispersed to or toward their traditional winter ranges. The remaining 5 deer moved to the extreme edges of their own subpopulations or into other subpopulations.

Female dispersal was examined for 20 yearlings, 7 with known and 13 with assumed birth ranges. Of those with assumed ranges, 8 deer had migrated to specific summer sites while 5 were nonmigratory. Of the 20 yearlings, all were followed to 1.5 years old; 14 to 2.0 years old; 11 to 2.5 years old; 6 through 3.0 years old; and one through 5 years old, for a total of 28 deer-years of tracking. During that time, only one of the 20 yearlings (5%) dispersed. When 1 year old, a migratory yearling moved 22 km to a new home range after a 1-month stay on an assumed birth range. She was followed until 2.9 years old, during which time she became nonmigratory and remained on her new range throughout the year. The remaining 19 females (95%) continued to utilize the summer and winter ranges they first used as fawns.

Breeding Movements by Adult Males

October–November locations ($N = 133$) of 15 bucks aged 3.5 to 6.5 years old were examined to measure movements during the breeding season. The mean straight-line distances from their locations to the center of their ranges averaged 3.4 km (SE = 0.6). Three bucks shifted their movements to areas 8 km to 22 km away from their summer ranges but remained within their own subpopulation. The only buck followed through two breeding seasons shifted to the same area each year. The 12 bucks not shifting their ranges were located an average of 2.6 km (SE = 0.3) from their home range centers.

## DISCUSSION

Our results suggest that the deer from each of the winter yards represent subpopulations that constitute genetic demes. Adult deer from each yard occupy summer ranges in largely exclusive areas that have little overlap with those of neighboring yards. This separation can last for years since adult movement patterns are traditional. Moreover, most yearling females establish home ranges on or near their birth ranges and continue the migration pattern of their mothers. This site tenacity could lead to inbreeding between daughters and their fathers since dominant bucks probably maintain breeding tenure on their ranges for more than one year. Female philopatry could also lead to inbreeding with brothers and other close kin that never disperse.

Outbreeding by adult bucks appears limited to deer on the edges of demes since average rutting movements are short relative to deme area. As such, most bucks must breed does within 3 km of their home ranges and, most likely, does using the same winter yards as themselves. In addition, some late breeding occurs in December when most deer are in their yards. Finally, while it appears that yearling male dispersal tends to promote outbreeding by removing males from their birth ranges, average dispersal distances would be insufficient to disperse many deer beyond the boundaries of our two larger demes. Eight of our 13 dispersering males established new ranges within their own demes. In fact, 4 of the 5 dispersers that moved to the edges of their demes or beyond were from small demes adjacent to larger ones, and therefore were more likely to enter an adjacent deme by chance alone.

Our deer spacing is surprisingly similar to that in New York, where ear-tagged deer from winter yards were distributed into contiguous and generally exclusive regions during summer (Gotie 1976) (fig. 2.4). Data from an additional 30 New York yards, not shown here, indicate a similar pattern, although some cases of summer range overlap are known to exist between deer from different winter ranges there (N. Dickinson, pers. comm.). Migration patterns and seasonal ranges of radio-tagged deer in northern New York are traditional, with female yearlings establishing ranges on or adjacent to their mothers' ranges (Tierson et al., 1985). One- and 2-year-old males also establish new ranges up to 28 km from their birth ranges. Thus, using our reasoning for deme recognition, we conclude that these New York deer are also genetically subdivided based on deeryard location. Similar patterns of deer distribution seem to exist in Michigan (Verme 1974) and Wisconsin; in the latter state, deer yarding 6 km and 18 km apart have little or no home range overlap on summer ranges (O'Brien 1976).

Pronghorn (*Antiolocapra americana*) herds in southeastern Idaho also winter and summer in adjacent areas with little or no exchange between herds (Hoskinson and Tester 1980). Their summer ranges are separated by

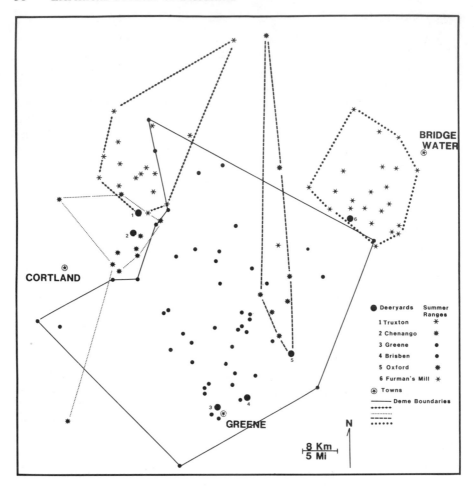

FIGURE 2.4. Apparent demes formed by deer in east-central New York. Symbols represent nonwinter locations of individual deer ear tagged in each winter yard. Data are from G. Mattfeld and R. W. Sage (pers. comm.); lines indicate Nelson and Mech's concept of the deme boundaries.

mountain ranges, but winter ranges are at the same elevation. Pronghorn herds use traditional ranges, and yearling females join their natal herds and drop fawns in the areas they were born (Pyrah 1970). Females breed almost exclusively with dominant males that establish traditional territories, which they defend from April to October (Kitchen 1974). Given this system of movement and breeding, it seems likely that pronghorns are also genetically subdivided.

Our hypothesis that deer populations consist of a conglomerate of demes is further supported by biochemical data from other populations and other species. Gene frequency differences have been found between adjacent upland and lowland deer in South Carolina (Manlove et al. 1976; Ramsey et al. 1979; Chesser et al. 1982) and between moose occupying areas within 2 km to 50 km of each other (Chesser, Reutherwall, and Ryman 1982). Genotypic differences have also been documented in subgroups of other large mammals (red deer, *Cervus elaphus*, McDougall and Lowe 1968; Bergman 1976; Gyllensten et al. 1980; reindeer, *Rangifer tarandus*, Braend 1964; and elephant, *Loxodonta africana*, Osterhoff et al. 1974).

Our results are also consistent with the model of Smith (1979) and Shields (1982), which considers dispersal as it occurs in most vertebrates to have evolved from selection for nonincestuous inbreeding. Adult home-range tenacity, female philopatry, relatively short male dispersal, short breeding movements, and little range overlap between deer from different yards suggest that deer populations composed of inbred groups are the rule.

Our study also raises several questions about the origin of demes, their sizes and configurations, and the mechanics of their separation. For instance, do demes originate from the infrequent female dispersal that eventually leads to a new migration pattern, or do they develop from gradual proliferation from an existing deme with a disruption in the traditional pattern? Conceivably a female fawn orphaned on its summer range could fail to make its mother's migration and thus become the progenitor of a new deme. Are the larger demes older than the smaller ones, thus reflecting the time each has had to develop? Our larger demes had approximately 8 to 16 times the number of deer occupying 6 to 9 times the area of our two smaller demes. Deer in our largest deme migrated 3 to 4 times as far as deer from the smaller demes. The two largest demes have histories of 40 years or more, but their origins and the origins of the smaller demes are unknown. Is deme size also related to the proportion of migrating versus nonmigrating deer comprising the deme? Our smallest two demes had 5 times the proportion of nonmigrants of the two larger demes. Is this difference somehow reflected in the relative survival of demes and their ultimate success? Are these small demes more inbred than the larger demes because of less migratory behavior or are females more apt to outbreed because they are closer to bucks from larger adjacent demes?

These latter questions lead to the subject of deme separation, since deme boundaries seem fairly well defined (fig. 2.2). Is there an active mechanism by which deer from one deme can recognize deer from another deme or recognize when they are in another deme? Or is the mechanism more passive in which boundaries simply reflect deme age and movement tradition? Conceivably, scent marking and aggression could be two mechanisms for active separation. Black-tailed deer (*O. hemionus columbianus*) frequently scent mark objects and tend to claim and defend localities (Muller-Schwarze 1971). Individual blacktails in coastal Oregon exclude others from their home ranges and form groups that exclude other groups from their preferred areas (Miller 1974). Female mule deer (*O. h. hemionus*) also occupy traditional ranges and form closed social groups that are aggressive toward outsiders (Bouckhout 1972). White-tailed deer do form "scrapes" and scent mark during, but only during, the rut (Moore and Marchinton 1974; Hirth 1977). However, 7 of our 13 yearlings dispersed in summer, so they would not have encountered "scrapes" until fall, yet most settled down in their own demes before then. Thus the role of scent marking in deme separation remains unclear.

Aggression remains a possible mechanism for deme separation. Hirth (1977) observed adult bucks chasing 2-year-old bucks and driving them away. He also observed a group of three bucks excluding 2-year-olds from their group. Unfortunately, dense forest cover in our study area inhibited our ability to observe deer as they interacted with other deer.

Whatever the case with bucks, there appears to be no active mechanism by which females could know about deme boundaries since most never explore or disperse beyond their birth ranges. In addition, there are several cases, in both our study and those from New York and Wisconsin, of deme overlap and interdeme dispersal, which suggests that deme boundaries are not totally exclusive or actively guarded. Thus evidence suggests that site familiarity and tradition are major factors in determining where deer establish home ranges (Nelson 1979; Nelson and Mech 1981, 1984), and that deme size, configuration, and separation are passive results of movement tradition, differential deme survival, and deme age.

## ACKNOWLEDGMENTS

This study was supported by the U.S. Fish and Wildlife Service, the U.S.D.A. North Central Forest Experiment Station, the Mardag Foundation (formerly Ober Foundation), and Wallace Dayton. We also thank Jeff Renneberg and several volunteer technicians who assisted with the live capture of the deer. George Mattfeld and Richard Sage graciously allowed us to use their unpublished data.

## REFERENCES

Bergman, F. 1976. Contributions to the study of infra-structures of red deer, pt. 2: First investigations of elucidation of the genetic structure of red deer populations employing protein polymorphism of serum. *Zeitschrift für Jagdwissenschaft* 22:28–35 (in German).

Bouckhout, L. W. 1972. The behavior of mule deer (*Odocoileus hemionus hemionus* Rafinesque) in winter in relation to the social and physical environment. M.S. thesis, University of Calgary, Alberta.

Braend, M. 1964. Genetic studies on serum transferrin in reindeer. *Hereditas* 52:181–88.

Chesser, R. K., C. Reuterwall, and N. Ryman. 1982. Genetic differentiation of Scandinavian moose *Alces alces* populations over short geographical distances. *Oikos* 39:125–30.

Chesser, R. K., M. H. Smith, P. E. Johns, M. N. Manlove, D. O. Straney, and R. Baccus. 1982. Spatial, temporal, and age-dependent heterozygosity of beta-hemoglobin in white-tailed deer. *Journal of Wildlife Management* 46(4):983–90.

Cothran, E. G., R. K. Chesser, M. H. Smith, and P. E. Johns. 1983. Influences of genetic variability and maternal factors on fetal growth in white-tailed deer. *Evolution* 37(2):282–91.

Floyd, T. J., L. D. Mech, and M. E. Nelson. 1979. An improved method of censusing deer in deciduous-coniferous forests. *Journal of Wildlife Management* 43(1):258–61.

Gilbert, F. F. 1966. Aging white-tailed deer by annuli in the cementum of the first incisor. *Journal of Wildlife Management* 30(1):200–202.

Gotie, R. F. 1976. Deer trap and tag, Region 6. New York Department of Environmental Conservation. Mimeo.

Gyllenstein, U., C. Reuterwall, N. Ryman, and G. Stahl. 1980. Geographical variation of transferrin allele frequencies in three deer species from Scandinavia. *Hereditas* 92:237–41.

Harris, M. J., T. H. J. Huisman, and F. A. Hayes. 1973. Geographic distribution of hemoglobin variants in the white-tailed deer. *Journal of Mammalogy* 54:270–74.

Hawkins, R. E., and W. D. Klimstra. 1970. A preliminary study of the social organization of white-tailed deer. *Journal of Wildlife Management* 34(2):407–19.

Hirth, D. H. 1977. Social behavior of white-tailed deer in relation to habitat. *Wildlife Monographs* 53:1–55.

Hoskinson, R. L., and L. D. Mech. 1976. White-tailed deer migration and its role in wolf predation. *Journal of Wildlife Management* 40(3):429–41.

Hoskinson, R. L., and J. R. Tester. 1980. Migration behavior of pronghorn in southeastern Idaho. *Journal of Wildlife Management* 44(1):132–44.

Kitchen, D. W. 1974. Social behavior and ecology of the pronghorn. *Wildlife Monographs* 38:1–96.

McDougall, E. I., and V. P. W. Lowe. 1968. Transferrin polymorphism and serum proteins of some British deer. *Journal of Zoology, London* 155:131–40.

Manlove, M. N., M. H. Smith, H. O. Hillestad, S. E. Fuller, P. E. Johns, and D. O. Straney. 1976. Genetic subdivision in a herd of white-tailed deer as demonstrated by spatial shifts in gene frequencies. *Proceedings of the Southeast Association Game and Wildlife Commission* 30:487–92.

Mech, L. D., and P. D. Karns. 1977. Role of the wolf in a deer decline in the Superior National Forest. USDA Forest Service Research Paper NC-148. North Central Forest Experiment Station, St. Paul, Minn.

Miller, F. L. 1974. Four types of territoriality observed in a herd of black-tailed deer. In *The behavior of ungulates and its relation to management*, ed. V. Geist and F. Walter, 2:644–60. IUCN New Series Publication No. 24. Morges, Switzerland: IUCN.

Mohr, C. O. 1947. Table of equivalent populations of North American small mammals. *American Midland Naturalist* 37(1):223–49.

Moore, W. G., and L. Marchinton. 1974. Marking behavior and its social function in white-tailed deer. In *The behavior of ungulates and its relation to management*, ed.

V. Geist, and F. R. Walther, 1:447–56. IUCN New Series Publication No. 24. Morges, Switzerland: IUCN.

Muller-Schwarze, D. 1971. Pheromones in black-tailed deer (*Odocoileus hemionus columbianus*). *Animal Behavior* 19:141–52.

Nelson, M. E. 1979. Home range location of white-tailed deer. USDA Forest Service Research Paper NC-173. North Central Forest Experiment Station, St. Paul, Minn.

Nelson, M. E., and L. D. Mech. 1981. Deer social organization and wolf predation in northeastern Minnesota. *Wildlife Monographs* 77:1–53.

———. 1984. Home range formation and dispersal of deer in northeastern Minnesota. *Journal of Mammalogy* 65:567–75.

———. 1986a. Deer population in the Superior National Forest, 1967–85. USDA Forest Service Research Paper NC-271. North Central Forest Experimental Station, St. Paul, Minn.

———. 1986b. Mortality of white-tailed deer in northeastern Minnesota. *Journal of Wildlife Management* 50:691–98.

O'Brien, T. F. 1976. Seasonal movement and mortality of white-tailed deer in Wisconsin. M.S. thesis, University of Wisconsin, Madison.

Osterhoff, D. R., S. Schoeman, J. Op't Hof, and E. Young. 1974. Genetic differentiation of the African elephant in the Kruger national park. *South African Journal of Science* 70:245–47.

Pyrah, D. 1970. Antelope herd ranges in Central Montana. *Proceedings of the Antelope States Workshop, Scottsbluff*, 16–20.

Ramsey, P. R., J. C. Avise, M. H. Smith, and D. Urbston. 1979. Genetics of white-tailed deer in South Carolina. *Journal of Wildlife Management* 43:136–42.

Rees, J. W. 1969. Morphologic variation in the cranium and mandible of the white-tailed deer (*Odocoileus virginianus*). A comparative study of geographical and four biological distances. *Journal of Morphology* 128:95–112.

Seal, U. S., A. W. Erickson, and J. G. Mayo. 1970. Drug immobilization of the carnivora. *International Zoological Yearbook* 10:157–70.

Seal, U. S., M. E. Nelson, L. D. Mech, and R. L. Hoskinson. 1978. Metabolic indicators of habitat differences in four Minnesota deer populations. *Journal of Wildlife Management* 42(4):746–54.

Shields, W. M. 1982. *Philopatry, inbreeding, and the evolution of sex*. Albany: State University of New York Press.

Smith, M. H., C. T. Garten, Jr., and P. R. Ramsey. 1975. Genic heterozygosity and population dynamics in small mammals. In *Isozymes*, vol. 4: *Genetics and evolution*, ed. C. L. Market, 85–102. New York: Academic Press.

Smith, R. H. 1979. On selection for inbreeding in polygynous animals. *Heredity* 43:205–11.

Tierson, W. C., G. F. Mattfeld, R. W. Sage, Jr., and D. F. Behrend. 1985. Seasonal movements and home ranges of white-tailed deer in the Adirondacks. *Journal of Wildlife Management* 49(3):760–69.

Verme, L. J. 1974. Movements of white-tailed deer in upper Michigan. *Journal of Wildlife Management* 37(4):545–52.

Wright, S. 1978. *Evolution and genetics of populations*, vol. 4: *Variability within and among natural populations*. Chicago: University of Chicago Press.

# 3. Reproductive Fates of Dispersers in a Harem-Dwelling Ungulate: The Wild Horse

*Joel Berger*

Charles Darwin observed his first wild horses (*Equus caballus*) on the Falkland Islands in 1834. It was unclear to him why animals that had been feral for at least 60 years had not colonized the other half of the island, especially since Darwin considered the island "not to appear stocked . . . [and] that part of the island is not more tempting than the rest." Gauchos "were unable to account for it excepting for the strong attachment which horses have to any locality to which they are accustomed" (Darwin 1845, 219). Darwin had, in fact, detected curious differences in patterns of land use in feral horses. For, unlike the Falkland horses, which apparently were poor dispersers, feral horses in other areas of South America had spread from Buenos Aires in 1537 to the Straits of Magellan by 1580. Even today little has been published about dispersal patterns in horses.

Relationships among social behavior, dispersal, and population structure have been topics of recent investigation, although further information is needed on a number of matters, including how far animals move, why, and what their reproductive fates might be. Such information is difficult to obtain since problems frequently arise in determining kinship patterns, in observing nocturnal species, and in sorting out why some members of a population leave but others do not (Bekoff 1977; Krebs 1978; Gaines and McClenaghan 1980). While no choice of species will circumvent all these problems, wild (i.e., feral) horses are relatively good subjects for study. They are large, diurnal, social animals of steppes and deserts. They are individually identifiable by natural colors and markings, and they mate and bear young in open areas. In this chapter, I present information on factors that influence dispersal in wild horses and on postdispersal breeding relationships. I also contrast patterns of emigration in horses with those of monogamous and polygynous ungulates and point out how parental associations may have influenced dispersal.

41

FIGURE 3.1. Overview of the Granite Range indicating the insularity of the habitat within the Great Basin Desert. Photo by Emory Kristoff, Courtesy of The National Geographical Society.

## METHODS
### Study Site and Methods of Observation

Horses were observed in the Granite Range of Nevada, a portion of the Great Basin Desert, from 1979 to 1984, during which time population size increased from 49 to 171. Although about 8,000 horses live in the northwestern portion of Nevada, the Granite Range is an insular, fault-blocked mountain where emigration is prevented on three sides by vast dry lake playas (fig. 3.1). Horses can and do migrate through a corridor, toward the northern end of the 40-km-long range, although most of the population (ca. 93%) occurs in the southern Granites (fig. 3.2). Over the 6-year study period, no animals emigrated from the study population and, except for a couple of domestic ranch horses, no new animals entered the population.

Intensive observations (ca. 8,100 hours) were made on known individuals during the first 5 years of the study. Primitive base camps were established through the horse's winter and summer ranges, and animals were observed throughout all portions of the day and during all months of the year except February (see Berger 1986 for specific details). During the sixth year (1984) about 100 hours were spent gathering additional information on movements, dispersal and puberty, and habitat relationships. Fixed-wing or helicopter surveys were made from 1979 through 1983.

Reproductive Status

Maternity was easily evaluated by direct observations of associations between mares and their newborn. Paternal relationships were estimated by reliance on 244 observed copulations, from which 92 foals were born. Stallions were considered the true fathers when the median date of copulation during a mare's estrous cycle fell within the gestation period of Granite Range foals ($\overline{X} = 345.1 \pm$ (SD) 8.5 days for females; $\overline{X} = 348.3 \pm 9.3$ days for males; Berger 1986). In cases where two or more males copulated with a single female during the critical period and a foal was born subsequently, each male was assigned a proportional representation of sire status, since I could not infer which fathered the offspring. Pubescent status was assigned to females when they were mounted successfully or when they lifted their tails and pranced toward unfamiliar males (Asa, Goldfoot, and Ginther 1979). For males, age at puberty could not be determined in any meaningful fashion.

## SOCIAL ORGANIZATION AND AGES AT POSTNATAL EMIGRATION

Wild horses live in year-round *bands*, each with one or more stallions and their *harems* (females and young). Males not associated with bands are *bachelors*; they form transient associations with other males or they some-

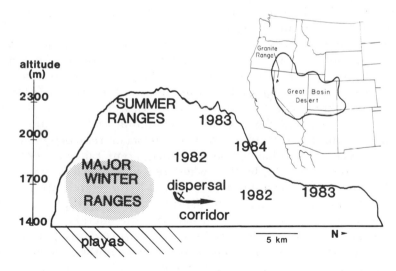

FIGURE 3.2. Overview of the study site within the Great Basin Desert and a schematic diagram of the major features of the Granite Range study site. The major winter range encompasses about 90% of the population, while summer ranges account for between 70% to 80%. Yearly figures represent locations of 8 or more animals. The X in the dispersal corridor is the location of the only ranch; it is at the eastern base of the Granite Range.

times are solitary (Keiper 1976; Miller and Denniston 1979; Rubenstein 1981). Emigration from natal bands occurs in both young males and females; most often the young depart from their natal groups voluntarily (Berger 1986).

The mean age at puberty in 32 females was 2.1 ( ± .77) years. Of these, 26 left their natal bands permanently at that time; 6 pubescent females remained. Mean age at which these latter 6 females emigrated permanently was 3.0 ( ± .94) years. The overall mean age at which all 32 females departed permanently from their natal bands was 2.26 years. For 31 males, mean postnatal dispersal age was 2.3 ( ± .88) years; the range was from 11 months to 4.3 years.

## ON EMIGRATION DISTANCES AND SAMPLING BIASES

How far do dispersers move from their natal bands? To address the question one must decide the criteria to be used, an important yet still unresolved issue. Should the criterion be the maximum distance or the average distance on a given date, in a given year, or at a specified age or density? If density is to be used, how is it to be calculated and over what period of time? Perhaps multiple criteria should be employed. The possibilities are endless. Depending on the answers chosen and the specific problems confronting researchers who study different species (e.g. nocturnal rodents, secretive carnivores, etc.), different findings with regard to dispersal distances will emerge. Some data will undoubtedly portray true movements while others may simply reflect sampling or criteria biases.

Consider the following example based on about 3,100 actual sightings of a single Granite bachelor, $A$, who was 3 years old when leaving his natal band. $A$ then spent about 6 months in a home range (90% of the observations) about 3 km² in size, located about 2.5 linear km from his natal band's home range. During the next 4 months most of his time (85%) was spent within 6 km of the natal band, but for two periods, one of 7 days and one for at least 12 days, he was a minimum of 9 km and 16 km respectively from that area. In the next year, he again returned to his birth area for 2 weeks, then inhabited another small home range area of about 3 km² for about 4 months; he then wandered widely, at least twice up to 5 km/day for an additional 2 months. Obviously there are different ways in which such movement data can be handled, but without specifying what the methods are, some very different patterns can emerge.

In assessing emigration distances, I used standardized time periods after individuals left their natal bands. These were 4 to 6 months after emigration and then as close to 1 year after, 2 years after, and so forth, as possible. Distances were straight-line map estimates between birth sites (or the center of a disperser's maternal home range) and the areas of subsequent residency. For more than 90% of the individuals, at least 150 sightings/individual were used at the specified dispersal times. Hence, if the dispersal distance of animal *B* was desired at the end of one year, a minimum of 75 observations immediately before the end of the year and 75 observations after the next year began would be tallied for the distance from the natal band site. An exception occurred in 1984 when I made sporadic reconnaissance visits to the study site.

## EMIGRATION DISTANCES AND POPULATION SIZE

While both sexes emigrated from natal bands, all females transferred to other bands or formed consortships with bachelors on the same day they left. In contrast, about 90% (28 of 31) of the males formed associations with other males within the first 3 days of emigration. The other 10% (3 of 31) remained solitary (one of the males for almost 4 months).

The distances of movement differed among individuals and were related to three factors. First was the sex of the individual. Males always moved farther than females, regardless of age or population size (fig. 3.3). Second was the period that followed emigration from the natal band. The amount of time that had elapsed after emigration (measured in number of years) was positively related to distance from natal birth site for males ($r_s = .66$; $N = 7$; $p < 0.001$), followed over 4 consecutive years, but not for females ($r_s = .16$; $N = 12$; NS) over the same time period. Third was population size. As the total number of animals increased, winter density remained constant, an indication that the population as a whole was enlarging its area

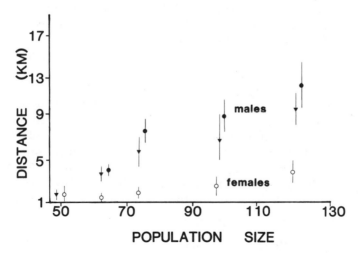

FIGURE 3.3. Relationships between population size (yearlings and older) in a year and emigration distances in males and females within 2 years of leaving their natal bands. Solid triangles and solid circles are male averages at 1 and 2 years post–natal band dispersal (respectively). Open circles are female averages for 1 and 2 years post–natal band dispersal. Emigration distances are measured as mean distances from the center of birth home ranges and are based on a minimum of 150 points per individual.

of use (fig. 3.2; see also Berger 1986). Coincident with increasing population size were greater emigration distances (fig. 3.3). In the last year of study it was difficult to know whether the increased distances were a result of the sporadic censusing or a real increase in travel distances by horses.

Overall, females moved much shorter distances from their natal bands than males, a result of the consortships that females rapidly formed with males once they left their bands. Only 4 of 32 females (12.5%) ever shifted to home ranges that were more than 4 km away from their maternal home ranges (e.g., 90% frequency use areas). Males moved farther, probably as a result of increased food availability and mate competition. About 30% (10 of 31) of the males moved between 10 km to 15 km from their natal home ranges, but 7 of these 10 males (70%) returned to areas near their birth sites at least once within 2 years of emigration. Such movements probably resulted from the distribution of Granite females located primarily at the southern end of the range (fig. 3.2).

## EMIGRANTS AND THEIR REPRODUCTIVE FATES
### Mortality

In polygynous ungulates, males are often expected to suffer greater dispersal-related mortality (Ralls, Brownell, and Ballou 1980) since they traverse unfamiliar landscapes and move greater distances than do females (Estes 1969; Jarman and Jarman 1973). In horses and other equids, the available data suggest that males move greater distances than females and that adult sex ratios are skewed to favor females (Berger 1983b). In insular populations, including the Granite Range, where the effects of dispersal can be separated from those of mortality because emigrants cannot move beyond fixed boundaries, adult males have been found to experience greater mortality than females. This occurs most often as a result of wounds incurred in intermale competition for mates (Berger 1986).

Yet young emigrant Granite Range horses who left their birth home ranges and moved over greater areas survived as well as those who remained within or around natal areas; there were no sex differences in emigrant-related mortality. Feral equids experience little, if any, predation pressure, and only very young animals are exposed to nonrandom mortality risks (Berger 1983a). Native equids would probably be better indicators of dispersal-related mortality because they seem more finely tuned to pressures within their natural environments.

### Sex Differences in Reproductive Success

While both sexes of horses left their natal bands at similar ages, a logical prediction is that females will be more successful at reproduction than males, at least at early ages. This prediction stems from skewed breeding sex ratios found in polygynous mammals where variance in reproductive success is greater in males than in females (Clutton-Brock, Guinness, and Albon 1982; Hoogland and Foltz 1982).

For Granite horses the prediction held, as young females were more likely to reproduce than young males. Of 32 females that left their natal bands and had been away for 2 years, 63% ($N = 20$) produced at least one foal. For those away from natal bands for 3 years, about 90% (17 of 19) had produced at least one foal. Males were much less successful. No males ($N = 20$) sired offspring within their first two years away from natal bands. A 3-year-old managed to copulate once with an emigrating female, but she subsequently recycled and later produced a foal fathered by an older male. Only about 13% (2 of 16) of the males that had been away for 3 years sired offspring.

## FACTORS INFLUENCING REPRODUCTIVE SUCCESS

Population density affected reproductive performance. As the Granite population increased in size, it was likely that a greater proportion of young males would become successful breeders simply because not all the females could be monopolized by older males. Although younger individuals failed to sire foals, in the last 3 years of the study younger animals did copulate, whereas only older animals mated during the earlier years (Berger 1986). In an analogous situation, LeBouef (1974) found that as an elephant seal (*Mirounga angustirostris*) population expanded, a greater proportion of younger males bred. Furthermore, young emigrant female seals that moved to lower density sites experienced greater reproductive success than those that remained at higher density sites (Reiter, Panken, and LeBouef 1981). For young emigrating female horses, this was not the case, probably because of resource limitations. Two emigrant Granite females that moved to sites of lower population density produced their first foals an average of 3 years after leaving their natal bands, while the mean for the remainder of the population was 1.7 years past postnatal dispersal. Delayed reproduction in these two females resulted from exposure to new physical and social environments and poor-quality food resources in the new home ranges (Berger 1986).

### Females

Female abilities to produce foals were affected by at least three factors.

*Age.* Older females had better reproductive success than younger females until the age of 5 years, presumably as a function of greater reserves that could be allocated to offspring production.

*Home Range Site.* Areas used by transferring females had marked influences on their abilities to produce foals. Foal production in 2- and 3-year-olds that moved to home ranges designated as "good" or "poor" quality (after Berger 1986) averaged 69% (11 of 16) and 27% (7 of 26), respectively. Home range location was, however, confounded with membership in stable bands (e.g., those with no membership changes).

*Band Stability.* Of the 16 females in better-quality home ranges, only one of 5 (20%) that had switched bands in the prior year produced a foal, whereas 10 of 11 (91%) in stable bands did so ($Z = 2.84$; $p < 0.005$). A similar but less dramatic effect occurred in mares from poor-quality areas; 37% (7 of 19) of the females in stable bands had foals, while none of the 7 females from unstable bands did ($Z = 1.88$; $p < 0.06$).

These differences in reproductive performances between females from stable and unstable bands are due in part to the physical difficulties associated with transferring. Both pregnant and nonpregnant females may be forced by unfamiliar males to copulate, and females that joined new bands experienced aggression from resident females (Berger 1986).

## Males

Opportunities for young males to mate were extremely limited. For males 5 years or older, home range locations, fighting skills, and age were all factors that affected reproductive success. Young emigrating males, however, possessed neither the skills nor the body weights needed to acquire harems, and their chances to mate were limited by their ability to locate young (e.g., emigrating) females. Mating opportunities for young emigrating males were improved by adopting one or several of the following tactics.

*Maximizing Body Weight While Young.* Of nine males whose natal bands and histories as bachelors were known, three of the four heaviest bachelors won females by fighting. The lightest three only secured their first three females by encountering them as they wandered; later they lost these females in combat, suggesting that weight (in association with combat) mediated retention of females (see Berger 1986).

*Ranging Widely to Improve Chances of Meeting Females.* Young bachelors had significantly larger home ranges, as well as daily and nightly movements, than did stallions (Berger 1986), a pattern leading to the prediction that males wandering more widely would encounter proportionately more female wanderers than did males that were more sedentary. This idea was not supported by the data. An average of 5.8 females/year wandered by themselves. When the number of bachelor males and stallions within 41 km² (4 × 4 mi) of 32 wandering solitary females was compared with expected encounter rates, there was no difference between categories of males ($X^2 = .29$; $N = 369$; NS). However, significantly more of the consortships that lasted at least an hour in duration were with stallions ($X^2 = 4.68$; $p < 0.05$), and 88% (23 of 26) of females' first copulations were with stallions. These data suggest that females may select stallions over bachelors as possible mates or, alternatively, that stallions are more adept at usurping young females from bachelors.

Still, two bachelors (a 4- and a 5-year-old) were successful at locating young females and then shifting these females to new home ranges in areas not previously used by other horses. This is one of the ways in which the population expanded its area of usage. Also, by 1983, 13 bachelors had moved about 24 km north of the rest of the population. Although none of these males mated with the study females, they possibly (1) scouted out new areas where they might move if (or when) females were acquired, as in the above cases; or (2) could meet other males and develop fighting skills or meet females and establish relationships.

*Joining Bachelor Groups to Improve Fighting Skills.* Numerous benefits may be derived by membership in bachelor groups (Rubenstein 1981); none of the nine bachelors (specified earlier) acquired females without first belonging to these groups. Fighting skills may be one of the benefits obtained, as bachelors playfight often. No data are available on whether bachelors that interact more also leave behind more offspring, but it is known that stallions that sired a relatively higher number of foals won a greater proportion of their fights ($p < 0.03$; $\tau = .48$; $N = 12$) and spent a greater proportion of their time interacting aggressively ($p < 0.02$; $\tau = .55$).

*Acquiring Females in Less Competitive Situations.* Among ungulates, including bighorn sheep (*Ovis canadensis*, Smith 1954), elk (*Cervus canadensis*, Portillo, pers. comm.), and horses (Azara 1838), cases exist where males leave the population and mate surreptitiously with domestic sheep, cattle (e.g., cows!), and domestic mares, respectively. These situations may occur because, in domestic populations of these particular species, intermale competition for mates may be less intense than in natural populations since humans ordinarily control which males breed. As a result of the distorted breeding sex ratios, domestic male animals might not have the fighting skills found in their wild counterparts. For example, a domestic stallion from a nearby ranch was killed in combat by a Granite bachelor that then appropriated the females.

## RECOGNITION, INBREEDING, AND EMIGRATION DISTANCES

Since bachelor males were sometimes successful at taking over harems, a natural experiment was created in which stallions were either true fathers (i.e., no stallion replacement) or stepfathers (i.e., a new stallion) of young that grew up in their band. Regardless of paternal status, both young females and males emigrated voluntarily. Of 5 stepdaughters and 19 true daughters whose first copulations were known, all occurred with nonband males even though fathers and stepfathers had mating opportunities (Berger and Cunningham 1987). Since young females emigrate from familiar members, it may be that an innate mechanism to avoid mating with familiar individuals exists (Duncan et al. 1984), an idea substantiated from experimental evidence for other vertebrates (see Bateson 1982). It is still unclear whether young emigrating females avoid mating with familiar males or vice versa.

The incidence of inbreeding due to father-daughter, mother-son, and brother-sister matings is low in horses under free-ranging conditions (Duncan et al. 1984). In the Granite Range, only 3.1% (4 of 129) of the copulations were between fathers and daughters when genealogies were completely known. None of these occurred when daughters attained puberty or were emigrating. The incestuous matings occurred later in life, when fathers reacquired daughters, a minimum of 19 months after initial associations had ended (Berger and Cunningham 1987).

Opportunities for extreme inbreeding in wild horses are reduced in ways other than through recognition in early life. For although both sexes emigrate, males whether young or old tend to move farther from natal areas than females. In many cases, brothers or even old emigrant fathers will be unavailable to mate with sisters or even mothers. Exceptions do occur. In the last year of the study a brother (a 5-year-old) was the stallion of a band in which his 3-year-old full sister produced his foal; the stallion had left the natal band before the sister was born and as a result he had never known her.

## CHARACTERISTICS OF THOSE UNGULATES OF WHICH BOTH SEXES DISPERSE
### Monogamous and Harem-Dwelling Species

There are only a few species of ungulates characterized by the dispersal of both male and female young (Greenwood 1980; Dobson 1982). Among these can be found both monogamous species, such as dik-diks (*Madoqua kirki*, Hendrichs 1975) and presumably klipspringers (*Oreotragus oreotragus*), oribis (*Ourebia ourebi*), and southern reedbucks (*Redunca redunca*) (Dunbar and Dunbar 1980; Wittenberger and Tilson 1980), and all of the polygynous ungulates that live in year-round harems. These latter include the two South American camelids, guanacos (*Lama guanicoe*) and vicunas (*Vicugna vicugna*) (Franklin 1983), and under "typical" conditions, four species of equids: mountain and plains zebras (*E. zebra* and *E. burchelli*), Przewalski's horses (*E. przewalski*), and feral horses (Klingel 1975; Penzhorn 1979).

An especially notable feature distinguishing the above monogamous and polygynous species from the more "generalized" polygynous ungulates, whose males emigrate (e.g., bighorn sheep, deer, large African antelope; see Owen-Smith 1977; Jarman 1974; Geist 1971, 1978; Clutton-Brock, Guinness, and Albon 1982), is the duration of parental associations with their young. When parental associations are long, as in either monogamous species or year-round harem dwellers, it is likely that periods of male tenure exceed the ages at which puberty is attained in the young of both sexes and that male and female young then emigrate. In contrast, in generalized ungulates that favor seasonal or other temporary male-female breeding associations, including tending bonds and leks, males tend to emigrate and move farther than females (Geist 1971; Nelson and Mech, chap. 2, this vol.; Robinette 1966). To verify the generality of these patterns, specific data, especially on male tenure length in haremic species and on turnover rates in monogamous ones (Tilson and Tilson 1986), will be needed.

## Functional Differences in Equid Emigration Because of Social Organization

Simply because some equids occur in year-round harems and others (feral asses, *E. asinus*; onagers, *E. hemionus*; Grevy's zebras, *E. grevyi*) are found in loose or more transient groupings is it fair to conclude that various emigration patterns have different effects on the genetic structure of the population? Probably not, since, despite the type of social organization, the young of both sexes of equids depart from their mothers (or their bands), with males moving farther than females.

Recent work on feral asses best illustrates this. In xeric areas such as Death Valley, asses are found in loose, casual associations; emigration from natal bands occurs in both sexes (Norment and Douglas 1977). In contrast, year-round harems of asses sometimes form when resources are adequate, such as on comparatively lush subtropical islands; still, both sexes of young emigrate from natal bands (McCort 1980). The result is that demes consist of more closely related females than males (Norment and Douglas 1977), as one also finds in generalized polygynous ungulates characterized by male dispersal (Nelson and Mech, chap. 2, this vol.). Thus, regardless of whether or not equids are in year-round harems, the pattern that males move farther than females suggests that the consequences of male and female emigration on the genetic structure of the population are probably no different than in the more generalized polygynous ungulates.

## Selection Pressures for Dispersal by Sex in Other Taxa

Describing dispersal patterns is clearly different from explaining why animals disperse, and many points require further elaboration. For instance, why corollaries do not occur between emigration patterns in ungulates and in other mammalian groups is not obvious. Red howler monkeys (*Alouatta palliata*), African lions (*Panthera leo*), black-tailed prairie dogs (*Cynomys ludovicianus*), and hyraxes (*Heterohyrax brucei*) represent four taxa, all with year-round social systems analogous to wild horses in that male tenure usually exceeds one year. Yet red howler females emigrate more than males (Crockett 1984); lion males generally leave, while females do not, except under very xeric conditions, when both sexes may emigrate (see Packer and Pusey 1983; Owens and Owens 1984). Young male prairie dogs usually leave, but when harem tenure of dominant males exceeds the period required for daughters to mature sexually, the father may emigrate as well (Hoogland 1982). Finally, both sexes of hyrax usually emigrate, although some females may remain in their fathers' groups when neither father or daughter leave (Hoeck 1982). Other empirical findings also require explanation. For example, resident males in some groups of harem-dwelling camelids drive out young of both sexes (Franklin 1983), but it is still unclear why this does not occur in the harem-dwelling equids. Additional research will be necessary to understand the reasons why dispersal patterns differ among the many species. As Darwin noted on the Falkland Islands and elsewhere, many kinds of diversity exist, and unitary explanations may be insufficient to account for all of them.

## REFERENCES

Asa, C. S., Goldfoot, D. A., and Ginther, O. J. 1979. Socio sexual behavior and the ovulatory cycle of ponies (*Equus caballus*) observed in harem groups. *Hormones and Behavior* 13:49–65.

Azara, Don F. de. 1838. *The natural history of the quadrupeds of Paraguay and the river La Plata.* Vol. 1. Edinburgh: Adam and Charles Black.

Bateson, P. 1982. *Mate choice.* Cambridge: Cambridge University Press.

Bekoff, M. 1977. Mammalian dispersal and the ontogeny of individual behavioral phenotypes. *American Naturalist* 111:715–32.

Berger, J. 1983a. Ecology and catastrophic mortality in wild horses: Implications for sociality in fossil assemblages. *Science* 220:1403–4.

———. 1983b. Predation, sex ratios, and male competition in equids (Mammalia: Persissodactyla). *Journal of Zoology, London* 201:205–16.

———. 1986 *Wild horses of the Great Basin: Social competition and population size.* Chicago: University of Chicago Press.

Berger, J., and Cunningham, C. 1987. Influence of familiarity on frequency of inbreeding in wild horses. *Evolution* 41: 229–31.

Clutton-Brock, T. H., Guinness, R. E., and Albon, S. D. 1982. *Red deer: Behavior and ecology of two sexes.* Chicago: University of Chicago Press.

Crockett, C. M. 1984. Emigration by female red howler monkeys and the case for female competition. In *Female primates: Studies by women primatologists,* ed. M. F. Small, 159–73. New York: A. R. Liss.

Darwin, C. 1845. *Journal of researches.* New York: P. F. Collier and Son.

Dobson, R. S. 1982. Competition for mates and predominant juvenile male dispersal in mammals. *Animal Behaviour* 30:1183–92.

Dunbar, R. I. M., and Dunbar, E. P. 1980. The pairbond in klipspringer. *Animal Behaviour* 28:219–29.

Duncan, P., C. Feh, J. C. Gleize, P. Malkas, and A. M. Scott. 1984. Reduction of inbreeding in a natural herd of horses. *Animal Behaviour* 32:520–27.

Estes, R. D. 1969. Territorial behavior of the wildebeest (*Connochaetes taurinus* Burchell, 1823). *Zeitschrift für Tierpsychologie* 26:284–370.

Franklin, W. L. 1983. Contrasting socioecologies of South America's wild camelids: The vicuna and guanaco. In *Advances in the study of mammalian behavior,* 573–629. Special Publication No. 7, American Society of Mammalogists, Shippensburg, Pa.

Gaines, M. S., and McClenaghan, L. R., Jr. 1980. Dispersal in small mammals. *Annual Review of Ecology and Systematics* 11:163–96.

Geist, V. 1971. *Mountain sheep.* Chicago: University of Chicago Press.

————. 1978. *Life strategies, human evolution, environmental design: Toward a biological theory of health.* New York: Springer-Verlag.

Greenwood, P. J. 1980. Mating systems, philopatry and dispersal in birds and mammals. *Animal Behaviour* 28:1140–62.

Hendrichs, H. 1975. Changes in a population of dikdik, *Madoqua* (*Rhynchotragus*) *kirki* (Gunther, 1880). *Zeitschrift für Tierpsychologie* 38:55–69.

Hoeck, H. N. 1982. Population dynamics, dispersal, and genetic isolation in two species of hyrax (*Heterohyrax brucei* and *Procavia johnstoni*) on habitat islands in the Serengeti. *Zeitschrift für Tierpsychologie* 59:177–210.

Hoogland, J. L. 1982. Prairie dogs avoid extreme inbreeding. *Science* 215:1639–41.

Hoogland, J. L., and Foltz, D. W. 1982. Variance in male and female reproductive success in a harem-polygynous mammal, the black-tailed prairie dog (Sciuridae: *Cynomys ludovicianus*). *Behavioral Ecology and Sociobiology* 11:155–63.

Jarman, P. J. 1974. The social organization of antelope in relation to their ecology. *Behaviour* 48:215–67.

Jarman, P. J., and Jarman, M. V. 1973. Social behaviour, population structure and reproduction in impala. *East African Wildlife Journal* 11:329–38.

Keiper, R. R. 1976. Social organization of feral ponies. *Proceedings of the Pennsylvania Academy of Science* 50:89–90.

Klingel, H. 1975. Social organization and reproduction in equids. *Journal of Reproduction and Fertility* (*supp.*) 23:7–11.

Krebs, C. J. 1978. A review of the Chitty hypothesis of population regulation. *Canadian Journal of Zoology* 56:2463–80.

LeBouef, B. J. 1974. Male-male competition and reproductive success in elephant seals. *American Zoologist* 14:167–95.

McCort, W. D. 1980. The behavior and social organization of feral asses (*Equus asinus*) on Ossabaw Island, Georgia. Ph.D. diss., Pennsylvania State University.

Miller, R., and Denniston, R. H. II. 1979. Interband dominance in feral horses. *Zeitschrift für Tierpsychologie* 51:41–47.

Norment, C., and Douglas, C. C. 1977. Ecological studies of feral burros in Death Valley. *University of Nevada Cooperative National Park Research Studies Unit Contribution* 17.

Owens, M., and Owens, D. 1984. *Cry of the Kalahari.* New York: Houghton Mifflin.

Owen-Smith, N. 1977. On territoriality in ungulates and an evolutionary model. *Quarterly Review of Biology* 52:1–38.

Packer, C., and Pusey, A. E. 1983. Adaptations of female lions to infanticide by incoming males. *American Naturalist* 121:91–113.

Penzhorn, B. L. 1979. Social organization of the Cape Mountain zebra, *Equus zebra zebra*, in the Mountain Zebra National Park. *Koedoe* 22:115–56.

Ralls, K., Brownell, R. L., Jr., and Ballou, J. 1980. Differential mortality by sex and age in mammals with specific reference to the sperm whale. *Report of the International Whaling Commission, Special Issue* 2:223–43.

Reiter, J., Panken, K. J., and LeBouef, B. J. 1981. Female competition and reproductive success in northern elephant seals. *Animal Behaviour* 29:670–87.

Robinette, W. L. 1966. Mule deer home range and dispersal in Utah. *Journal of Wildlife Management* 30:335–49.

Rubenstein, D. I. 1981. Behavioural ecology of island feral horses. *Equine Veterinary Journal* 13:27–34.

Smith, D. R. 1954. *The bighorn sheep of Idaho.* Idaho Department of Fish and Game Bulletin No. 1. Boise, Idaho.

Tilson, R. L., and Tilson, J. 1986. Adult mortality and population turnover in a monogamous antelope, *Madoqua kirki*, in Namibia. *Journal of Mammalogy* 67: 610–13.

Wittenberger, J. F., and Tilson, R. L. 1980. The evolution of monogamy: Hypotheses and evidence. *Annual Review of Ecology and Systematics* 11:197–232.

# 4. Age, Season, Distance, Direction, and Social Aspects of Wolf Dispersal from a Minnesota Pack

*L. David Mech*

Although theories about mammalian dispersal have recently evoked considerable scientific interest (Bekoff 1977; Frame et al. 1979; Gaines and McClenaghan 1980; Greenwood 1980; Dobson 1982; Bekoff, Daniels, and Gittleman 1984; Moore and Ali 1984), little detail is known about dispersal in most carnivores (Bekoff and Wells 1985). The literature on dispersal in wolves (*Canis lupus*), one of the most studied carnivores, consists primarily of scattered records of long movements by a few individuals (Pimlott, Shannon, and Kolenosky 1969; Mech and Frenzel 1971; Van Camp and Gluckie 1979; Berg and Kuehn 1982; Stephenson and James 1982; Van Ballenberghe 1983; Fritts 1983). In addition some wolves have been followed as they dispersed from natal territories and founded their own packs (Fritts and Mech 1981; Peterson, Woolington, and Bailey 1984).

The available reports indicate that considerable variation occurs in many aspects of wolf dispersal and that more detailed information about dispersal is necessary before generalizations can be made about this phenomenon. This chapter provides more such information and attempts to draw some preliminary generalizations. It describes the founding of the Perch Lake wolf pack (PLP) in northeastern Minnesota, the breeding history of pack members, and the dispersal and pack tenure of up to 18 offspring from 1973 through 1984.

## STUDY AREA

The PLP was part of a wolf population in the central Superior National Forest (SNF) (48° N latitude, 92° W longitude). Wolf packs inhabit territories in every direction from the PLP, although a J-shaped lake about 0.5 km wide—Birch Lake—separates the PLP from its neighbors to the east and south. The nearest frontier of the wolf population is at least 125 km to the southwest, and the only extensive physical impediment to wolf travel

55

is the Lake Superior shore running northeast-southwest, 75 km to the southeast (fig. 4.1).

The study area is generally flat, with numerous small rocky ridges, hills, swamps, lakes, and rivers, and the vegetation is transitional between the boreal forest and deciduous types. Summers are hot (up to 36° C) and winters cold (to − 50° C); snowfall averages over 150 cm during 5 months, and rain about 60 cm. Wolves prey primarily on white-tailed deer (*Odocoileus virginianus*) but supplement their diet with beavers (*Castor canadensis*), moose (*Alces alces*), and snowshoe hares (*Lepus americanus*).

The wolf population in the central SNF of which the PLP is part has been studied continually since 1966 (Mech 1979). Generally the central-SNF population has been declining since about 1970 because of a decline in white-tailed deer beginning in 1968 (Mech 1977b, Mech and Karns 1977). The PLP itself, however, has flourished because of its reliance upon a garbage landfill (Hertel 1984). Wolf hunting and trapping have been illegal in the study area since 1970 but were legal in Ontario to which some wolves dispersed. Illegal killing of wolves in Minnesota persisted throughout the

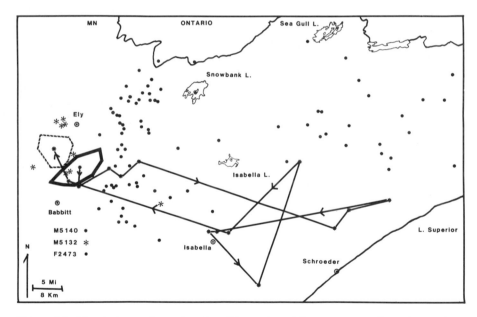

FIGURE 4.1.  The study area in northeastern Minnesota and the movements of certain members of the Perch Lake pack (PLP) of wolves. The black dots indicate where female 2473, one progenitor of the pack, was found as a dispersed lone wolf for a year before settling into the PLP territory (heavy black encircled area). Large asterisks represent locations of dispersed lone male 5132 before he took over the alpha male role from the unknown male progenitor of the pack. Lines and arrows connecting small asterisks represent the dispersal of 1-year-old male 5140 and his eventual settlement (broken encircled area) northwest of his natal territory. (Only selected towns and lakes are shown.)

study, usually from October through January during the general hunting/trapping period.

## METHODS

Wolves were live-trapped, anesthetized, radio tagged, ear tagged, blood sampled, dye marked (1981–83), and aerially radio tracked and observed in the central-SNF population from 1968 through December 1984 (Mech 1974, 1979, and unpub. data). Each wolf generally was located at least weekly year-round and several times per week during winter. Most actual observations were made from October through March, but for part of summers 1981, 1982, and 1983, members of the PLP were observed at the landfill during the night and were identifiable by dye-mark and radio-signal combinations (Hertel 1984). Individual radio collars functioned up to 3 years, with most of them lasting approximately 1 year. Attempts were made to recapture wolves and reradio them as many times as necessary, although generally the recapture rate was low due to the extreme wariness of trap-experienced animals.

Wolves were considered to have dispersed (table 4.1) if (1) radio signals indicated they had left the PLP and territory, or (2) their radio signals disappeared while still in the territory but the animals were reported killed outside the territory.

Study animals considered probable dispersers (table 4.1) were those individuals (1) whose signals disappeared sometime other than during October through January, the period when the animals were most likely to have been killed illegally, and (2) who were never recaptured afterward

TABLE 4.1  Background Information about Perch Lake Pack Wolves

| Wolf No. | Sex | Year Born | Date of Dispersal | Age of Dispersal | Probability of Having Dispersed |
|----------|-----|-----------|-------------------|------------------|--------------------------------|
| 5139 | F | 1974 | Apr 1975 | 1 yr | Definitely |
| 5140 | M | 1974 | Mar 1975 | 1 yr | Definitely |
| 5488 | F | 1977 | Jun 1978 | 1 yr | Definitely |
| 5497 | M | 1977 | > Oct 1978 | > 1 1/2 yr | Possibly |
| 5777 | M | 1977 | > Apr 1978 | > 1 yr | Probably |
| 5778 | F | 1977 | > Sep 1978 | > 1 1/2 yr | Possibly |
| 5941 | M | 1978 | > Mar 1980 | > 2 yr | Probably |
| 6029 | F | 1979 | > Sep 1980 | > 1 1/2 yr | Possibly |
| 6073 | F | 1980 | > Feb 1981 | > 10 mo | Definitely |
| 6075 | F | 1980 | > Feb 1981 | > 10 mo | Probably |
| 6091 | F | 1980 | > Mar 1981 | > 11 mo | Probably |
| 6121 | M | 1980 | Feb 1983 | 33 mo | Definitely |
| 5331 | M | 1981 | > Oct 1982 | > 1 1/2 yr | Definitely |
| 6313 | M | 1981 | Feb 1984 | 33 mo | Definitely |
| 6429 | M | 1981 | > Apr 1983 | > 2 yr | Probably |
| 6433 | F | 1982 | Nov 1983 | 1 1/2 yr | Definitely |
| 6441 | M | 1982 | > May 1983 | > 1 yr | Definitely |
| 6530 | M | 1982 | Apr 1984 | 2 yr | Definitely |

despite considerable effort and recapture of other pack members, and (3) who could not be accounted for in observations of the pack. Probable dispersers were used only for supporting information, and no conclusion was based exclusively on their data.

A third category of animals that provided data relevant to dispersal consisted of PLP offspring remaining with the pack for a known minimum period, followed by disappearance of their signals during the general fall hunting/trapping season. They were considered possible dispersers (table 4.1), and their data were useful for inferences about minimum tenure with the pack.

When possible, dispersing wolves were followed at least weekly until they moved too far from the study area or until their signals were lost. Information from some dispersed animals was obtained via reports from trappers or hunters who had killed the animals in Ontario. (Collars were inscribed with messages requesting the finder to call the author collect and obtain a reward.)

## RESULTS

Some 300 wolves from approximately 25 contiguous packs were captured, radio tagged, and radio tracked. This sample included 25 members of the PLP, whose territory lay toward the western edge of the study population; 3 other members were ear tagged only. Throughout the study PLP members were consistently in the best condition and were the most productive of any of the packs studied (Hertel 1984). Individual members were captured up to seven times and their radios replaced, so a more complete history was obtained for this pack than for any other (Mech and Hertel 1983).

The PLP was founded by adult female 2473, who had traveled nomadically around an area of some 4,117 km² (111 locations) of the study area alone (59 observations) from October 1972 to December 1973. About 20 December 1973 this wolf met an untagged male, and the two set up a territory of approximately 69 km² (71 locations) (fig. 4.1). The pair produced at least four pups in spring 1974. The adult male remained with the pack at least through August 1974. However, apparently he was killed sometime thereafter, probably during November. Two female pups, 5139 and 5176, and male pup 5140 were radio tagged during autumn (table 4.1; fig. 4.2).

Meanwhile, lone male wolf 5132 had been radio tagged some 27 km away on 14 August 1974. He traveled around an area of some 294 km² (11 locations) near the PLP territory (fig. 4.1). On 15 November 1974, 5132 was found in the PLP territory. When next located a week later, he was with female 2473 and the pups; from then on he remained with the pack. When the PLP was observed that winter, there were no extra animals, indicating that the original, unknown alpha male was gone. Two members, presumably wolves 2473 and 5132, were seen copulating on 23 February 1975. Pups 5139 and 5140 dispersed in early spring, and only 5176 remained (fig. 4.2). At least three pups were produced in 1975; one of them was radio tagged, but all had disappeared by winter.

In spring 1976, 2-year-old female 5176 apparently usurped the breeding role from her mother. She remained with 5132 (her stepfather) much more often and attended the den more than 2473. In addition, 5176's nipple sizes during her recapture on 26 June 1976 confirmed that she had produced pups (Meier, Mech, and Seal, n.d.). However, her nipple sizes, her premature termination of den attendance, and the lack of pups in the PLP the following winter indicated that her pups had died before summer. Meanwhile, 5176's mother, 2473, was found away from both 5176 and 5132 during all of the six times she was located between 23 April and 17 August 1976. Because her radio failed, her fate remains unknown. A third wolf, which could have been 2473, was observed with 5132 and 5176 on 6 October 1976, but from 13 December 1976 through 4 April 1977, no wolves other than 5132 and 5176 were seen in this pack during 35 observations. In spring 1977 they bore a litter of pups. From then through 1984, these two wolves produced all the offspring of the pack (Mech and Hertel 1983; Hertel 1984).

## Dispersal Case Histories

Ten radio-tagged members of the PLP were known to have dispersed, and 5 others were considered probable dispersers. The dispersal histories of each known disperser are detailed here.

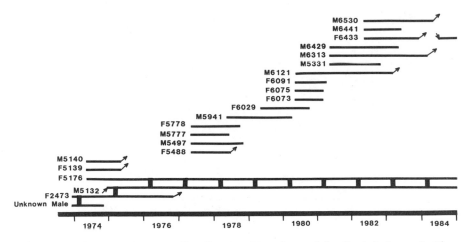

FIGURE 4.2. History and tenure of radio-tagged members of the Perch Lake pack. Lines represent known minimum tenure of each individual with the pack; arrows show dates of known dispersal from the pack or ingress into the pack. Crossbars indicate pair bonds. Additional offspring were produced but were not radio tagged.

*Female 5139.* This wolf was born in April 1974 and dispersed in April 1975. She left the PLP at a younger age than any other wolf we have followed in any pack: she was only found with 2473 and 2473's new mate 5132 the first time we located the two animals together on 22 November 1974. After that 5139 was always located away from the PLP (44 locations), although she remained within its territory. Both littermates 5140 and 5176 remained with the pack during the same period.

Then between 2 and 8 April 1975, 5139 left the PLP territory at the age of 11 months and traveled nomadically over 68 km$^2$ (54 locations) immediately southeast of the territory (fig. 4.3). The wolf remained there for more than a year. In winter 1975–76 she was observed with another wolf during 30 of 33 sightings, and in April and May, the usual denning

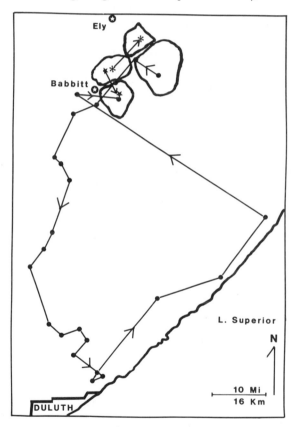

FIGURE 4.3. Directional dispersal of male 2489 (black dots) from the Harris Lake pack territory (east encircled area) to a new territory (south encircled area) just south of the Perch Lake pack (PLP) territory. Dispersed female 5139 (black star) from the PLP territory (west encircled area) mated with 2489 to form the new territory. Asterisk shows dispersal of males 6121 and 6313 one year apart from the PLP territory to the Birch Lake pack territory. (Only selected towns are shown.)

period, her movements were localized as though she were attending a den. When recaptured on 29 July 1976, her nipple measurements (Meier, Mech, and Seal, n.d.) indicated that she had produced pups.

Wolf 5139's mate was not known, but there were strong indications that it was male wolf 2489. This animal, who had been studied earlier, was recaptured on 28 July 1976 in 5139's territory, wearing an expired radio collar. During seven of eight times we located him after his recapture until 29 September 1976, he was located in or near wolf 5139's territory (fig. 4.3), although we only found the two wolves together once during that time. They both then left the territory, probably having lost their pups. Female 5139 returned to her natal territory and was found once near her stepfather 5132 on 29 September 1976 at the landfill. Her supposed mate 2489 rejoined 5139 in 5139's natal territory on 27 October, after which 2489's signal disappeared. Female 5139 then returned to her own territory within a few weeks where she was alone all 16 times she was observed until 10 March 1977 when her signal was lost. On 5 November 1977, wolf 5139 was killed by a human 75 km WNW of her natal territory.

Because male 2489 probably was 5139's mate, his origin and dispersal are of interest. The wolf was born in spring 1972 to the Harris Lake pack (Mech 1977a), whose territory lay a kilometer east across Birch Lake (fig. 4.3). He was last observed with his natal pack on 25 January 1974. During his next eight locations he remained within his pack's territory but away from the pack. On 8 February 1974, when 21 months old, 2489 left his natal territory and dispersed southerly toward Duluth some 50 km away. His further southward movement apparently was deterred by the city, and he returned northward where his signal was lost after 5 April 1974 (fig. 4.3). On 15 November 1974 the animal was captured accidentally by a trapper and released just west of where 5139 dispersed to the next spring. He was not observed again until his recapture on 28 July 1976.

*Male 5140.* Born in April 1974, this wolf remained with the PLP until 17 March 1975 when he was found about 150 m away. The next day and during his next two locations he was alone but in the territory. On 25 March 1975 when 10 months old, he began a series of moves southward and eastward that took him at least 80 km from the territory (fig. 4.1). Some 10 weeks after dispersing, wolf 5140 returned to the PLP territory and was even found once within about 400 meters of the PLP on 25 July 1975. He then settled in an area of 70 km² (15 locations) just northwest of the PLP territory (fig. 4.1) until at least 21 November 1975, after which his signal was lost.

*Male 5331.* Born in spring 1981, male 5331 remained with the PLP until at least 22 October 1982 when his signal was last heard. Despite intensive trapping and observations during summer 1983 through a night-vision scope at the garbage landfill the pack frequented (Hertel 1984), this animal was not found. On 7 May 1984 his collar and remains were discovered in Ontario some 290 km northeast of the PLP and some 34 km from where wolf 6073 had been killed (fig. 4.4). Evidence indicated that the animal probably had perished by summer or fall 1983.

*Female 5488.* Wolf 5488 was born to the PLP in spring 1977 and remained with the pack until 26 May 1978. On 3 June she was away from the pack and at the southeast end of the territory within a kilometer of Birch Lake. Two days later she was back in the center of the territory and in 4 more days was 6.4 km northwest of the territory. A week later she was back at the east edge of the territory within a kilometer of the lake, and in 3 more days was 3 km west of the territory. She then headed north and later veered westward to a point some 53 km west of her natal territory on 23 July 1978 (fig. 4.4). Her signal was then lost.

*Female 6073.* Female 6073 was born in spring 1980, and her signal was last heard in the PLP territory on 24 February 1981. She was killed in December 1983 by a trapper in Ontario, about 290 km northeast of the PLP (fig. 4.4).

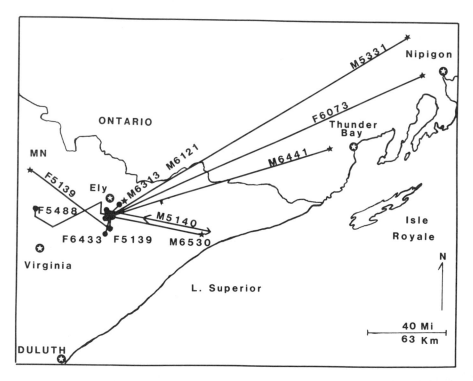

Figure 4.4. Composite map of 10 known dispersers from the Perch Lake pack. Letters preceding wolf identification numbers represent sex. Black stars indicate known deaths, and black dots show locations where individuals were last located. (Only selected towns are shown.)

The trapper reported that the wolf had been traveling with a pack, a strong indication she had mated and helped found the pack.

*Male 6121.* Born in spring 1980, wolf 6121 remained with the PLP until 11 March 1982. For the next 20 months he alternated between living in his natal pack territory and the neighboring (Birch Lake pack—BLP) territory (fig. 4.3). There lone female 6119 had recently been estranged by her father, wolf 5809. (Male 5809 was at least 6 years old, had been found with his yearling daughter 6119 only twice out of 58 locations from 4 July 1981 through 28 January 1982, 17 of 25 times from 30 January through 22 March 1982, and none of 28 times from 27 March to 9 November 1982, when he was last heard from. No other pack members inhabited this territory during that period.)

From 19 March through 10 June 1982, wolf 6121 spent about equal time in both PLP and BLP territories and associated with members of his natal pack and at least twice with 6119 (fig. 4.5). Throughout summer and most of fall, he remained with his natal pack, but between 29 November and 13 December he returned to the BLP territory and joined 6119 through 5 January 1983. He then rejoined his natal pack through 10 February but returned to 6119 and spent most of his time with her from 16 February through 22 November when he was killed illegally. Because 6119's radio failed in May, we do not know whether the pair produced pups, although there is indirect evidence that they did.

*Male 6313.* This wolf was a littermate of wolf 5331, born in spring 1981. However 6313 remained with the alpha pair of the PLP for some 33 months through 1 February 1984. On 9 February he was found in the BLP territory

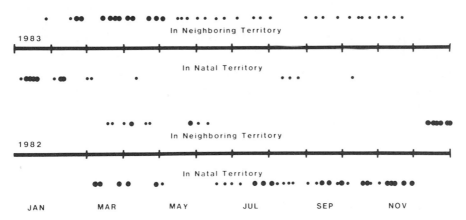

FIGURE 4.5. History of dispersing male 6121 as he intermittently shifted between his natal pack and territory to neighboring female 6119 and her territory. Small dots represent radio locations only. Large dots indicate that 6121 was known to be with other wolves, while small dots show only his known presence in either territory.

(fig. 4.3), where he remained at least through September 1984 when his radio expired. On five occasions when he was observed, he was seen with another wolf, and twice it was confirmed that his companion was wearing a defunct radio collar. Presumably this animal was 6119, the wolf that 6313's older brother 6121 had paired with the year before. Wolf 6313 and mate produced at least four pups.

*Female 6433.* Wolf 6433 was born in spring 1982 and remained with the PLP until 15 November 1983. The only PLP wolf known to have dispersed during fall, she moved to an area southeast of the PLP territory, including part of the region where 5139 had settled 8 years earlier (fig. 4.3). From 11 January 1984 on, 6433 usually associated with another wolf eventually determined to be wolf 6496, a nomadic, adult male we had followed intermittently since 4 July 1983. The male had been found once with members of the Sawbill pack whose territory was about 11 km east of 6433's, although his relationship to the Sawbill pack remains unknown. Female 6433 settled in an area of a few km² for about 3 weeks during the 1984 denning season, but she then resumed more widespread movements, indicating that she probably had not produced pups or had lost them if she did. By 25 June 1984, 6433 had split from her mate and returned to her natal territory. She remained there throughout November 1984 and was usually found with her mother. She dispersed southward for a few kilometers again in December 1984, after which she remained alone within the territory or just outside its south and west edges through 12 February 1986, when she was killed (Mech and Seal 1987).

*Male 6441.* This wolf was born in spring 1982, and his signal was last heard from on 3 May 1983 in the PLP territory. Despite intensive trapping and night observations referred to earlier, this animal was not found in summer 1983. On 27 December 1983 he was killed by a trapper in Ontario, some 191 km to the northeast (fig. 4.4). Tracks indicated to the trapper that the wolf had been alone.

*Male 6530.* Male 6530, born in spring 1980, remained with the PLP through 28 March 1984, but on 5 April when almost 2 years old, he was found 11 km east of the territory. He then became nomadic over an area of about 1,867 km² (135 locations), ranging up to 83 km northeast of his natal territory (fig. 4.4) and including a 3-week return to his natal territory in March and April 1985. On 8 December 1985 he died of natural causes (Thiel et al. 1987).

## Age and Season of Dispersal

Two of three female dispersers, whose ages at dispersal were known, left in April and early June when 11 to 13 months old. In addition, wolf 6073's signal disappeared in February when she was 10 months old, and she was later found to have dispersed. Two other female probable dispersers disappeared in February and March at 10 and 11 months of age. Only wolf 6433 dispersed at about 18 months of age in November; two other females remained with the PLP at least until September of their second year (table 4.1).

The males dispersed from 1 February through early April when 10 to 33 months old, based on six known dispersers and two probable dispersers; two other males remained with the pack at least until October of their second year (table 4.1).

## Dispersal Distance and Direction

Dispersal distances and directions each fell into two main classes (fig. 4.4). Two members of each sex appropriated territories adjacent to their natal territory, with the two males dispersing to the northeast and the two females to the south and southeast. At the other extreme two males and one female dispersed northeastward over 190 km. Between these two extremes was male 5140, who dispersed eastward at least some 80 km, returned to his natal territory, and then dispersed to an area immediately northwest of his natal territory; female 5488, who dispersed westward a minimum of 53 km; and male 6530, who became nomadic in an area within 70 km east-northeast of his natal territory.

## Social Aspects of Wolf Dispersal

A few of the known dispersers, both male and female, first separated from other pack members but remained in the territory for 1 to 2 weeks before dispersing; one female inhabited the territory as a loner for several months before leaving. Other dispersers could have separated from the pack for several days preceding dispersal, but the frequency of radio tracking might have been insufficient to show it. On the other hand, male 6121 maintained his social bonds with his natal pack for a year while he intermittently courted a neighboring female.

In all four cases where male or female dispersers moved immediately to adjacent territories and their radios continued to function, the animals found mates by the first breeding season. Only one or two of the four produced pups that survived through their first autumn, however. Two of the pairs broke up after failure of pup production or survival, and in each case the PLP disperser returned to the natal territory. Wolf 6433 reintegrated into the natal pack, but 5139 dispersed again to her own territory within a few weeks.

At least one of the long-distance dispersers apparently also had mated and produced pups within her first breeding season away from her natal pack. None of the five remaining known dispersers were radio tracked through their first post dispersal breeding season.

Although the data are sparse, the age of dispersal seems to have increased with time in the PLP (fig. 4.2). Of the seven known dispersers of known dispersal age, the three born before 1978 left at about 1 year of age, whereas three of the four born after 1979 remained with the pack for at least 2 years; the fourth dispersed at 18 months of age and returned to the pack when 2 years old (table 4.1). A male born in 1980 and one in 1981 remained with the pack until 33 months old.

## DISCUSSION

This study emphasizes the extreme degree of variability intrinsic in the dispersal behavior of wolves, which in turn reflects the adaptability of the species' ability to propagate. All wolves produced in the PLP after 1977 were siblings, yet their propagation strategies seemed to cover most of the possibilities. Although data from the packs adjacent to the PLP were not analyzed thoroughly, indications are that the dispersal behavior of members of those packs generally fit within the variability noted in the PLP (Mech, unpub. data) and within most of the published information on wolf dispersal. Even male 6121's maintenance of bonds with his natal pack while he intermittently courted a neighboring female is similar to an observation from Alaska (Peterson, Woolington, and Bailey 1984).

About the only clear difference between data from the PLP and from previous reports is that the PLP data showed no tendency for males to disperse farther than females. However, males predominated in a sample of wolves colonizing Finland (Pulliainen 1982), implying either that a disproportionate number of males were produced or survived in the adjacent USSR or that males disperse farther; males also predominate in the reports of the longest dispersing wolves in North America (Mech and Frenzel 1971; Van Camp and Gluckie 1979; Fritts and Mech 1981; Berg and Kuehn 1982; Stephenson and James 1982; Fritts 1983; Peterson, Woolington, and Bailey 1984).

It is clear from the present study that the genetic basis for a wide variety of dispersal strategies can reside in as few as two wolves. It also appears that even single individuals have enough flexibility to utilize various strategies in order to adapt to environmental opportunities. Each individual strives to seek a mate and enough resources to propagate, but the particular strategy employed at a given time seems to depend on the opportunity available.

The minimal, although not sufficient, physiological state necessary for dispersal in wolves seems to be puberty, as in most other species (Howard 1960) except bears (*Ursus americanus*; Rogers, chap. 5, this vol.). Wolves can reach puberty as early as 10 months of age and produce offspring when 1 year old (Medjo and Mech 1976; Zimen 1976). However they probably do not attain full sexual and physical maturity until about 5 years of age (Packard, Mech, and Seal 1983). At 8 to 10 months old, tame, free-ranging pups show considerable independence from the pack (Zimen 1982). Social interactions intensify as the breeding season (January and February) approaches (Rabb, Woolpy, and Ginsburg 1967), and frequency of aggression elevates from November through June (Zimen 1975), probably reflecting similar changes in secretion of reproductive hormones (Seal et al. 1979, and unpub. data).

The specific behavior that results in dispersal of wolves from a pack has only been studied in captivity, but miscellaneous observations from the field tend to support the conclusions from captives. According to Zimen's (1975, 1976) detailed study, dispersers are generally former alphas and maturing, low-ranking individuals, and aggression from other pack members is one of the primary factors forcing dispersal. In the present study, the original alpha female of the PLP, wolf 2473, fits the former category, apparently having been ousted by her 2-year-old daughter 5176. Peterson, Woolington, and Bailey (1984) reported dispersal by an alpha male after the animal was captured during December and became subordinate upon return to the pack; 2 weeks later he dispersed. Breeding within the natal pack after ousting a parent (Packard, Seal, and Mech 1985), while necessarily less common than dispersal, probably is the most advantageous strategy because both a territory and a mate are readily available (Packard and Mech 1983).

The remaining PLP dispersers were all maturing individuals. In four cases, including wolf 2489 from an adjacent pack, separation from the pack within the territory for 1 to 2 weeks preceded dispersal from the territory, supporting the contention that some type of alienation from the pack was involved. However, it should be noted that wolf 6121 maintained bonds with his natal pack while dispersing.

Observations similar to all of these have also been reported from other areas (Murie 1944; Mech 1966; Van Ballenberghe 1983; Peterson, Woolington, and Bailey 1984; Messier 1985). Probably the reason for the apparent diversity in dispersal behavior among various wolves is the complex of factors affecting the probability that an individual will disperse: age, sex, social rank, season; age, sex, social rank of the other pack members; pack size; and food availability (Zimen 1976). In addition, some wolves, probably high-ranking ones that may enjoy positive relations with members of their natal packs, may still take advantage of breeding opportunities that present themselves. Examples are wolf 6121 (fig. 4.5) and an individual described by Peterson, Woolington, and Bailey (1984).

Upon dispersal, a continuum of possibilities for settling, courting, mating, and forming a new pack confronts a wolf. Two needs must first be met—locating a large enough area with sufficient prey not yet claimed by other territorial wolves, and encountering a mate (Rothman and Mech 1979). Depending on the status of the local population, the disperser might find cither or both in one corner of its natal territory, immediately adjacent to its natal territory, at some greater distance, or at the very frontier of the wolf population.

Although only weakly demonstrated in the present study, by female 6433 during her second dispersal attempt, it is not uncommon for maturing wolves to begin concentrating their activity at one corner of the natal territory, especially in an increasing population (Fritts and Mech 1981; Mech, unpub. data). Eventually they may find a mate there, produce pups, and carve out a territory of their own that includes part of their natal pack's original territory—a clear example of "natal philopatry" (Waser and Jones 1983).

The next step in the progression of possibilities is finding a vacancy next to the natal territory. If a suitable area exists there, chances are good that potential dispersers learn that quickly, for the packs cover their territories frequently, and each advertises regularly via scent marking (Peters and Mech 1975; Rothman and Mech 1979) and howling (Harrington and Mech 1979). Thus individuals such as wolves 6121 and 6313 could readily be aware of neighboring vacancies and probably even of potential mates there. When, through the social shuffling that accompanies the annual courtship (Schenkel 1947; Rabb, Woolpy, and Ginsburg 1967; Zimen 1975), an individual learns that its chances of breeding are low within its natal pack, it can immediately disperse to the adjacent vacancy and possibly mate.

Wolf 6121 attempted this strategy near the end of the 1982 breeding season but continued to maintain ties with his natal pack. For some reason, the pairing did not work, so 6121 returned to the pack in June (fig. 4.5), soon after the reproductive hormones had diminished (Seal et al. 1979, unpub. data). In December, when both testosterone and estradiol levels are increasing (Seal et al. 1979, unpub. data), 6121 dispersed to the same adjacent territory and again began pairing with 6119. The breeding season in this area covers at least 28 January through 2 March (Mech and Knick 1978). Although 6121 also returned to the PLP intermittently in January and early February (to check the possibility of breeding with his mother?), he rejoined 6119 by 21 February and remained with her primarily from then on (fig. 4.5). Wolf 6121 was killed in fall 1983, but between 1 and 9 February 1984 his 33-month-old brother 6313 suddenly dispersed to the same territory, paired with a wolf that most probably was 6119, and formed a new pack.

However, if vacancies are unavailable within or next to a natal territory, any vacancies one territory or more beyond probably would be difficult to detect without temporary or permanent extraterritorial moves. Wolves make such temporary extraterritorial travels (Van Ballenberghe 1983; Peterson, Woolington, and Bailey 1984), but they are much more common in food-stressed populations (Messier 1985).

Permanent extraterritorial travels, or dispersal, can also vary considerably in direction, distance, and pattern, as indicated by this study and those cited previously. If a dispersed wolf's goal is to locate a vacancy and a mate, a standard search pattern involving a spiraling outward from the natal territory would logically be the quickest method of finding either. No such dispersal pattern has been described in wolves. Instead, wolves show two main types of long-distance dispersal—directional dispersal and nomadism.

Directional dispersal involves a series of more or less straight movements that carry an animal in one general direction. Although this type of movement was first described for a dispersed wolf (Mech and Frenzel 1971), it has been documented best for red foxes (*Vulpes vulpes*, Storm et al. 1976). Wolf 2489 in the present study typifies such a move, even though he doubled back, once having encountered Duluth (fig. 4.2). Van Camp and Gluckie (1979), Fritts and Mech (1981), Berg and Kuehn (1982), Stephenson and James (1982), Fritts (1983), and Peterson, Woolington, and Bailey (1984) also documented such moves. The advantage of directional dispersal over other types of movements in locating vacancies and mates is not apparent. However, such dispersal would most quickly carry an individual to, or beyond, the frontier of the population's range. There it might help found a new population and thus gain a disproportionate genetic representation.

Nomadism is also a well-known behavior for dispersing wolves (Mech and Frenzel 1971; Fritts and Mech 1981; Van Ballenberghe 1983), with individuals sometimes wandering over areas as large as 4,000 km$^2$ (fig. 4.1). Vacancies over a large region are visited frequently by these nomadic wolves (Mech and Frenzel 1971; Van Ballenberghe, Erickson, and Byman 1975; Rothman and Mech 1979; Fritts and Mech 1981; Peterson, Woolington, and Bailey 1984), presumably checking for suitable mates. In some regions, a mate may be found first, and the pair wanders over large areas searching for vacancies (Fritts and Mech 1981).

With all these strategies available, wolves find vacancies and mates reasonably quickly as documented by the previously cited studies and by the present investigation. Even if attempted colonization fails, some wolves can return to their natal pack and try again later (wolf 6433). Or if directional dispersal is blocked (wolves 2489 and 5140 in the present study, and wolf 5173 of Fritts and Mech 1981), wolves can resort to settling close to their natal territory. Such flexibility implies that, at least in wolves, the expression of any innate tendency for dispersal (Howard 1960) can be modified by circumstances adverse to that strategy. Given this variability, population voids tend to fill quickly, and from a population perspective, the wolf's dispersal strategy is highly effective in allowing the species to fully exploit available resources.

From a genetic standpoint, the wolf's dispersal strategy seems to insure both a high level of inbreeding as well as extreme outcrossing. Although incest has not been well documented in wild wolves, it is known in captives (Packard, Seal, and Mech 1985). In the present study, wolf 5176's breeding with her stepfather 5132 implies that there is no immediate social taboo in mating with the pack leader or with the mate of one's mother; the possibility remains, however, that kin recognition (reviewed in Moore and Ali 1984), if it exists in wolves, might have prevented a similar type of mating with a natural father.

Weak evidence against incestuous matings is found in the present study when adult wolf 5809 dispersed instead of breeding with his daughter 6119.

However, even 6121 did not pair-bond with 6119 the year that 5809 left her, so perhaps 6119 was not physiologically ready to mate. The apparent tendency of PLP males to disperse east to northeast and females southeast to west might be additional evidence of incest avoidance. Nevertheless Peterson, Woolington, and Bailey (1984) presented strong circumstantial evidence of a female-grandfather mating in Alaskan wolves.

Regardless of whether incest occurs in wild wolves, the fact that offspring "biders" (Packard and Mech 1983) may remain in a pack and breed (Fritts and Mech 1981; Mech, unpub. data), and the frequency with which dispersers move to adjacent territories (e.g., wolves 2489, 5139, 6121, 6313, and 6433), strongly implies that close, although not necessarily incestuous, inbreeding is common in wolves. At least one computer simulation of wolf genetics concludes that wolves must be strongly inbred (Woolpy and Eckstrand 1979), and empirical testing of the effects of incestuous mating in captive wolves has shown few birth defects compared with the same in coyotes (*Canis latrans*) and dingoes (*Canis* sp.) (Woolpy and Eckstrand 1979; Woolpy, pers. comm.). Furthermore, the Isle Royale wolves that have prospered for 35 years (Mech 1966; Peterson and Page 1983), with very little possible outcrossing, demonstrate that a natural wolf population can withstand deleterious inbreeding effects, at least in the short term. These conclusions tend to support Shields' (1982, 1983) theory of an inbreeding optimum and Templeton's (chap 17, this vol.) contention that little inbreeding depression should occur in populations that are historically inbred.

On the other hand, the occurrence of distant, directional dispersal in wolves (fig. 4.4), which has reached as far as 886 km (Fritts 1983), also insures a certain degree of extreme outcrossing in wolf populations. Thus, one ready explanation for directional dispersal (rather than a spiral search pattern covering the same travel distance for example) is that it promotes outcrossing. The adaptive advantage of directional dispersal, in fact, would be unclear unless outcrossing resulted in at least some increased chance of offspring survival. None of the alternative explanations posed by Moore and Ali (1984) for dispersal seem to apply to distant, directionally dispersing wolves.

One might argue that directional dispersal would carry a disperser as fast as possible to a new population where there might be better resources. However, two pieces of evidence tend to counter that argument: (1) wolves from the PLP were in better condition than most members of the population because of their reliance on a garbage landfill, yet some dispersed directionally, and (2) members of the surrounding population were dispersing directionally even when the prey base was much better and the wolf population in much better condition than at present (Mech and Frenzel 1971).

Whatever the theoretical problems, the conclusion can be made from the present study that the best strategy of an alpha wolf pair under good conditions seems to be to disperse some of its propagules nearby and some to new populations, much as Howard (1960) found for many other vertebrates. This strategy implies that the population will be constantly balancing the effects of inbreeding with those of outcrossing (cf. Theberge 1983; Shields 1983), and that whether greater homozygosity or heterozygosity is favored under any given conditions, the potential is present for the population to shift quickly either way.

## ACKNOWLEDGMENTS

This study was supported by the U.S. Fish and Wildlife Service, the U.S.D.A. North Central Forest Experiment Station, and a generous individual who wishes to remain anonymous. I am also grateful to H. Hertel, M. Nelson, J. Renneberg, and numerous technicians and bush pilots for field assistance, and to B. Wall and G. Koistinen for reporting radioed wolves killed in Ontario.

## REFERENCES

Bekoff, M. 1977. Mammalian dispersal and the ontogeny of individual behavioral phenotypes. *American Naturalist* 111:715–32.

Bekoff, M., T. J. Daniels, and J. L. Gittleman. 1984. Life history patterns and the comparative social ecology of carnivores. *Annual Review of Ecology and Systematics* 15:191–232.

Bekoff, M., and M. C. Wells. 1985. Social ecology and behavior of coyotes. *Advances in the Study of Behavior* 16:251–338.

Berg, W. E., and D. W. Kuehn. 1982. Ecology of wolves in north-central Minnesota. In *Wolves of the world*, ed. F. H. Harrington and P. C. Paquet, 4–11. Park Ridge, N.J.: Noyes Publishing.

Dobson, F. S. 1982. Competition for mates and predominant juvenile male dispersal in mammals. *Animal Behaviour* 30:1183–92.

Frame, L. H., J. R. Malcolm, G. W. Frame, and H. van Lawick. 1979. Social organization of African wild dogs (*Lycaon pictus*) on the Serengeti Plains, Tanzania, 1967–1978. *Zeitschrift für Tierpsychologie* 50:225–49.

Fritts, S. H. 1983. Record dispersal of a wolf from Minnesota. *Journal of Mammalogy* 64(1):166–67.

Fritts, S. H., and L. D. Mech. 1981. Dynamics, movements, and feeding ecology of a newly protected wolf population in northwestern Minnesota. *Wildlife Monographs* 80:1–79.

Gaines, M. S., and L. R. McClenaghan, Jr. 1980. Dispersal in small mammals. *Annual Review of Ecology and Systematics* 11:163–96.

Greenwood, P. J. 1980. Mating systems, philopatry and dispersal in birds and mammals. *Animal Behaviour* 28:1140–62.

Harrington, F. H., and L. D. Mech. 1979. Wolf howling and its role in territory maintenance. *Behaviour* 68:207–49.

Hertel, H. H. 1984. The role of social niche in juvenile wolves: Puppyhood to independence. Master's thesis, University of Minnesota.

Howard, W. E. 1960. Innate and environmental dispersal of individual vertebrates. *American Midland Naturalist* 63:152–61.

Mech, L. D. 1966. *The wolves of Isle Royale.* National Park Fauna Series No. 7. Washington, D.C.: U.S. Government Printing Office.

———. 1974. Current techniques in the study of elusive wilderness carnivores. *Proceedings of the Eleventh International Congress of Game Biologists,* 315–22.

———. 1977a. Population trend and winter deer consumption in a Minnesota wolf pack. In *Proceedings of the 1975 Predator Symposium,* ed. R. L. Philips and C. Jonkel, 55–83. Montana Forest and Conservation Experimental Station, Missoula, Mont.

———. 1977b. Productivity, mortality and population trends of wolves from northeastern Minnesota. *Journal of Mammalogy* 58:559–74.

———. 1979. Making the most of radio-tracking. In *A handbook on biotelemetry and radio-tracking,* ed. C. J. Amlaner, Jr., and D. W. MacDonald, 85–95. Oxford: Pergamon Press.

Mech, L. D., and L. D. Frenzel, Jr. 1971. Ecological studies of the timber wolf in northeastern Minnesota. U.S.D.A. Forest Service Research Paper NC-52. North Central Forest Experiment Station, St. Paul, Minn.

Mech, L. D., and H. H. Hertel. 1983. An eight-year demography of a Minnesota wolf pack. *Acta Zoologica Fennica* 174:249–50.

Mech, L. D., and P. D. Karns. 1977. Role of the wolf in a deer decline in the Superior National Forest. USDA Forest Service Research Report NC-148. North Central Forest Experiment Station, St. Paul, Minn.

Mech, L. D., and S. T. Knick. 1978. Sleeping distances in wolf pairs in relation to breeding season. *Behavioral Biology* 23:521–25.

Mech, L. D., and U. S. Seal. 1987. Premature reproductive activity in wild wolves. *Journal of Mammalogy* 65 (in press).

Medjo, D. C., and L. D. Mech. 1976. Reproductive activity in nine-and-ten-month-old wolves. *Journal of Mammalogy* 54:406–8.

Meier, T. J., L. D. Mech, and U. S. Seal. N.d. Wolf nipple measurements as indices of age and breeding status. Typescript.

Messier, F. 1985. Solitary living and extraterritorial movements of wolves in relation to social status and prey abundance. *Canadian Journal of Zoology* 63:239–45.

Moore, J., and R. Ali. 1984. Are dispersal and inbreeding avoidance related? *Animal Behaviour* 32:94–112.

Murie, A. 1944. *The wolves of Mount McKinley.* National Park Fauna Series No. 5. Washington, D.C.: U.S. Government Printing Office.

Packard, J. M., and L. D. Mech. 1983. Population regulation in wolves. In *Symposium on Natural Regulation of Wildlife Populations,* ed. F. L. Bunnell, D. S. Eastman, and J. M. Peek, 151–74. Proceedings No. 14. Forest Wildlife and Range Experimental Station, Moscow, Idaho.

Packard, J. M., L. D. Mech, and U. S. Seal. 1983. Social influences on reproduction in wolves. In *Wolves in Canada and Alaska,* ed. L. N. Carbyn, 78–85. Ottawa: Canadian Wildlife Service Report Series, 45.

Packard, J. M., U. S. Seal, and L. D. Mech. 1985. Causes of reproductive failure in two family groups of wolves (*Canis lupus*). *Zeitschrift für Tierpsychologie* 68:24–40.

Peters, R., and L. D. Mech. 1975. Scent-marking in wolves: A field study. *American Scientist* 63:628–37.

Peterson, R. O., and R. E. Page. 1983. Wolf-moose fluctuations at Isle Royale National Park, Michigan, U.S.A. *Acta Zoologica Fennica* 174:251–53.

Peterson, R. O., J. D. Woolington, and T. N. Bailey. 1984. Wolves of the Kenai Peninsula, Alaska. *Wildlife Monographs* 88:1–52.

Pimlott, D. H., J. A. Shannon, and G. B. Kolenosky. 1969. The ecology of the timber wolf in Algonquin Park. Ontario Department of Lands and Forests Research Paper (Wildlife), no. 87.

Pulliainen, E. 1982. Behavior and structure of an expanding wolf population in Karelia, northern Europe. In *Wolves of the world*, ed. F. H. Harrington and P. C. Paquet, 134–45. Park Ridge, N.J.: Noyes Publishing.

Rabb, G. B., J. H. Woolpy, and B. E. Ginsburg. 1967. Social relationships in a group of captive wolves. *American Zoologist* 7:305–11.

Rothman, R. J., and L. D. Mech. 1979. Scent-marking in lone wolves and newly formed pairs. *Animal Behaviour* 27:750–60.

Schenkel, R. 1947. Expression studies of wolves. *Behaviour* 1:81–129.

Seal, U. S., E. D. Plotka, J. M. Packard, and L. D. Mech. 1979. Endocrine correlates of reproduction in the wolf, pt. 1: Serum progesterone, estradiol and LH during the estrous cycle. *Biology of Reproduction* 21:1057–66.

Shields, W. M. 1982. *Philopatry, inbreeding, and the evolution of sex.* Albany: State University of New York Press.

———. 1983. Genetic considerations in the management of the wolf and other large vertebrates: An alternative view. In *Wolves in Canada and Alaska: Their status, biology and management,* ed. L. N. Carbyn, 90–92. Ottawa: Canadian Wildlife Service Report Series, 45.

Stephenson, R. O., and D. James. 1982. Wolf movements and food habits in northwest Alaska. In *Wolves of the world,* ed. F. H. Harrington and P. C. Paquet, 26–42. Park Ridge, N.J.: Noyes Publishing.

Storm, G. L., R. D. Andrews, R. L. Phillips, R. A. Bishop, D. B. Siniff, and J. R. Tester 1976. Morphology, reproduction, dispersal, and mortality of midwestern red fox populations. *Wildlife Monographs* 49:1–82.

Theberge, J. B. 1983. Considerations in wolf management related to genetic variability and adaptive change. In *Wolves in Canada and Alaska: Their status, biology and management,* ed. L. N. Carbyn, 86–89. Ottawa: Canadian Wildlife Service Report Series, 45.

Thiel, R. P., L. D. Mech, G. R. Ruth, J. R. Archer, and L. Kaufman. 1987. Blastomycosis in wild wolves. *Journal of Wildlife Diseases* 23 (in press).

Van Ballenberghe, V. 1983. Extraterritorial movements and dispersal of wolves in southcentral Alaska. *Journal of Mammalogy* 64:168–71.

Van Ballenberghe, V., A. W. Erickson, and D. Byman. 1975. Ecology of the timber wolf in northeastern Minnesota. *Wildlife Monograph* 43:1–43.

Van Camp, J., and R. Gluckie. 1979. A record long-distance move by a wolf (*Canis lupus*). *Journal of Mammalogy* 60:236.

Waser, P. M., and W. T. Jones. 1983. Natal philopatry and solitary mammals. *Quarterly Review of Biology* 58:355–90.

Woolpy, J. H., and I. Eckstrand. 1979. Wolf pack genetics: A computer simulation with theory. In *The behavior and ecology of wolves,* ed. E. Klinghammer, 206–24. New York: Garland STPM Press.

Zimen, E. 1975. Social dynamics of the wolf pack. In *The wild canids,* ed. M. W. Fox, 336–62. New York: Van Nostrand Reinhold.

———. 1976. On the regulation of pack size in wolves. *Zeitschrift für Tierpsychologie* 40:300–341.

———. 1982. A wolf pack sociogram. In *Wolves of the world,* ed. F. H. Harrington and P. C. Paquet, 282–322. Park Ridge, N.J.: Noyes Publishing.

# 5. Factors Influencing Dispersal in the Black Bear

*Lynn L. Rogers*

Black bear dispersal was studied from 1969 to 1982 as part of a study of social behavior, habitat use, and population dynamics in northeastern Minnesota (Rogers 1987). Objectives were to determine (1) age and sex of dispersers, (2) factors that promote dispersal, and (3) factors that induce dispersing bears to settle. For this study, dispersal is defined as movement from the mother's territory to a nonadjacent, more or less permanent breeding area. This definition differs from that of Shields (chap. 1, this vol.) in that bears that did not disperse to a nonadjacent territory were not considered to have dispersed; these nondispersing bears commonly continued to use part of their natal range, including their birthplace, while their mothers shifted slightly away.

## STUDY AREA AND METHODS

Bears were captured in a 300 km² portion of the Superior National Forest in northeastern Minnesota. The capture area and surrounding region had gently rolling terrain with small rock outcrops. There were few towns, farms, large highways, large waterways, or other physiographic barriers except Lake Superior to hinder dispersal. Forest habitat was nearly continuous for more than 500 km to the north and for more than 150 km to the west. Lake Superior was 30 km to 60 km to the east and south. The human population was low over most of the forested habitat, so human selection against dispersal was probably minimal.

Bears studied for dispersal were born to radio-collared mothers in dens in January and ear tagged as 2-month-old cubs in March, shortly before the families emerged. The cubs remained with their mothers through the next winter and were radio collared in dens as yearlings. Young females were then radio tracked for the next several years as they matured, established territories, and produced cubs of their own. Radio collars were replaced each winter during hibernation. However, radio collars were removed from some males at 2 years of age after it became apparent that the majority dispersed at that age and were likely to move beyond signal range. Dispersal data from males without radio collars were obtained through ear-tag returns from Wisconsin, Ontario, and various parts of Minnesota. Radio-collared individuals were monitored up to 200 km outside the capture area.

## MATING SYSTEM, SOCIAL ORGANIZATION, AND DEMOGRAPHY OF THE STUDY POPULATION

Characteristics of the capture area population are summarized here and detailed elsewhere (Rogers 1976, 1977, 1983, 1987). Social organization depended upon the distribution and abundance of food. In the few places where food was clumped, the bears formed hierarchies. In most places, food was dispersed, and females held territories averaging 3.5 km in diameter, while males used mating ranges averaging 12.25 km in diameter (average length plus average width of mating ranges divided by 2). Each mating range contained 7 to 15 female territories. However, the mating ranges were indefensibly large and overlapped to the extent that no male had exclusive access to any female. Both sexes were observed to be promiscuous. Mating occurred in June or early July. After that, some members of each sex foraged outside their usual ranges but returned for denning. They used approximately the same areas for mating each year. Minor shifts in territory locations are discussed later. Social organization of the bears differed in several respects from that of the more social carnivores discussed in this volume (Rood, chap. 6, and Mech, chap. 4). The differences may stem in part from the fact that bear foods are usually too small and scattered to support group feeding except by mothers and cubs (Rogers 1987).

Males began mating as early as 4.4 years of age (Rogers 1987). Females produced their first litters at 4 to 8 years of age (mode 7 years, average 6.3 years), depending on food supply (Rogers 1987). Sex ratios among cubs, yearlings, and 2-year-olds were even or slightly male biased (Rogers 1977). The proportion of males decreased with age; for bears over 4 years there was 1 male to 3 to 4 females (Rogers 1977). Population density was approximately 1 bear per 4.5 km², including cubs, or 1 adult ($\geq 4$ years old) per 12.37 km² during the years of most intensive study (Rogers 1987). This density is lower than has been reported for most other regions (Lindzey and Meslow 1977b).

## FEMALE DISPERSAL
### Ages and Movements

Of 31 females whose birthplaces and adult ranges were known, only 3 (10%) dispersed. The 3 included 1 of 22 that were radio tracked from the time they left their mothers and 2 of 9 that were repeatedly captured or observed. Dispersal distances were only 3 km for the radio-collared bear and 8 km and 11 km for the other 2. These distances may not have removed the females from the mating ranges of their fathers. Three females dispersed unknown distances and settled in the study area. Four dispersed or immigrated at 3 or 4 years of age. All 4 probably dispersed at maturity because 3 of the 4 produced their first litters the subsequent winter and the fourth was in estrus when captured as an immigrant. No older female dispersed.

### Relationships between Mothers and Independent Daughters

The low dispersal rate among females probably stemmed from benefits of remaining near their mothers' territories. Yearlings of both sexes changed their movement patterns after family breakup and began using small portions of their mothers' territories (Rogers 1987). These yearling ranges then were avoided by the mothers, giving the yearlings more or less exclusive feeding areas (Rogers 1987), as has been reported for various primates (Tilson 1981; Waser and Jones 1983). Two young bears that had exclusive feeding areas gained weight more rapidly than their same-sex siblings that used larger areas but fed in competition with their mothers (Rogers 1987). As the young bears grew, males dispersed and females increased their range size. Nine mothers shifted their territories away from 12 maturing daughters, thereby aiding daughters in obtaining territories. No mother shifted her territory toward a daughter. Of 22 subadult females that were radio tracked, 15 expanded yearling ranges, 6 left their yearling ranges and opportunistically established territories in adjacent areas, and 1 dispersed 3 km and usurped part of the territory of an older female whose weight of 45 kg was only 41% of her peak weight. The tolerance mothers showed daughters and the aggression mothers displayed toward nonkin (Rogers 1987) suggest that the shifting or spacing behavior probably resulted from differential aggression against nonkin when young females began expanding their ranges.

Without parental aid, young females may have had difficulty obtaining space. Two of the 3 immigrants had problems not found among the 31 philopatric females. One of them incurred lacerations on her head and neck as if from fighting. The other did not obtain an exclusive area as a pregnant young adult; she denned and gave birth in the territory of another bear. The next spring she obtained for herself and her cub the smallest exclusive area used by an adult that year. In late July, she permanently abandoned the exclusive area and the study area (Rogers 1977, 1987). The cub's fate after 18 July is unknown.

For territorial mothers, competition with philopatric daughters and the effort required to shift territories could conceivably decrease subsequent reproductive success. However, if mothers' efforts on behalf of their daughters enhance the reproductive success of the daughters by at least twice the amount that the mothers' reproductive success is decreased, those efforts will be favored by natural selection (Hamilton 1964; Wilson 1975). The 2:1 ratio holds because daughters possess half their mothers' genes; therefore, two grandprogeny carry the same amount of an individual's genetic material, on the average, as does one progeny.

## MALE DISPERSAL
### Ages and Movements

In contrast with the largely nondispersing females, all 20 males that were studied dispersed: 13 as 2-year-olds, 5 as 3-year-olds, and 2 as 4-year-olds. Strongly male-biased dispersal is common in species with polygynous mating systems and female defense of resources (Greenwood 1980; Dobson 1982; Waser and Jones 1983).

Two males were radio tracked extensively before they dispersed. Both showed marked increases in travel outside their mothers' territories in the 1 to 5 weeks before leaving permanently. Three that were followed during dispersal included the males that made the longest (219 km) and shortest (13 km) movements. The first moved 145 km in 12 months and then 74 km farther in only 15 days. The fact that he moved more than half as far in 15 days as he did in the previous 12 months suggests that the bears could have dispersed farther than they did. Another bear moved in an essentially straight line for 42 km in the first 5 days of dispersal, moved 47 km farther the next 55 days, then reversed direction and moved 133 km back past his birthplace to settle approximately 48 km from it. The bear that dispersed the shortest distance moved only 13 km to the area he would use for mating as an adult but then roamed more than 40 km away from it while foraging. The main areas he used as a 2-year-old, after dispersing, were reused for at least the next 2 years at approximately the same times each year. The average distance that 18 dispersing males were recovered from their birthplaces was 61 km (median 49 km, range 13 km to 219 km). This distance represented less than 5 mating range diameters (range 1–18 diameters), which probably was not far enough to disrupt genetic adaptations to regional conditions (Shields 1982, 1983; Rogers 1987).

### Food Shortage and Aggression

Dispersal was not prompted proximally by local food shortages at the observed population densities. Males showed no more likelihood to disperse at the minimum age of 2 years in years of fruit and nut crop failures (7 of 11 2-year-olds dispersed) than in years of abundant food (6 of 9 2-year-olds dispersed). Five males that ate supplemental garbage all dispersed at 2 years of age. For comparison, bears in another part of Minnesota, where natural food was more abundant and growth was more rapid, dispersed as yearlings in 4 of 7 instances (D. Garshelis and K. Noyce, pers. comm.). Similarly, black bears in Pennsylvania showed unusually rapid growth (Alt 1980) and dispersed mainly as yearlings (Alt 1978). On Long Island, Washington, where black bear density was more than 5 times that of northeastern Minnesota, 3 of 4 males delayed dispersal until 4 years of age (Lindzey and Meslow 1977a, 1977b). The fact that dispersal was delayed rather than hastened at high density suggests, further, that any aggression associated with high density did not initiate dispersal. If there is differential aggression against male nonkin in the natal range, as was observed among females (Rogers 1987), the observed pattern of delayed dispersal at high population densities would be expected because risks of dispersal by bears with small body size probably would be greater under those conditions. Under the same logic, dispersal by males would be expected to occur at an earlier age where abundant food accelerates growth and sexual development. Aggression is known to deter dispersing bears from settling in new areas.

Aggression did not appear to initiate male dispersal in northeastern Minnesota. For example, after a mother died in winter, her radio-collared son nevertheless left in spring at the usual age of 2 years. His abandoned range and that of another 2-year-old that dispersed that spring were not immediately used by territorial neighbors or siblings, all or nearly all of which were radio collared. Both dispersed in a year (1972) when the potential for aggression by adult males was unusually low: 2 of 3 adult males that had overlapped the ranges of those subadults the year before were dead, and there was evidence that the third was incapacitated for several weeks by injuries from a fight in mid-June (Rogers 1987). The loss of influence of those 3 adults opened the capture area to immigration by 7 subadult males, which was nearly as many as immigrated in all of the remaining 8 years in which immigration was assessed (8 bears). Thus, although space was available locally in 1972, both males that reached 2 years of age that year dispersed. Further, movements of 32 km, 74 km, and 80 km were recorded for dispersing subadult males in September and October when aggression was unlikely. Aggression and testosterone levels are particularly low at that time of year (McMillin et al. 1976), and many bears of both sexes are either lethargic or in dens. In three other movements not readily explained by aggression, 3 subadult males moved 75 km, 80 km, and 100 km outside the bear range in southern Minnesota. These data and the fact that all males dispersed led to a conclusion that the initial dispersal movements and some subsequent movements by males were voluntary. This conclusion suggests that dispersal confers advantages on the dispersing individual.

Inbreeding Avoidance

All males dispersed before mating, and all males dispersed farther than any female. These facts are consistent with the hypothesis that dispersal evolved as a mechanism for avoiding inbreeding. However, this explanation is weakened by three observations. First, dispersal was mainly by males even though in a polygynous system this sex would stand to lose less with inbreeding than would females, due to males' smaller parental investment and their larger reproductive potential (Smith 1979; Dawkins 1979). Second, the few females that did disperse probably did not move outside their fathers' mating ranges. Third, males did not disperse a second time when daughters began to mature in their mating ranges (Rogers 1987). Twenty-five percent of the adult males were over the minimum age of 8 years, at which father-daughter matings would become possible if males and females began mating at the minimum ages of 4 years and 3 years, respectively (Rogers 1987). Fifteen percent of the adult males were over the age of 10 years, at which father-daughter matings would become possible if males began mating at 4 years and females began mating at 5 years. Males did not avoid pairing with females young enough to be their daughters (Rogers 1987), but paternity was not certain for any female.

For females, the benefits of remaining in or adjacent to their mothers' territories apparently outweighed any potential disadvantages of remaining in their fathers' ranges. No deleterious effects of inbreeding were noted when a sibling pair was mated in captivity. They produced 34 cubs in 11 litters of normal number (2 to 4 cubs) with no mortality or obvious birth defects (D. Eggleston, pers. comm.). This could imply that bears may be at least mildly inbred in nature (Shields 1982; Templeton, chap. 17, this vol.). Avoidance of close inbreeding is logically a factor promoting dispersal, but it may be of limited importance relative to other factors influencing reproductive success and dispersal in black bears.

## Food, Females, and Dominant Males

With few exceptions, the foods of black bears are small items that cannot efficiently be carried to offspring (Rogers 1987). Therefore, males cannot efficiently provision their offspring, and they do not directly aid in raising them (Rogers 1987). Instead, they attempt to maximize reproductive success by inseminating as many females as possible. At population densities observed in northeastern Minnesota, reproductive success of males appeared to depend upon ability to find receptive females before other males do and upon ability to defeat other males that find the same females. Females became attractive before they became receptive, thereby heightening competition among males (Rogers 1977). Mating privileges appeared to be obtained primarily through male-male competition rather than as an obvious result of female choice. The outcome of mating battles depended heavily upon body size. Where contestants differed significantly in size, the larger simply chased the smaller away (Rogers 1987).

Consequently, dispersing males might be expected to establish mating ranges where there are few dominant males, many mature females, and sufficient food for rapid attainment of large body size. Costs of dispersing to look for such an area might be small for three reasons. Black bears have few predators (Rogers and Mech 1981). Mating ranges are so much larger than natal ranges that most of the mating range would initially be unfamiliar whether males dispersed or not. And the large size of mating ranges precludes males from obtaining meaningful amounts of space from their mothers (as do daughters).

Evidence that dispersing males tended to establish mating ranges in areas with reduced numbers of dominant males was obtained in this study and in a study in Alberta. In this study, the loss of influence of 3 adult males in the study area resulted in a great increase in immigration in 1972. In that year, 12 new males remained in the study area long enough to be caught, and 7 of them established permanent ranges there. By comparison, in the 8 other years of intensive trapping, an average of only 5 new males per year remained in the study area long enough to be caught, and only 1 new male established a range there per year (Rogers 1987). In Alberta, 26 adult males were experimentally removed from a population, with the result that 95 new bears, mostly subadult males, immigrated (Kemp 1976; Ruff, Young, and Pelchat 1976).

However, areas with few or no dominant males may not be attractive to dispersing males if the areas lack females. In the wooded vicinities of towns there were few bears of either sex due to people shooting them (Rogers 1976). Dispersing males commonly stopped to feed in those areas (Rogers 1976), but none was known to settle permanently. Two that were radio tracked fed around towns for 23 and 39 days, respectively, before moving farther (Rogers 1987). Food supply influenced movements to the extent that dispersing males stopped to feed at garbage dumps for up to 72 days, despite the presence of numerous dominant males (Rogers 1987). Conversely, food, females, and reduced numbers of dominant males did not induce young males to remain in their natal ranges (Rogers 1987). Possible explanations for this behavior, in addition to avoidance of inbreeding, are presented later in this chapter.

Inclusive Fitness

The majority of males dispersed 2 years before reaching sexual maturity. They typically dispersed at 2 years of age, while the youngest male to pair with a female was 4 years of age (Rogers 1987). Testicle size and body size of Minnesota bears, when compared with those of Michigan bears (Erickson, Nellor, and Petrides 1964), indicate that Minnesota males less than 4 years of age were probably sexually immature in most cases (Rogers, unpub. data). By dispersing prior to maturity, males might increase the time available for finding areas favorable to reproductive success. However, the fact that the males were immature suggests that the dispersals involved factors not directly related to obtaining mates. Early dispersal may increase inclusive fitness by reducing competition among the kin left behind (Rogers 1977; Shields 1983, and chap. 1, this vol.). Males in Minnesota dispersed when they weighed 29 kg to 59 kg ($N = 17$). At that size, due to sexual dimorphism, they usually outweighed their sisters and were approaching the weights of their mothers. If they remained, they probably could have displaced female kin from preferred feeding sites, which might have interfered with the females' growth and reproductive success (Rogers 1976, 1987). If they remained and showed deference to female kin, they might have reduced their own growth, which could have reduced their potential for winning future mating fights. In a new area, males could compete vigorously without reducing their inclusive fitness. Any resulting gains in reproduction by female kin would cumulatively increase the male's inclusive fitness. Although the proximal mechanism for the pattern is unclear, 9 of 10 males that had living sisters dispersed at 2 years of age, whereas only 4 of 9 without living sisters dispersed that early (difference significant; $p < 0.05$; $X^2 = 4.55$).

Early dispersal can reduce competition for the nondispersers only if the latter prevent immigration by other potential competitors. The effectiveness of resident adult males in deterring immigration was discussed earlier. The effectiveness of territorial females in deterring immigration by young males is not as well documented, but during 1,480 hours of radio tracking a territorial female, 3 transient subadults (males by circumstantial evidence) were observed fleeing from her (Rogers 1987).

If resident adults can prevent or reduce immigration, and if dispersing males establish mating ranges where chances of mating are no worse, on the average, than in their fathers' ranges, dispersal may further improve the inclusive fitness of young males by reducing mate competition with fathers and brothers. Males benefit more by taking matings away from nonkin than from kin. Where a male's competitors share half his genes, he can achieve, by dispersing, up to a 50% increase in gene copies passed to the next generation. For example, if by dispersing, a young male enables his father to mate with a female that the young male otherwise would have mated with, while the young male mates elsewhere, 50% more of the young male's gene copies are passed to the next generation than if he had taken the mating away from his father. All males dispersed at least 1 mating range diameter from their birthplaces. By dispersing long distances, males may reduce the degree of genetic relatedness of their competitors.

CONCLUSIONS

Despite their solitary habits, bears behaved in accordance with kinship theory. Movements of mothers in relation to daughters and of brothers in relation to sisters were consistent with a hypothesis that individuals recognize their independent offspring and littermates and behave in manners beneficial to them within limits dictated by the degree of genetic relatedness (Hamilton 1964; Wilson 1975). Differential aggression by mothers toward nonkin may partly explain delayed dispersal by males where growth rates were slow or population densities were high. This differential aggression by mothers also aided daughters in establishing adjacent territories, which may explain the low dispersal rate among females. If the parent most involved in resource defense shows differential aggression against nonkin in other mammal and bird species, thereby aiding same-sex offspring in establishing adjacent territories, such behavior may partly explain the fact that the philopatric sex in polygynous species tends to be the sex most involved in resource defense (Greenwood 1980; Waser and Jones 1983).

Evidence indicated that initial dispersal movements by males were voluntary and not forced by aggression or food shortage. Costs of male dispersal probably were low because of low predation risk and small natal: adult range-size ratios. Dispersal may enable males to find more favorable locations for establishing mating ranges. Dispersal also may increase males' inclusive fitness by reducing mate competition with male kin and feeding competition with female kin left behind. The increase in inclusive fitness depends upon the ability of the dispersers' kin to prevent immigration by other competitors. Given that demonstrated ability of kin, dispersing males can achieve up to a 50% increase in gene copies passed to the next generation if they mate where they will take mates away from nonkin rather than kin. Inbreeding avoidance may not have been the primary factor promoting dispersal.

## ACKNOWLEDGMENTS

I thank W. M. Shields, L. D. Mech, and B. D. Chepko-Sade for helpful suggestions.

## REFERENCES

Alt, G. L. 1978. Dispersal patterns of black bears in northeastern Pennsylvania: A preliminary report. In *Proceedings of the Fourth Eastern Black Bear Workshop*, ed. R. D. Hugie, 186–99. Maine Department of Inland Fisheries and Wildlife, Orono.

———. 1980. Rate of growth and size of Pennsylvania black bears. *Pennsylvania Game News* 51(12):7–17.

Dawkins, R. 1979. Twelve misunderstandings of kin selection. *Zeitschrift für Tierpsychologie* 51:184–200.

Dobson, F. S. 1982. Competition for mates and predominant juvenile dispersal in mammals. *Animal Behaviour* 30:1183–92.

Erickson, A. W., J. Nellor, and G. A. Petrides. 1964. *The black bear in Michigan.* Michigan State University Agricultural Experiment Station Bulletin No. 4. East Lansing.

Greenwood, P. J. 1980. Mating system, philopatry, and dispersal in birds and mammals. *Animal Behaviour* 28:1140–62.

Hamilton, W. D. 1964. The genetical theory of social behaviour, pts. 1 and 2. *Journal of Theoretical Biology* 7:1–52.

Kemp, G. A. 1976. The dynamics and regulation of black bear *Ursus americanus* populations in northern Alberta. In *Bears: Their biology and management*, ed. M. R. Pelton, J. W. Lentfer, and G. E. Folk, Jr., 191–97. IUCN Publications, New Series No. 40. Morges, Switzerland: IUCN.

Lindzey, F. G., and E. C. Meslow. 1977a. Home range and habitat use by black bears in southwestern Washington. *Journal of Wildlife Management* 41:413–25.

———. 1977b. Population characteristics of black bears on an island in Washington. *Journal of Wildlife Management* 41:408–12.

McMillin, J. M., U. S. Seal, L. L. Rogers, and A. W. Erickson. 1976. Annual testosterone rhythm in the black bear (*Ursus americanus*). *Biology of Reproduction* 15:163–67.

Rogers, L. L. 1976. Effects of mast and berry crop failures on survival, growth, and reproductive success of black bears. *North American Wildlife and Natural Resources Conference* 41:431–38.

———. 1977. Social relationships, movements, and population dynamics of black bears in northeastern Minnesota. Ph.D. diss., University of Minnesota, Minneapolis.

———. 1983. Effects of food supply, predation, cannibalism, parasites, and other health problems on black bear populations. In *Symposium on Natural Regulation of Wildlife Populations*, ed. F. Bunnell, D. S. Eastman, and J. M. Peek, 194–211. Forest, Wildlife, and Range Experiment Station Proceedings No. 14. University of Idaho, Moscow.

———. 1987. Effects of food supply and kinship on social behavior, movements, and population dynamics of black bears in northeastern Minnesota. *Wildlife Monograph* 97.

Rogers, L. L., and L. D. Mech. 1981. Interactions of wolves and black bears in northeastern Minnesota. *Journal of Mammalogy* 62:434–36.

Ruff, R., B. F. Young, and B. O. Pelchat. 1976. A study of the natural regulatory mechanisms acting on an unhunted population of black bears near Cold Lake, Alberta. Progress Report. Alberta Department of Recreation, Parks, Wildlife, and Fish. Wildlife Division, Edmonton, Alberta.

Shields, W. M. 1982. *Philopatry, inbreeding, and the evolution of sex*. Albany: State University of New York Press.

———. 1983. Optimal inbreeding and the evolution of philopatry. In *The ecology of animal movement*, ed. I. R. Swingeland and P. J. Greenwood, 132–59. Oxford: Oxford Clarendon Press.

Smith, R. H. 1979. On selection for inbreeding in polygynous animals. *Heredity* 43:205–11.

Tilson, R. 1981. Family formation strategies of Kloss' gibbon. *Folia Primatologica* 35:259–87.

Waser, P. M., and W. T. Jones. 1983. Natal philopatry among solitary mammals. *Quarterly Review of Biology* 58:355–90.

Wilson, E. O. 1975. *Sociobiology.* Cambridge: Harvard University Press.

# 6. Dispersal and Intergroup Transfer in the Dwarf Mongoose

*Jon P. Rood*

In most group-living mammals, males are the mobile members of the society (Greenwood 1980). Group-living mammals are primarily polygynous (Eisenberg 1981), and thus subordinate males are often excluded from breeding and may only be able to increase their reproductive potential by intergroup transfer. Females, however, are able to breed in their natal groups and may remain in them for their entire lives. In monogamous species, females as well as males are normally excluded from breeding by a dominant pair and may need to disperse if they are to reproduce successfully.

In this chapter I present dispersal data on an essentially monogamous, group-living carnivore, the dwarf mongoose (*Helogale parvula*). Like most small carnivores, the majority of the 36 species of mongooses (family Herpestidae) are solitary. However, 8 species in four genera forage, travel, and sleep in communal groups. Of these, the breeding systems of 3 species have been studied in the field. The banded mongoose (*Mungos mungo*) and meerkat (Suricata suricatta) are polygamous (Rood 1974, 1975; Macdonald, pers. comm.), and groups normally contain several breeding males and females. The dwarf mongoose differs in that the groups contain a dominant breeding pair, usually the oldest individuals, which are the parents of young raised in the group (Rood 1980).

Dwarf mongooses show many characteristics of a monogamous species (see Kleiman 1977), such as reduced sexual dimorphism, delayed reproduction in the presence of the parents, and aid in the rearing of young by the father and older siblings (Rasa 1977; Rood 1978). Monogamy is rare among mammals and has been reported in less than 3% of mammalian species (Kleiman 1977). Among these, only the dwarf mongoose and a few other species travel and sleep in cohesive social groups that, besides the reproductive pair, may consist of more than one generation of offspring,

85

siblings of the breeders, and sometimes other kin. Two other carnivores, the wolf, *Canis lupus*, (Mech 1970) and African wild dog, *Lycaon pictus* (Frame et al. 1979), share these characteristics, as do the marmosets and tamarins in the primate family Callitrichidae (Epple 1975; Dawson 1977; Neyman 1977). In the dwarf mongoose, some group members may be unrelated immigrants of either sex (Rood 1983).

The dwarf mongoose is the smallest member of the Herpestidae, weighing approximately 300 g to 350 g when adult (Rood 1983). Like the other group-living mongooses—and in contrast to many solitary species—dwarfs feed primarily on insects, particularly dung beetles. Although subject to predation from a wide variety of aerial and ground predators, dwarf mongooses have a well developed antipredator system (Rood 1983, 1986), and some individuals live longer than 10 years. For the investigator interested in dispersal, this species offers several advantages. It frequently occurs at high densities, is diurnal, which facilitates field observations, and has a relatively high reproductive rate and short generation time. The data presented here were obtained by following the histories of mongoose groups and individuals over a 10-year period.

## METHODS

Fieldwork was carried out at the Serengeti Research Institute (SRI), Serengeti National Park, Tanzania, beginning in 1974. Full-time observations were made until 1977, and I have subsequently visited the area for 6 weeks to 4 months each year. In the early years of the study, most observations were made in a 2.2 km² area of open *Acacia, Commiphora* woodland, bordering the Sangere River, containing a high density of *Macrotermes* termite mounds that are used by the mongooses as dens (Rood 1983). Additional observations were made on packs living near houses of investigators at the Serengeti Research Institute and Seronera, and on mongooses inhabiting a heavily grazed area at Kirawira in the western corridor of the park. In the Sangere River area, mongoose packs used overlapping home ranges averaging 34.6 ha and attained densities at the start of the birth season (October–November) of up to 31 individuals/km². In the more open Kirawira habitat, densities were much lower (3/km²) and ranges larger (up to 160 ha). The Sangere River study area was enlarged to 25 km² in 1980, incorporating the area used by the SRI packs. This enlarged area consists of *Acacia, Commiphora* woodland interspersed with areas of open grassland, which are avoided by the mongooses; the area currently contains the ranges of 13 packs. Population density at the start of the birth season has varied since 1980 from 3.9/km² to 4.6/km².

Dwarf mongooses were live-trapped at their dens, sexed, aged by tooth wear, and permanently freeze-marked with Freon 12 or Quik Freeze for individual recognition (Rood and Nellis 1980). A total of 478 mongooses have been marked to date. At the start of the birth season each year, all dwarf mongooses on the study area were censused, and young of the year and immigrants were live-trapped and marked. In addition, suitable mongoose habitat surrounding the area was searched for emigrants.

Most censuses and observations of social interactions were made when the mongooses were at dens in the early morning or late afternoon, but observations were also made opportunistically throughout the day. I normally observed the mongooses with 10x binoculars from a Land Rover parked 40 m to 50 m from the den. Thirty-two packs in which all individuals could be identified were monitored for periods of up to 10 years, and all observed changes in pack composition were recorded. The number of packs under observation increased as the study progressed and also varied because of the disappearance of some packs and the formation of new ones. The total monitoring time was 97 pack-years (one pack-year = one pack monitored for one year).

In the Serengeti, births have been recorded in all months from October to June, but most (86% of 78 litters) were born between November and April—the months of greatest rainfall (Rood 1983). Breeding females normally produce three litters of from one to six young during the breeding season. Mongooses born in the same birth season have been placed in the same age class. In the Serengeti, the dry season extends from June to October. Animals that have not yet survived a dry season are termed juveniles. Yearlings have survived one dry season; 2-year-olds, two dry seasons; and so forth. Dwarf mongooses reach puberty as yearlings but are typically unable to breed successfully because of reproductive suppression by their parents. The youngest alpha individuals (breeders) in my study were 2-year-olds. These and older animals are termed adults.

## RESULTS
### Group Composition

Dwarf mongoose packs are potential reproductive units and contain at least one breeding male and female. A typical pack at the start of the birth season might consist of the dominant breeding pair, one or two subordinate adult males, two or three subordinate adult females, and two or three yearlings born the previous breeding season (table 6.1). Average pack size at this time was 8.4 and median size was 7. The maximum pack size recorded during the study (including juveniles) was 26.

Both males and females may remain and breed in their natal packs, but, as table 6.2 shows, females are considerably more likely to do so than males. Of adults of known origin at the start of the birth season, 78% of the females and 36% of the males were natal animals. Among breeders, 43% of the females, but only 5% of the males, were in their natal packs.

### Emigration

Emigration was recorded if an individual or group was seen either outside the pack range, or in a separate part of the range from the main pack,

TABLE 6.1 Pack Composition in Dwarf Mongoose Packs at the Onset of the Birth Season

|  | NUMBER | MEAN ± (SD) | RANGE |
|---|---|---|---|
| Alpha males | 115 | 1 | |
| Alpha females | 115 | 1 | |
| Subordinate adults | | | |
|    Males | 175 | 1.5 ± (1.1) | 0–6 |
|    Females | 260 | 2.3 ± (2.1) | 0–7 |
|    Unsexed | 7 | 0.1 ± (0.4) | 0–3 |
| Yearlings | | | |
|    Males | 128 | 1.1 ± (1.3) | 0–5 |
|    Females | 136 | 1.2 ± (1.2) | 0–7 |
|    Unsexed | 26 | 0.2 ± (0.6) | 0–3 |
| Packs | 115 | 8.4 ± (4.1) | 2–19 |

Note: These figures represent 32 mongoose packs at the onset of 115 birth seasons.

before disappearance. Mongooses emigrated from three main types of packs:

1. *Intact packs.* These individuals left undisturbed packs that were viable (contained a breeding pair), without observed aggression from conspecifics.

2. *Invaded packs.* These individuals left packs after they were invaded by groups of same-sex intruders. Aggression by the intruders was sometimes observed before dispersal.

3. *Disintegrating packs.* These individuals left packs following the death of one of the breeders, usually the alpha female. Pack failure occurred when all pack members died or emigrated from the pack range.

Emigration from intact packs appeared to be, and is here referred to as, voluntary emigration. Emigration from invaded or disintegrating packs was instigated by some external event, that is, aggression from intruders or the death of an important pack member.

Greenwood (1980) distinguished natal dispersal (the permanent emigration of young animals from their birthplaces) from breeding dispersal

TABLE 6.2 Proportion of Adult Natal Animals in Dwarf Mongoose Packs at the Onset of the Birth Season

|  | NO. OF KNOWN ORIGIN | NO. NATAL ANIMALS |
|---|---|---|
| Adult males | | |
|    Alphas | 59 | 3   (5%) |
|    Subordinates | 128 | 65 (51%) |
|    Total | 187 | 68 (36%) |
| Adult females | | |
|    Alphas | 60 | 26 (43%) |
|    Subordinates | 195 | 173 (89%) |
|    Total | 255 | 199 (78%) |

Note: See table 6.1.

(the movement between successive breeding sites of individuals that have reproduced). However, because mongooses may transfer between groups several times in their lives, and may disperse for the second or third time without having reproduced, I refer to animals dispersing from their natal pack as natal dispersers, and those dispersing subsequently as secondary dispersers.

Table 6.3 shows the number of natal mongooses that have disappeared, and the number of these that are known to have emigrated, in each age-group from yearlings to 7-year-olds. Most natal dispersers were young, recently sexually mature mongooses. Significantly more males than females disappeared from their natal packs as yearlings and 2-year-olds. It is likely that this finding represents a sexual bias in dispersal rather than in mortality

TABLE 6.3 Disappearance of Male and Female Dwarf Mongooses from Natal Packs

|  | MALES | FEMALES | $\chi^2$ | $p$ |
|---|---|---|---|---|
| Yearlings |  |  |  |  |
| No. monitored | 110 | 144 |  |  |
| No. disappeared | 61 (55%) | 42 (29%) | 17.9 | <.001 |
| No. known emigrated | 18 (16%) | 8 (6%) |  |  |
| 2-year-olds |  |  |  |  |
| No. monitored | 38 | 91 |  |  |
| No. disappeared | 23 (61%) | 35 (38%) | 5.3 | <.05 |
| No. known emigrated | 5 (13%) | 14 (15%) |  |  |
| 3-year-olds |  |  |  |  |
| No. monitored | 13 | 51 |  |  |
| No. disappeared | 4 (31%) | 13 (25%) |  | NS |
| No. known emigrated | 3 (23%) | 3 (6%) |  |  |
| 4-year-olds |  |  |  |  |
| No. monitored | 7 | 31 |  |  |
| No. disappeared | 1 (14%) | 8 (26%) |  | NS |
| No. known emigrated | 0 | 0 |  |  |
| 5-year-olds |  |  |  |  |
| No. monitored | 4 | 18 |  |  |
| No. disappeared | 2 (50%) | 5 (28%) |  | NS |
| No. known emigrated | 0 | 1 (6%) |  |  |
| 6-year-olds |  |  |  |  |
| No. monitored | 1 | 10 |  |  |
| No. disappeared | 0 | 2 (20%) |  | NS |
| No. known emigrated | 0 | 0 |  |  |
| 7-year-olds |  |  |  |  |
| No. monitored | 1 | 5 |  |  |
| No. disappeared | 0 | 2 (40%) |  | NS |
| No. known emigrated | 0 | 0 |  |  |
| Total 3-year-olds to |  |  |  |  |
| 7-year-olds |  |  |  |  |
| No. monitored | 26 | 115 |  |  |
| No. disappeared | 7 (27%) | 30 (26%) |  | NS |
| No. known emigrated | 3 (12%) | 4 (3%) |  |  |

Note: Animals that disappeared may have either died in the pack or emigrated. Of these, the numbers of known emigrants are given. $\chi^2$ values test for differences in the disappearance of males and females.

while in the natal pack. This was difficult to confirm, however, because dispersers that left the study area were seldom located. Natal animals in older age classes showed no sexual bias in rate of disappearance. In animals 4 or more years old there was only one case of natal emigration; most of these older mongooses that disappeared probably died in their natal packs. The oldest known natal animal is a 9-year-old female who is currently the alpha female in the pack of her birth.

Approximately half the 74 mongooses known to have emigrated during the study dispersed as a consequence of pack disintegration or invasion (table 6.4). These externally influenced emigrations occurred in animals of all ages from juveniles to 9-year-olds. Aggression from within the pack was not observed before emigration, but six males and one female were ousted by same-sex intruders during pack invasions (table 6.4). Emigration from disintegrating small packs frequently resulted in pack failure. Of the 15 packs that failed during the study, 13 contained less than seven pack members when last seen, and 11 contained only one adult female. Pack failure occurred most commonly during the dry season when I was not on the study area, but I suspect the typical sequence is death of the alpha female followed by dispersal of other pack members. The events surrounding the failure of the H1 pack are shown in the appendix (a).

What was the fate of known emigrants from the monitored packs? As table 6.4 shows, many joined intact packs or pack remnants. Some formed new packs by joining with opposite-sex emigrants, either singly or in groups, and establishing a territory in an area unoccupied by conspecifics. In most instances the packs had already formed when first located so that the sequence of events could not be determined. Examples of new pack formation are given in the appendix (a and d).

Mortality is higher in small than in large packs (Rood 1986), and dispersing mongooses are at greater risk from predators than residents. Mortality through predation undoubtedly accounted for some of the emigrants

TABLE 6.4  Types of Emigration and the Fates of the Emigrants

|  | MALES | FEMALES | TOTAL |
| --- | --- | --- | --- |
| Structure of pack left |  |  |  |
| Intact pack | 19 | 20 | 39 |
| Disintegrating pack | 18 | 10 | 28 |
| Invaded pack | 6 | 1 | 7 |
| Total emigrants | 43 | 31 | 74 |
| Fate of emigrants |  |  |  |
| Joined intact pack or pack remnant | 26 | 14 | 40 |
| Unknown | 15 | 9 | 24 |
| Formed new pack | 1 | 7 | 8 |
| Joined transient group | 1 | 0 | 1 |
| Established solitary territory | 0 | 1 | 1 |

that disappeared from the study area. Others may have left the area and joined or formed packs elsewhere.

A sexual difference occurs in the response of breeders to the death of their mates. As in the H1 pack (appendix, a), males frequently leave the pack range and search for breeding openings. Four of the 15 alpha males whose packs failed were found subsequently—3 in other packs and 1 alone. Alpha females remained in their ranges after the deaths of male breeders. In one instance, the alpha female of a disintegrating pack remained in her home area for over a year before she was joined by an adult male, thereby rejuvenating the pack (appendix, b).

In attempting to determine the causes of voluntary dispersal, I compared natal mongooses that emigrated voluntarily with those that stayed in their natal packs, for pack size, number of older same-sex animals, and the degree of relationship of the opposite-sex breeder in their packs at the start of the birth season. Natal animals that dispersed during the ensuing year were living in significantly larger packs at the start of the birth season than those that remained in their natal packs (table 6.5). Emigrants also tended to come from packs with more older same-sex individuals than did nonemigrants, but the difference was not significant. While many mongooses stayed in packs in which the opposite-sex breeder was a close relative (a parent in 80% of the cases), only two individuals emigrated from packs in which the opposite-sex breeder was unrelated (table 6.6). Combining males and females to increase sample size indicates that mongooses are significantly more likely to leave packs in which the opposite-sex breeder is a close relative ($\chi^2 = 4.10$; $p < .05$).

## Immigration and Transfer

I distinguished three types of immigration:

1. *Joining an intact pack as a subordinate.* These immigrants joined packs containing an alpha male and female and accepted subordinate status.

TABLE 6.5  Relationship of Pack Size and Composition to Emigration

| | MALES | | | FEMALES | | |
|---|---|---|---|---|---|---|
| | Emigrants | Nonemigrants | $p$ | Emigrants | Nonemigrants | $p$ |
| No. individuals | 14 | 61 | | 17 | 195 | |
| Mean pack size | 14.3 | 11.3 | <0.01 | 14.4 | 12.2 | <0.03 |
| ± (SD) | ± (4.3) | ± (3.8) | | ± (3.8) | ± (3.7) | |
| Mean no. older same-sex animals | 2.6 | 2.5 | <0.40 | 4.2 | 3.8 | <0.20 |
| ± (SD) | ± (1.6) | ± (1.3) | | ± (1.5) | ± (2.3) | |

*Note:* Emigrant and nonemigrant males and females are shown with mean pack size and mean number of older same-sex animals in the pack at the onset of the birth season. Emigrants are natal animals that voluntarily left their natal pack during the subsequent year, while nonemigrants remained in their natal packs. Mann Whitney tested differences between packs in which the mongooses remained and those from which they emigrated.

TABLE 6.6 Voluntary Emigration and Relationship to the Opposite-sex Breeder

| | OPPOSITE-SEX BREEDER | |
| --- | --- | --- |
| | Close Relative | Unrelated |
| Males | | |
| Emigrants | 13 (93%) | 1 (7%) |
| Nonemigrants | 44 (83%) | 9 (17%) |
| Females | | |
| Emigrants | 13 (93%) | 1 (7%) |
| Nonemigrants | 108 (73%) | 39 (27%) |
| Both sexes | | |
| Emigrants | 26 (93%) | 2 (7%) |
| Nonemigrants | 152 (76%) | 48 (24%) |

2. *Takeover of a pack.* This occurred when a group of immigrants joined a pack and ousted same-sex pack members. In all cases the number of immigrants exceeded the number of mongooses that emigrated. Male take-overs resulted in the emigration of breeding males. Females that had previously bred never lost their status to immigrants, but in one instance a yearling female was ousted by two older immigrant females following the death of the alpha female.

3. *Joining a pack remnant.* These immigrants joined and rejuvenated packs that, usually because of mortality, were lacking a breeder. An example is given in the appendix (a). In another case a yearling female remained alone in her natal range for over a year before being joined by a male (appendix, b).

Immigration rates were considerably higher in males than in females. Fifty-five males (25 of known origin) and 21 females (11 of known origin) joined monitored packs during the study. The study packs received an average of one immigrant per 15.3 months (one male per 21.2 months and one female per 55.4 months).

Groups of immigrants always consisted of mongooses of the same sex. Although up to six individuals transferred together, most joined or formed packs singly or in same-sex pairs (table 6.7). Most moved into breeding openings or joined packs with one older same-sex individual. Packs with at least two adult males and females were the least open to immigration. Females never joined such packs, but seven males joined packs with two older males and one joined a pack with at least three. Males that attempted to join packs with more than one adult male encountered frequent aggression from the beta male. They sometimes trailed the pack for as long as a month before becoming accepted, and one was probably killed by the beta male of the pack he was attempting to join (Rood 1983).

Most females that leave their natal packs disperse only once, but secondary dispersal was common in males. Among dispersers whose packs of origin and destination were known, 17 of 21 females were natal dispersers,

TABLE 6.7 Group Sizes of Immigrants and Pack Founders

| GROUP SIZE | MALE GROUPS | FEMALE GROUPS | TOTAL |
|---|---|---|---|
| 1 | 25 | 14 | 39 |
| 2 | 14 | 8 | 22 |
| 3 | 3 | | 3 |
| 6 | 1 | | 1 |
| TOTAL | 43 | 22 | 65 |

2 were secondary dispersers, and the type of dispersal was unknown in 2. Fourteen of 27 dispersing males were natal dispersers, and 13 were secondary dispersers. Two males were known to have lived in at least four packs.

Most recorded transfers between monitored packs were between packs with overlapping home ranges (77% of 26 males, and 54% of 13 females). As a consequence, median dispersal distances were not great (0.5 km for males and 1.0 km for females). Maximum distances were 1.8 km for males and 5.0 km for females. In the absence of telemetry, individuals that left the study area were unlikely to be found and many may have dispersed farther than this. It is likely that the frequent scent marking that occurs at termite mound dens enables potential immigrants to assess the composition of resident packs and move into breeding openings when these occur. Young adult mongooses sometimes monitor the situation in neighboring packs for long periods before transferring permanently. For example, IF9 transferred repeatedly between the O and I packs for at least four months (fig. 6.1). When the alpha female in the 0 pack died, IF9 replaced her. Of the three C-pack males that joined the G pack (fig. 6.2), one had been trailing it periodically for over 18 months, and another for over 6 months. These males interacted peacefully with the G-pack females and aggressively with the males, while retaining their position in the C pack—before becoming permanent G-pack members.

Mongooses that transferred between monitored packs of known composition increased their reproductive potential by moving from packs with more older, same-sex individuals to packs with fewer (figs. 6.1 and 6.2). The only known exceptions were an alpha male whose pack had failed, and one that had lost his breeding status in a takeover. These joined packs containing a male breeder and accepted subordinate status. Such males are acting in their own best interest by attaining the advantages of group life, particularly in decreasing their vulnerability to predators. In two cases, EM20 (fig. 6.2) and MF3 (fig. 6.1), individuals assisted another pack member to effect a takeover and subsequently returned to their pack of origin. In both, they returned to larger packs in which adult mortality is lower (Rood 1986), and in which their reproductive potential was at least as high as it would have been in the new pack.

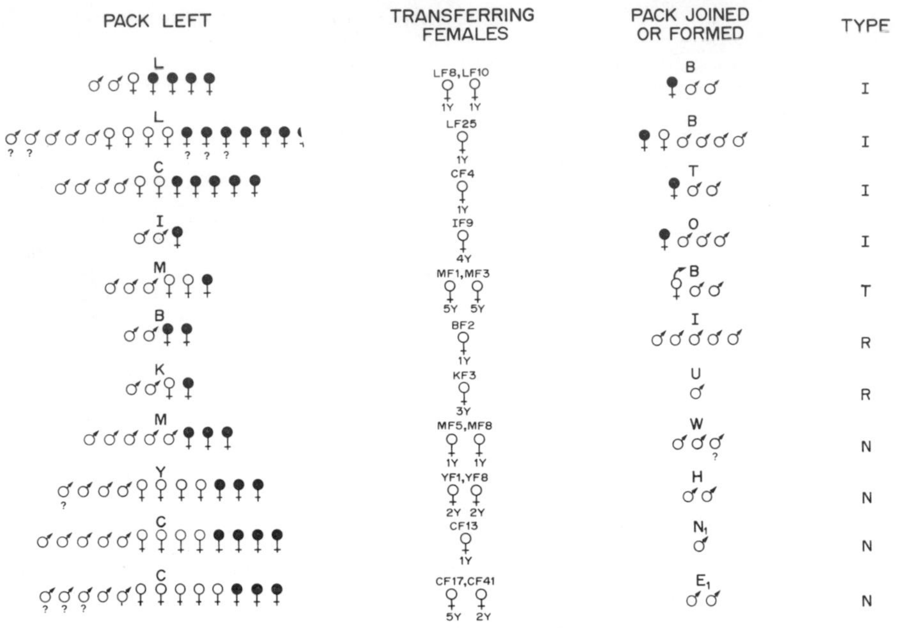

| PACK LEFT | TRANSFERRING FEMALES | PACK JOINED OR FORMED | TYPE |
|---|---|---|---|

FIGURE 6.1. Interpack transfer by females. "Pack Left" and "Pack Joined" give the number of male and female adults in these packs at the time of transfer; solid symbols indicate adults of the same sex and older than the transferring individual. "Transferring Females" lists the names and ages (in years) of transferees.

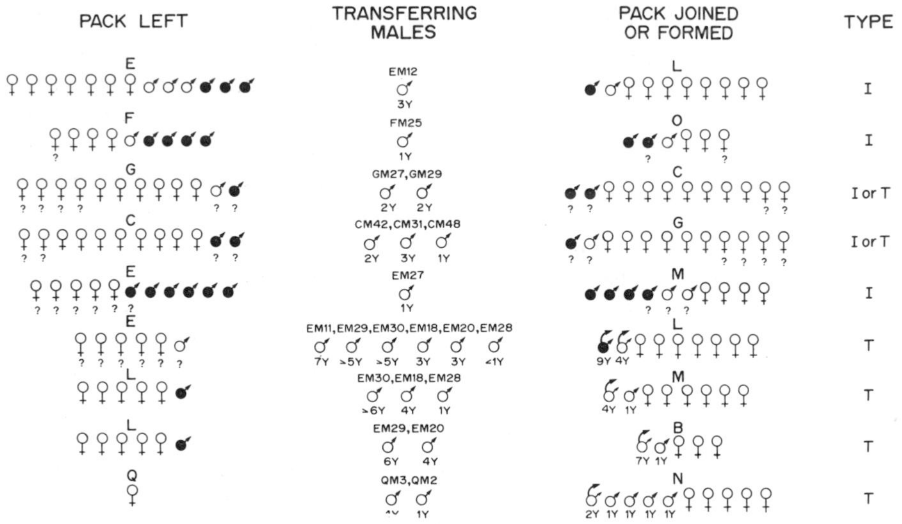

| PACK LEFT | TRANSFERRING MALES | PACK JOINED OR FORMED | TYPE |
|---|---|---|---|

FIGURE 6.2. Interpack transfer by males. See fig. 6.1 for conventions.

I- Joined intact pack.
T- Takeover.
R- Joined opposite sexed remnants of pack thereby rejuvenating it.

N- Formed new pack on unoccupied area.
⌐- Ousted in takeover.
?- Individuals which were gone when transfer was first observed but could still have been present at time of transfer.

Most adult males in dwarf mongoose packs are immigrants, while most females are natal animals. As a consequence, and because secondary transfer occurs more commonly in males than in females, dispersing groups of females are more likely to be close relatives than dispersing groups of males. Kinship relationships of dispersers were known in six groups of females and seven groups of males. The female groups were composed of close relatives (five of sisters and one of half sisters). Three of the male groups consisted of brothers, but four contained at least one unrelated individual. In addition, an emigrant from a monitored pack was observed traveling with two unrelated males, suggesting that membership in male transient groups may be comparatively open.

## DISCUSSION

Dwarf mongooses differ from many group-living mammals in that dispersal commonly occurs in both males and females (but see Berger, chap. 3, this vol.). Individuals of either sex may remain and breed in their natal pack or emigrate and search for breeding opportunities elsewhere. As in most mammals, however, dispersal is sex biased, with males being the principal dispersers. Thus, (a) significantly more males than females disappeared from their natal packs as yearlings and young adults; (b) the immigration rate into monitored packs of animals of unknown origin was considerably higher for males than for females; and (c) most transient groups consisted of males, suggesting that males spend more time as transients than females (Rood 1983, unpub. data). The movements of a pair of transient males are shown in the appendix (d).

Most males that survive their year of birth leave their natal packs before they are 3 years old and may transfer several times during their lives. Males seldom breed in their natal packs. In contrast, many females that survive to yearling status breed and spend their entire lives in their natal packs, and those females that do transfer usually do so only once. Among breeders, some males are ousted in takeovers and others disperse following the deaths of their mates. Females that have bred successfully retain their status and stay within their ranges for life.

Neither the proximate nor the ultimate causes of dispersal have been determined for any species of mammal (Gaines and McClenaghan 1980; Lidicker 1975). Aggression has frequently been correlated with dispersal, particularly in rodents, in which dominant individuals have been found to or are hypothesized to drive subordinate individuals into suboptimal habitats (Bronson 1964; Broadbrooks 1970; Christian 1970; Healy 1967; Steiner 1972). In monogamous group-living mammals, Kleiman (1981) observes that competition or aggression among females has been reported to be more intense than that among males in tamarins, wolves, and African wild dogs; she suggests that these females may be competing for groups of males that exhibit good paternal care, and that dispersal of young in monogamous group-living mammals may be initiated mainly through intrasexual sibling aggression. I found no evidence, however, that aggression by same-sex siblings or other pack members was instrumental in causing dispersal in dwarf mongooses. Overt aggression was not observed immediately prior to dispersal, and dispersers appeared healthy and in good condition. Intrapack aggression occurred most frequently during mating periods, when the alpha male frequently chased subordinate males from the alpha female, but these attacks did not lead to dispersal.

It is evident that no single factor can explain all cases of dispersal in dwarf mongooses. As Hollekamp (1984) points out, causal mechanisms for dispersal need not be mutually exclusive, and two or more variables may be necessary to initiate it. In a recent review of female transfer in primates, Moore (1984) concludes that different females transfer for different reasons. Most of these are apparently tactical, that is, for short-term or proximate motives (e.g., avoiding an infanticidal male) rather than strategic or ultimate (e.g., avoiding inbreeding). In mongooses, dispersal is affected by demographic and environmental factors such as pack size, population density, and predation rates. Some individuals disperse for tactical reasons following a pack failure or takeover, but often there is no readily apparent precipitating factor. What might cause voluntary dispersal in dwarf mongooses?

1. *Range choice hypothesis.* Mongooses could disperse to search for areas where food and cover are more abundant. The finding that voluntary emigrants come from larger packs than individuals that do not disperse suggests that competition for food resources could be important. Dwarf mongooses feed primarily upon insects, however, which they find individually, and little competition occurs during foraging behavior. Individuals frequently emigrated from packs that were receiving supplementary feeding from scientists at the research institute, and dispersal also frequently occurred during the rainy season when food resources are most abundant (Rood 1983).

2. *Social cohesion hypothesis.* In canids, Bekoff (1977) reports that relationships formed as pups influence later dispersal behavior, indicating that interactions at the time of dispersal may be less important than the developmental history of the disperser. He suggests that individuals showing social avoidance, and that are least likely to interact with littermates, will be the most likely individuals to disperse of their own accord. This hypothesis could be tested by recording frequencies of social-bonding mechanisms, such as allogrooming and play, in young mongooses and determining if these differ between individuals that subsequently disperse and those that do not. At present, insufficient data on individual behavioral profiles are available.

3. *Inbreeding avoidance hypothesis.* Some investigators have argued that dispersal is an adaptation for the avoidance of inbreeding depression, and

that many individuals disperse in order to avoid inbreeding (Demarest 1977; Greenwood 1980; Harcourt 1978; Packer 1979; Pusey 1980). Inbreeding avoidance can help explain sex differences in dispersal; once one sex has evolved mechanisms of dispersal that reduce the risk of inbreeding, pressures on the other sex to do likewise will be lessened (Clutton-Brock and Harvey 1976). Much of the evidence relating dispersal to inbreeding avoidance has been recently questioned (Moore and Ali 1984). These authors believe that "emigration and dispersal generally do not 'function' to minimize inbreeding" (p. 95) and that "observed dispersal patterns simply reflect sex differences in the balance between the advantages of philopatry and the costs of intrasexual competition" (p. 94). Dwarf mongooses provide some support for the inbreeding avoidance hypothesis, since they appear to be more likely to emigrate when the opposite-sex breeder in their pack is a close relative. In two cases, males emigrated in the absence of older, dominant resident males (appendix, b and c). The three yearling males that emigrated from the A pack after the death of the alpha male (appendix, b) were brothers of the alpha female and may have left to avoid breeding with her.

A low rate of inbreeding is a major consequence of the dwarf mongoose system of intergroup transfer. The origins of the members of 73 breeding pairs at the start of the birth season were known. In 69 pairs (95%) the males and females were born in different packs and were unlikely to be related. Inbreeding depression was not apparent in the four related pairs (two father-daughter, one mother-son, and one brother-sister). The mean number of offspring raised to yearling status by these related pairs was 2.25, slightly higher than the mean (1.92) raised by the 55 unrelated pairs for which the number of young raised was known.

4. *Search for breeding opportunities hypothesis.* I suggest that most dwarf mongooses that leave their groups voluntarily do so to fill breeding openings or search for breeding opportunities. Most voluntary dispersers were young, sexually mature animals, and they usually joined or formed new packs early in the breeding season (Rood 1983), when reproductive success would be most likely to be maximized. In essentially monogamous group-living mammals such as the dwarf mongoose, both males and females must attain alpha status to reproduce successfully. For many individuals, dispersal and a search for a breeding opening provide the best reproductive option. Some fortunate individuals are able to attain immediate alpha status by replacing breeders that have died. If they are able to do this in packs whose ranges overlap their own, they avoid the risks of predation as a transient. Many dispersers in this study moved into immediate breeding openings or joined packs with one older same-sex animal in which their chances of becoming breeders were high. Recorded cases of intergroup transfer increased the reproductive potential of the transferring mongooses.

The causes of dispersal in dwarf mongooses are complex. The previously mentioned hypotheses are not mutually exclusive, and many factors such as pack size and composition, an individual's status, degree of integration with and relatedness to other pack members, and the opportunities for breeding in neighboring packs may affect a mongoose's decision to disperse or remain in its natal pack. Both adult and juvenile survival increase with pack size (Rood 1986). It may frequently be advantageous for a mongoose to remain in a large pack and wait for a later breeding opportunity rather than transfer to a small one in which it may be able to breed immediately, but in which its chances of survival and those of its offspring will be low.

## ACKNOWLEDGMENTS

I thank the government of Tanzania for permission to conduct research and the Serengeti Wildlife Research Institute for allowing me to use their facilities. My wife, daughter, and the many scientists who have worked in the Serengeti since 1974 aided the project in numerous ways. I am also grateful to J. Berger, B. D. Chepko-Sade, J. A. King, H. Rood, P. Smouse, and P. M. Waser for their constructive criticism of an earlier draft of this chapter. Fieldwork was supported by grants from the National Geographic Society, the Harry Frank Guggenheim Foundation, and the Max-Planck-Institut fur Verhaltensphysiologie.

## APPENDIX

This appendix contains case histories demonstrating events surrounding the failure, disintegration, rejuvenation, and formation of dwarf mongoose packs. Alpha individuals were never known to voluntarily leave viable packs and are here assumed to have died if they disappeared.

Key

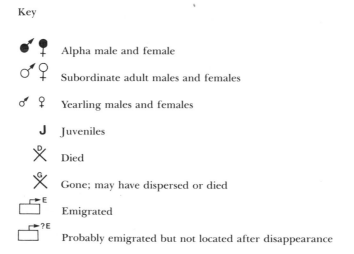

♂ ♀    Alpha male and female

♂ ♀    Subordinate adult males and females

♂ ♀    Yearling males and females

J    Juveniles

✕ᴰ    Died

✕ᴳ    Gone; may have dispersed or died

⬜→ᴱ    Emigrated

⬜→?ᴱ    Probably emigrated but not located after disappearance

a. Failure of the H1 Pack and formation of a new pack. Following the death of the alpha female, the alpha male attempted to recruit a female from an adjacent pack. He then joined a solitary female, forming a new pack.

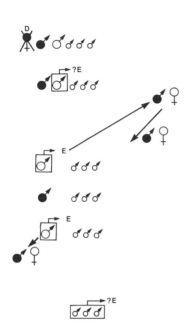

9 December 1980. Alpha female in H1 pack dies.

26–31 December. Subordinate adult male disappears (probably emigrates).

1 January 1981. Alpha male disperses .8 km to east and joins a subordinate female from the X1 pack. By 4 January the alpha male and new female have moved within .1 km of the three H1-pack males.

6 January. Alpha male has rejoined H1 pack. X1 female has returned to her pack.

7–10 January. Alpha male reemigrates and joins single female 1.8 km to west, forming new pack.

19–31 January. The three yearling males disappear (probably emigrate), leaving H1-pack range devoid of mongooses.

b. Disintegration and rejuvenation of the A Pack. Following the death of the alpha female, and subsequently the alpha male, three yearling males emigrated, leaving only females and juveniles. The entire pack left the area, and small groups returned and then reemigrated during a period of instability that lasted about 6 weeks. Eventually, three females (including the new alpha female) stayed in the original range. One year later only the alpha female remained. She was joined by an adult male, rejuvenating the pact.

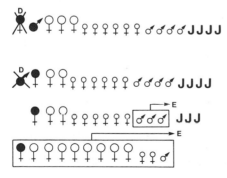

Ca. 15 March 1975. Alpha female in A pack dies.

August. Alpha male dies. New alpha female (the oldest daughter of the previous female breeder) produced one young in April.

25 September. Three yearling males emigrate.

5–12 November. Entire A pack emigrates. Several seen .5 km from usual range.

20 November. Seven return to usual range.

23 November. All reemigrate.

8 December. Three females return to usual range.

21 December. Three more females return.

25–31 December. Three females leave again.

24 January 1976. Alpha female gives birth to at least three young. She mated with an unknown male while away from natal range.

April. Two juveniles die.

September. Yearling female and last juvenile disappear.

October–November. Subordinate adult female disappears.

January 1977. Alpha female joined by new adult male, rejuvenating pack.

c. Rejuvenation of the Q Pack. Two males emigrated, leaving a yearling female. She remained alone in her natal range for over a year until she was joined by an adult male.

8 November 1981. Q pack consists of adult male and two yearlings. Alpha female from previous breeding season apparently has died.

25 November. Two males disperse 1.2 km to west and take over the N pack.

November 1982. Q-pack female still alone in natal range.

Ca. February 1983. Female joined by adult male, rejuvenating Q pack.

d. Formation of the D1 Pack. Two females moved into an unoccupied area. More than a year later the one remaining female was joined by an adult male. Subsequently they were joined by two transient males.

October 1980–January 1981. Two young adult females move in and establish a range in an area previously occupied by a pack that failed.

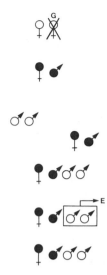

February–December 1981. One female disappears.

January–October 1982. Remaining female joined by a young adult male, forming D1 pack.

6 November. Two young adult transient males observed 1.8 km northwest of D1 pack range.

11 November. Two transient males join D1 pack.

13 November. Two transient males move 1.9 km to east.

8 December. Two transient males rejoin D1 pack.

# REFERENCES

Bekoff, M. 1977. Mammalian dispersal and the ontogeny of individual behavioral phenotypes. *American Naturalist* 111:715–32.

Broadbrooks, H. E. 1970. Home ranges and territorial behavior of the yellow-pine chipmunk, *Eutamias amoenus*. *Journal of Mammalogy* 53:310–26.

Bronson, F. H. 1964. Agonistic behavior in woodchucks. *Animal Behaviour* 12:470–78.

Christian, J. H. 1970. Social subordination, population density, and mammalian evolution. *Science* 168:84–90.

Clutton-Brock, T. H., and P. H. Harvey. 1976. Evolutionary rules and primate societies. In *Growing points in ethology*, ed. P. P. G. Bateson, 195–237. Cambridge: Cambridge University Press.

Dawson, G. 1977. Composition and stability of social groups of the tamarin, *Saguinus oedipus geoffroyi*, in Panama: Ecological and behavioral implications. In *The biology and conservation of the callitrichidae*, ed. D. G. Kleiman, 23–37. Washington, D.C.: Smithsonian Institution Press.

Demarest, W. J. 1977. Incest avoidance among human and nonhuman primates. In *Primate bio-social development*, ed. S. Chevalier-Skolnikoff and F. E. Poirier, 323–42. New York: Garland.

Eisenberg, J. F. 1981. *The mammalian radiations: An analysis of trends in evolution, adaptation, and behavior.* Chicago: University of Chicago Press.

Epple, G. 1975. The behavior of marmoset monkeys (callitrichidae). In *Primate behavior*, ed. L. A. Rosenblum, 4:195–239. New York: Academic Press.

Frame, L. H., J. R. Malcolm, G. W. Frame, and H. van Lawick. 1979. Social organization of African wild dogs (*Lycaon pictus*) on the Serengeti Plains, Tanzania, 1967–1978. *Zeitschrift für Tierpsychologie* 50:225–49.

Gaines, M. S., and L. R. McClenaghan. 1980. Dispersal in small mammals. *Annual Review of Ecology and Systematics* 11:163–96.

Greenwood, P. J. 1980. Mating systems, philopatry and dispersal in birds and mammals. *Animal Behaviour* 28:1140–62.

Harcourt, A. H. 1978. Strategies of emigration and transfer by primates with particular reference to gorillas. *Zeitschrift für Tierpsychologie* 48:401–20.

Healy, M. C. 1967. Aggression and self-regulation of population size in deermice. *Ecology* 48:377–92.

Hollekamp, K. E. 1984. Dispersal in ground-dwelling sciurids. In *The biology of ground-dwelling squirrels*, ed. J. O. Murie and G. R. Michener. Lincoln: University of Nebraska Press.

Kleiman, D. G. 1977. Monogamy in mammals. *Quarterly Review of Biology* 52:39–69.

———. 1981. Correlations among life history characteristics of mammalian species exhibiting two extreme forms of monogamy. In *Natural selection and social behaviour: Recent research and new theory*, ed. R. D. Alexander and D. W. Tinkle. New York: Chiron Press.

Lidicker, W. Z. 1975. The role of dispersal in the demography of small mammals. In *Small mammals: Their productivity and population dynamics*, ed. F. B. Golley, K. Petrusewicz, and L. Ryszkowski. New York: Cambridge University Press.

Mech, L. D. 1970. *The wolf: Ecology and social behavior of an endangered species.* New York: Natural History Press.

Moore, J. 1984. Female transfer in primates. *International Journal of Primatology* 5:537–89.

Moore, J., and R. Ali. 1984. Are dispersal and inbreeding avoidance related? *Animal Behaviour* 32:94–112.

Neyman, P. F. 1977. Aspects of the ecology and social organization of free-ranging cotton-top tamarins (*Saquinus oedipus*) and the conservation status of the species. In *The biology and conservation of the callitrichidae*, ed. D. G. Kleiman, 39–71. Washington, D.C.: Smithsonian Institution Press.

Packer, C. 1979. Inter-troop transfer and inbreeding avoidance in *Papio anubis*. *Animal Behaviour* 27:1–36.

Pusey, A. E. 1980. Inbreeding avoidance in chimpanzees. *Animal Behaviour* 28:543–52.

Rasa, O. A. E. 1977. The ethology and sociology of the dwarf mongoose (*Helogale undulata rufula*). *Zeitschrift für Tierpsychologie* 43:337–406.

Rood, J. P. 1974. Banded mongoose males guard young. *Nature* 248:176.

———. 1975. Population dynamics and food habits of the banded mongoose. *East African Wildlife Journal* 13:89–111.

———. 1978. Dwarf mongoose helpers at the den. *Zeitschrift für Tierpsychologie* 48:277–87.

———. 1980. Mating relationships and breeding suppression in the dwarf mongoose. *Animal Behaviour* 28:143–50.

———. 1983. The social system of the dwarf mongoose. In *Recent advances in the study of mammalian behavior*, ed. J. F. Eisenberg and D. G. Kleiman. Special Publication No. 7, American Society of Mammalogists, Shippensburg, Pa.

———. 1986. Ecology and social evolution in the mongooses. In *Ecological aspects of social evolution*, ed. D. Rubenstein and R. Wrangham. Princeton: Princeton University Press.

Rood, J. P., and D. Nellis. 1980. Freeze marking mongooses. *Journal of Wildlife Management* 44:500–502.

Steiner, A. L. 1972. Mortality resulting from intraspecific fighting in some ground squirrel populations. *Journal of Mammalogy* 53:601–3.

# 7. Natal Dispersal and the Formation of New Social Groups in a Newly Established Town of Black-tailed Prairie Dogs (*Cynomys ludovicianus*)

*Zuleyma Tang Halpin*

Among the terrestrial Sciuridae, prairie dogs (*Cynomys*), marmots (*Marmota*), and ground squirrels (*Spermophilus*) are characterized by complex forms of social organization (Michener 1983) and relatively little variation in patterns of dispersal (Dobson 1982). The female kin cluster is the basic social unit (Armitage 1981, 1984; Michener 1983), and the characteristic pattern of male dispersal, accompanied by female philopatry (Dobson 1982), reinforces this type of social organization. Even among solitary species, males are reported to disperse longer distances than do females, which usually settle in home ranges near those of their mothers and sisters (Haggerty 1968; Holekamp 1984; Vestal and McCarley 1984). Although under certain conditions male terrestrial sciurids may remain in or near their natal area (Michener and Michener 1973; Dunford 1977; Vestal and McCarley 1984; Rayor 1985), this appears to be relatively rare.

The lack of dispersal among female ground-dwelling squirrels results in the formation of the female kin groups, while male dispersal decreases the likelihood of inbreeding among close genetic relatives. The breeding males associated with the female kin clusters are not related to the females, and their association with each female group is generally transitory and impermanent. Thus, the sexually dimorphic dispersal patterns of these species clearly influence both the polygynous mating systems typical of the group and the genetic structure of terrestrial sciurid populations.

The black-tailed prairie dog, *Cynomys ludovicianus,* a species endemic to North America, is considered one of the most social species of ground-dwelling squirrels. King (1955) first described this species as living in large colonies or "towns" that are subdivided into distinct geographical areas called wards. The ward is further divided into a series of territories defended by distinct social groups or family units known as coteries. Typically, each coterie consists of one adult, breeding male, several adult, breeding females, and their yearling or juvenile offspring of both sexes; occasionally a coterie may contain more than one adult male (King 1955; Foltz and Hoogland 1981; Hoogland 1982; Hoogland and Foltz 1982; Halpin, pers. obs.) but this is relatively unusual. All members of a coterie, male and female, adult and young, actively defend the coterie territory from intruders on a year-round basis. Thus, interactions are generally aggressive between members of different coteries but amicable among all members of the same coterie. Mating usually occurs below ground, inside the burrows, during late winter, and breeding females defend a natal burrow against all other coterie members until their young are weaned and have emerged above ground (Hoogland 1982, pers. comm.). At all other times, coterie members are allowed equal access to all areas and burrows within the territory. The typical social structure of the black-tailed prairie dog, in which usually only one breeding male is associated with a group of females, results in a strongly biased adult sex ratio in favor of females, a situation that is the rule among the terrestrial sciurids (Dobson 1982).

In considering how dispersal can affect the genetic structure of prairie dog populations, three distinct types of movements or processes can be distinguished:

1. *Long distance dispersal.* This involves the movement of animals between towns. Emigration before the age of sexual maturity would tend to reduce the chances of inbreeding, while immigration would promote gene flow between towns and would result in increased genetic variability within populations.

2. *Dispersal between coteries.* These are movements in which a prairie dog leaves its natal coterie and breeds in another, previously established coterie within the same town. Such movements result in the establishment of new breeding assemblages, thereby contributing to gene exchange within the prairie dog town. Dispersal away from the natal coterie before breeding also reduces the probability of extreme inbreeding.

3. *Formation of new coteries.* In an established prairie dog town, new coteries most frequently form by either fission or fusion. The process of fission, in which one large coterie splits into two, is more common. Fusion occurs when two small coteries, or remnants of coteries, join together to form a new coterie; fusion may be rare and may occur only under special circumstances.

Until recently there had been relatively little information available on dispersal patterns of black-tailed prairie dogs or on the processes by which new coteries are formed. Additionally, most previous studies (King 1955; Tileston and Lechleitner 1966; Hoogland 1981) had been performed on prairie dog towns that were relatively large and stable (but see Garret, Hoogland, and Franklin 1982). The present study supplements the earlier findings and reports on long-distance dispersal, dispersal between coteries, and the formation of new social groups in a recently established, small, and fluctuating colony of black-tailed prairie dogs.

Although breeding dispersal occurs among prairie dogs (King 1955), this chapter is limited to natal dispersal—the movement of prereproductive animals away from the area in which they were born and into a new area where they may reproduce.

## MATERIALS AND METHODS

This study was conducted at a small *Cynomys ludovicianus* town, consisting of only one ward, located at the Quivira National Wildlife Refuge in Stafford County, Kansas. Two other small prairie dog towns were located within approximately 3 km and 6.5 km of the study area. There was relatively little human traffic or disturbance near the study site.

According to refuge records, the town was established in 1975. My study began in 1977, and during the first 3 years of study the prairie dog population expanded rapidly. However, as a result of a shooting incident in 1980, the population was drastically reduced in size, and by midsummer of 1982 the town had virtually disappeared. The area occupied by the town increased from 0.6 ha in 1977 to 1.7 ha in 1979; after the shooting in 1980 the size dropped to 1.1 ha and continued to decline until the end of the study. The number of coteries fluctuated between a minimum of four in 1977 to a maximum of nine in late 1978. The town was bounded on the east and west by a salt marsh and a creek, but there was ample room for expansion in other directions.

Observation periods in 1977, 1978, and 1979 consisted of approximately 2 weeks per month in March, May, July, and August; in 1980 and 1981, animals were watched from mid-May to mid-August, and in 1982 observations were conducted in May and June. Animals were live-trapped, sexed, weighed, ear tagged with individually numbered metal tags, and marked with individual patterns of black hair dye (Lady Clairol).

The presence of individuals within the prairie dog town was determined by massive trapping at the beginning of each year's observation period and by continued trapping at regular intervals throughout the rest of the summer. Data on each animal's movements and behaviors were collected by using three different sampling techniques (Altmann 1974). Once the animals had been trapped and individually marked, scan samples of the prairie dog town, taken at 30-minute intervals, were used to record on a map of the town the location and behavior of every individual that was visible. This technique allowed us to determine the size and composition of coteries, and the size and boundaries of coterie territories; it also provided a temporal record of each animal's movements and social interactions not only throughout each summer's observation period but also from one year to the next. In addition to the scan samples, any time an animal was observed in an unusual location or engaging in unusual behaviors, I recorded this observation as part of ad libitum data collection. Finally, 30-minute focal animal samples provided additional information on movements and on social interactions between and among animals.

During each observation period animals were observed for approximately 7 to 9 hours each day. On relatively cool days (<33° C) this observation time was distributed throughout the day, while on hot days (>33° C) the animals were generally observed for 4 to 5 hours in the morning and for 3 to 4 hours in the late afternoon or evening.

These methods also allowed us to determine with some certainty the familial relationships of the animals within the prairie dog town. Beginning in the summer of 1978, the young born each year were trapped and marked soon (generally within 3 weeks or less) after they emerged from their natal burrows. Since normally there is only one breeding male within a prairie dog coterie (Hoogland 1982; Hoogland and Foltz 1982), two assumptions were made: if only one breeding female was present in a coterie, all the young born in that coterie were assumed to be full siblings; if more than one breeding female was present, and since young from different litters within the same coterie mix almost immediately after emergence, it was assumed that all young in the coterie were at least half sibs. Although in most cases there is a high probability that both of these assumptions are correct, it should be noted that Foltz and Hoogland (1981), Hoogland and Foltz (1982), and Hoogland (1982) have found evidence that in some cases more than one male may father young within the same coterie.

Most of the following data are based on observations made between 1978 and 1982. In 1977 I attempted to use plastic, colored ear tags to mark the animals; all of these tags were removed by the prairie dogs and, therefore, I have only limited data for that year.

## RESULTS

A total of 193 prairie dogs were trapped and marked during this study; of these, 156 were juveniles trapped soon after emergence from their natal burrows. As in other studies, the prairie dogs in my area were strongly polygynous, with a mean of 1.2 adult males (range 1–3) and 2.9 adult females (range 1–7) per coterie. The modal situation was for there to be only one adult male in a coterie. When more than one adult male was present, it is possible they were related; for example, in one case in which three males simultaneously occupied the same coterie, they were known to be brothers. Among the animals trapped as juveniles, the sex ratio approximated 1.0 (75 males and 81 females). The sex ratio of breeding adults, however, was quite different, with 2.5 females for every 1 male (this figure includes breeding yearling females; nonbreeding yearling females and yearling males, which as a class do not normally breed, are not included). Thus, whether one considers coterie composition or the total adult population of the town, there was a strong sex bias in favor of females. Although

most other studies have reported that breeding by yearling females is uncommon, in my study 76% (range 40% to 100%) of the yearling females
remaining in their natal coteries bred, as evidenced by their enlarged (lactating) nipples. Normally, yearling males did not breed, although it is possible that a few did so in 1981, the year following the shooting incident in
which most adult males were killed.

In discussing the effects of long-distance dispersal on the genetic structure of prairie dog populations, two types of movements must be considered: (1) the emigration of animals away from the town in which they were
born, and (2) the immigration of new animals into already established
towns. In my prairie dog town, 73% of males trapped as juveniles disappeared before or during the summer of their yearling year, while the same
was true for only 31% of the females (table 7.1). Although these percentages
undoubtedly include animals that died (either before or after the initiation
of dispersal) as well as animals that successfully dispersed and reproduced
elsewhere, the effects on the natal town are effectively the same. Whether
an animal dies or disperses from its natal town, it is no longer available as
a potential mate, and as a genetic unit it is lost from the population. However, since not all animals that disperse make a genetic contribution to other
populations, the distinction between animals that die and animals that disperse successfully is important when one considers the effects of dispersal
on gene flow among prairie dog towns.

In the present study, the number of new animals that appeared and successfully bred in the study area was used as an indication of the frequency
of successful migrations between towns. During the 5 years of study only
three males of unknown origin (and no females) were trapped in the town
(table 7.2), and of these three males, only one gave evidence of being an
outsider. In the summer of 1979, this animal, which had an unusual number
of scars on its face and head, established a territory by himself on the pe-
riphery of the town (fig. 7.1, territory E) and proceeded to venture into
neighboring territories until he successfully joined a group of females in territory A. The other two males were also first trapped in 1979, but it is likely
that they were not true immigrants, but rather animals that had escaped

TABLE 7.1 Percentage of Yearling Animals Trapped during Their First Summer That Disappeared as Yearlings, 1978–1982

|  | N | % DISAPPEARED (exclud. 1980) | % DISAPPEARED (includ. 1980) | RANGE |
|---|---|---|---|---|
| Males | 75 | 73% | 82% | 64–95% |
| Females | 81 | 31% | 44% | 22–65% |

Notes: Animals known to have died are not included in these figures. In 1980 a shooting incident in the
prairie dog town is thought to have killed many young, and the figures for this year are probably unusually
high; for this reason, data are given including and excluding 1980.

TABLE 7.2 Males of Unknown Origin, Possibly Long-Distance Migrants, That Appeared in the Prairie Dog Town, 1979–1982

|  | 1979 | 1980 | 1981 | 1982 |
|---|---|---|---|---|
| No. of animals | 3 | 0 | 0 | 0 |
| % of adult and yearling male pop. | 13.6% | 0% | 0% | 0% |
| % of total adult pop. | 6.4% | 0% | 0% | 0% |

trapping during the previous year. All three males stayed and bred in the town. Regardless of whether there was one immigrant or three, it is clear that the rate of immigration into my study area during the period of study was low. In summary, the results suggest that in terms of long-distance dispersal, more males disperse than females, with the majority of males disappearing before or during their yearling year; the fate of most long-distance dispersers is unknown but, at least in the present study, the rate of successful immigrations appears to be low.

The sexual dimorphism observed in long-distance dispersal was also evident in dispersal between coteries (table 7.3). A total of 10 males remained in the prairie dog town after their yearling year, but of these none stayed in its natal coterie. Some of these males moved to and bred in neighboring coteries, but others moved to more distant coteries (fig. 7.2).

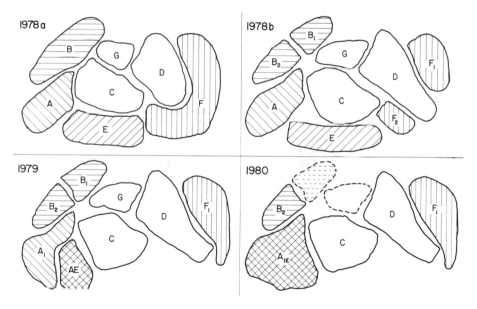

FIGURE 7.1. Changes in coterie territories during the period of study. Lines and cross-hatching indicate coteries that underwent fission or fusion. Dashed borders indicate coteries that disappeared. After 1980, territories remained stable.

TABLE 7.3  Percentage of Yearling Animals That Remained in the Prairie Dog Town and Later Bred in Their Natal Coterie, 1978–1981

|  | N | NO. MOVING | % REMAINING |
|---|---|---|---|
| Males | 10 | 10 | 0% |
| Females | 51 | 1 | 98% |

*Note:* In 1982 there were no signs of reproduction or breeding activity in the colony, so data for this year are not available.

In contrast to the males, of the 51 yearling females remaining in the town, only 1 changed coteries. All females remaining in the prairie dog town eventually bred. These results suggest that even when yearling males remain in the prairie dog town, they rarely, if ever, breed in their coterie of origin, while females almost never disperse to other groups.

Both fission and fusion of coteries were observed during the period of study. In July 1978, coterie B, originally consisting of three males and nine adult and yearling females, split into two coteries, B1 and B2 (fig. 7.1a, b). B1 was composed of two males and four females, while B2 contained one male and four females. One female, present before the coterie divided, could not be accounted for afterward. The final split of this coterie was preceded by a period of intense aggressive interactions between the animals from the two new coteries. Unfortunately, since this split occurred during the early part of the study, the previous history and genetic relationships of the animals involved were not known. Coterie B2 persisted until the end of the study in 1982, but coterie B1 disappeared after the shooting incident in 1980.

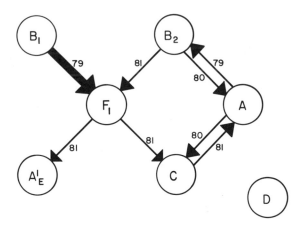

FIGURE 7.2.  Transfer of males between coteries during 1979–81. With the exception of the thick arrow, each arrow represents only one animal. The thick arrow represents 3 sibling males. The number above each arrow refers to the year the transfer occurred.

In May 1978 the animals from coterie D invaded and took over part of the territory of coterie F. Most of the animals from F shifted their range and continued to occupy the area designated as F1 (fig. 7.1a, b). However, one lactating female, whose young had not yet emerged above ground, remained and continued to defend her burrow in area F2. This female and her young had disappeared from the prairie dog town by the spring of 1979.

Coterie A was involved in a complicated series of events that included both fissions and fusions. During the summer of 1978, the resident male in coterie A was killed in a trapping accident. By March 1979 the immigrant male from E had joined three yearling females from A and formed a new coterie designated as AE (fig. 7.1c). The remaining adult and yearling females from A formed coterie A1 (fig. 7.1c) with another of the males of unknown origin. After the two males from these two coteries were killed in the shooting incident of 1980, the females remaining from both coteries were joined by a yearling male from B2, thus forming coterie A1E (fig. 7.1d). This male, a first cousin to the females in A1E, disappeared in late summer of 1980; in 1981 the females bred with a male that had transferred from F1. Coterie A1E remained in the study area until its disappearance in 1982.

Joint movements by related individuals are also of interest, and it appears that related males sometimes disperse together. In one case, three sibling males moved from coterie B1 to F1. One of these males initially moved as a yearling in 1978 from B1 to A, but by March 1979 he was in F1 where he had apparently bred. His two brothers (either full or half siblings) which had stayed in B1 also moved to F1. A second case involved three yearling brothers from A that in 1979 dispersed away from the prairie dog town. During the weeks preceding their final departure, these three animals wandered increasing distances away from the prairie dog town and eventually constructed a series of burrows approximately 60 meters from the main town. In July 1979 the three males disappeared simultaneously. A sister that had frequently accompanied them on their wanderings returned to her coterie in the prairie dog town.

## DISCUSSION

The results of this study are in accordance with those of earlier studies in showing that (1) black-tailed prairie dogs are strongly polygynous, with breeding units typically composed of one male and several females (King 1955; Hoogland 1981; Hoogland and Foltz 1982); (2) there is a sexually dimorphic pattern of dispersal, with males being more likely to disperse than females (King 1955; Hoogland 1981); (3) young males that stay in the prairie dog town move away from their natal coterie and breed in a different coterie, while females almost always stay and breed in their coterie of origin (Hoogland 1981, 1982); and (4) under certain conditions, large coteries may fission into smaller coteries (Hoogland 1981).

The low rate of immigration into my study area suggests that dispersal and gene flow between prairie dog towns may be low. Chesser (1983) found the same for a series of 21 prairie dog populations in New Mexico, but Foltz and Hoogland (1983) reported a high rate of male migration among their colonies. These differences may be partially due to different characteristics of the prairie dog populations studied by Foltz and Hoogland (1983) and those studied by Chesser (1983) and myself. The Wind Cave populations (Foltz and Hoogland 1983) appear to be relatively large and stable, with many towns or colonies in close proximity. In my area, there were only two other small prairie dog towns in the vicinity. The closest of these towns (3 km) disappeared after the second year of the study. The only other known prairie dog town that could have provided immigrants was 6.5 km away. In general, prairie dog towns in this area of Kansas tend to be small, unstable, and relatively isolated by distances and by unsuitable habitat (primarily farms, roads, human towns, etc.). Out of six prairie dog towns located in 1977, for example, only two were still known to be in existence by the summer of 1982. Shooting and poisoning activities are common and are known to have contributed to the extinction of several of these towns. Although Chesser (1983) did not provide data on size and stability of colonies, his report suggests barriers to intertown movements comparable to the ones I observed.

Some of the genetic effects of prairie dog social organization and dispersal patterns are clear. Males do not normally breed until they reach 2 years of age, and the pattern of yearling male dispersal accompanied by female philopatry reduces the likelihood of brother-sister and mother-son matings, thus reducing extreme inbreeding. It is possible, nonetheless, that a breeding male, especially if he stays in the same coterie for several years, could mate with his own daughters. Hoogland (1982) argues that this is unlikely because (1) most males stay in the same coterie for only 1 or 2 years, (2) yearling females are less likely to come into estrus if their father is present, and (3) estrous females will avoid mating with their father if he remains in the coterie. Although the shooting incident at my town makes it impossible to say how long most males would have stayed in the same coterie, at least some breeding males occupied the same coterie for more than 2 years, and one did so for 4 years. Furthermore, there was no evidence that yearling females were less likely to come into estrus when their father was present; the majority of yearling females (76%) came into estrus and produced young regardless of whether their father was present or not. It is possible, nonetheless, that Hoogland's (1982) third condition also operated in my town. Since I was not present during the breeding season, I have no definitive information on the yearling females' choice of a mating partner; my yearling females, like Hoogland's, may, in fact, have bred with neighboring, unrelated males. In any event, and regardless of the exact details, it seems clear that male dispersal and female philopatry greatly reduce the frequency of inbreeding with close relatives within prairie dog coteries.

Female philopatry is also important in that it results in the formation of coteries in which groups of closely related females are associated with (usually) an unrelated male. This close genetic relationship among the females can be expected to result in a higher degree of genetic similarity among coterie members than among individuals chosen at random from different coteries.

The movement of males between coteries may promote gene flow and prevent the extreme genetic differentiation of coteries. In the present study, dispersal between coteries was common; only one major coterie, D (fig. 7.2), was not involved in intercoterie dispersal either as a source or recipient of migrants. Although it is not clear if this high rate of intercoterie movements is also typical of older, larger, and more stable towns, reports by Hoogland (1981) and Foltz and Hoogland (1983) suggest that this may be the case.

Recent studies on the genetic structure of prairie dog populations provide conflicting information. Chesser (1983) analyzed genetic variation for seven polymorphic protein loci in 21 populations and found significant genetic differentiation even for prairie dogs from populations in relatively close proximity (5–15 km apart). Furthermore, individual towns showed a high degree of local differentiation. These results are consistent with the low levels of intertown dispersal observed in the present study. Chesser, however, also found higher than expected levels of homozygosity within coteries and suggests that this is probably due to inbreeding, or possibly to the high level of relatedness among coterie members. Also, he found the near fixation of rare alleles in certain towns and a lack of correlation between genetic and linear distances; on these bases he concluded that founder effects and genetic drift are important factors in prairie dog population genetics.

In a similar study, Foltz and Hoogland (1983) reported on the genetic structure of a prairie dog population as determined by electrophoretic analysis of four polymorphic protein loci and by pedigree analyses based on long-term behavioral observations. They found an overall excess of heterozygotes and a negative coefficient of nonrandom mating, indicating an avoidance of inbreeding. Additionally, electrophoretic paternity analyses suggested that inbreeding between close relatives is extremely rare and may have occurred only once (a father-daughter mating) during the 4 years of the study. Foltz and Hoogland also compared their main study colony with another located at 10 km distance and found only moderate genetic differentiation between the towns. This result is consistent with the higher rates of male migrations between towns observed in their study.

As discussed earlier the differences between the results of Foltz and Hoogland (1983) and those of Chesser (1983) may be due, at least in part, to geographical variations in the characteristics of prairie dog populations. The patterns of male dispersal and intercoterie movements observed in my study are similar to those reported by Hoogland (1981, 1982) and Foltz and Hoogland (1983), and can be expected to result in lower levels of inbreeding and increased gene flow within the town. The low rate of immigration into my town suggests low levels of intertown dispersal and, perhaps, greater local differentiation of the prairie dog populations in this area; such a result would be compatible with the findings of Chesser (1983). Finally, the high degree of relatedness of the females in each coterie can be expected to result in higher levels of genetic similarity within coteries, such as was reported by Chesser (1983).

Three major functions or ultimate causes of dispersal have been suggested for terrestrial sciurids (Holekamp 1984): the avoidance of competition for limited environmental resources, avoidance of competition for mates, and avoidance of inbreeding. Although data are limited, there is, at present, little evidence that the typical pattern of black-tailed prairie dog natal dispersal is related to the availability of environmental resources. In all populations so far studied, virtually all yearling males disperse from their natal coteries, regardless of the quality or quantity of resources in any given year (but see Rayor's 1985 study on Gunnison's prairie dogs).

Dobson (1982) argues that if avoidance of inbreeding were the primary function of sciurid dispersal, then one would expect *either* males or females to disperse. However, there is little variation in the dispersal patterns of terrestrial sciurids, and in all species it is the males that constitute the predominantly dispersing sex. Such a pattern is consistent with the hypothesis that male sciurids disperse in order to avoid competition over mates. Given the polygynous mating systems of ground-dwelling squirrels and the adult sex ratio bias in favor of females, males are the limited sex, and male competition for females can be expected to occur. These conditions strongly favor male dispersal (Dobson 1982).

This hypothesis is appealing, but on closer inspection several problems become apparent, at least in regard to black-tailed prairie dogs. Dispersal as a mechanism to reduce competition over mates should be favored only as long as dispersing males have a reasonable probability of finding a mate elsewhere. Among prairie dogs, the fates of most long-distance dispersers are unknown and the evidence on successful immigration into already established towns is equivocal. In some areas migrations between towns may be relatively common (Hoogland 1981; Foltz and Hoogland 1983), while in others (Chesser 1983; this study) they may be rare. All yearling males staying in my town eventually bred, either in a coterie in which there was no resident male or in a coterie from which the resident male eventually disappeared. Thus, one possible alternative to long-distance dispersal is to stay in the prairie dog town and wait for the opportunity to take over another coterie once the breeding, resident male has disappeared or weakened. Furthermore, if inbreeding considerations are not important, one would expect at least some males to stay in their natal coterie and take it over from their father, once the latter has left or died. This was never observed in my prairie dog town; even when breeding males were killed during the shooting incident, their surviving male offspring either left the prairie dog town or moved to a different coterie. The inbreeding avoidance hypothesis is further strengthened by Hoogland's (1982) observations that even when fathers and daughters are present in the same coterie, they rarely inbreed. On the other hand, it is also interesting that, over time, high rates of male dispersal between coteries probably result in matings among genetic relatives, such as when the male dispersing into a coterie is the cousin or uncle of resident females. Thus, while extreme inbreeding is probably rare, matings between more distant genetic relatives may not be uncommon.

These findings, however, do not obviate the fact that male-male competition may be an important factor in prairie dog coteries. The large number of scars on the immigrant male discussed in "Results" attests to his involvement in aggressive interactions. Furthermore, the behavioral profiles of yearling males that stayed in the prairie dog town showed that, as juveniles, they had been involved in many more agonistic encounters than had the yearling males that disappeared from the town (Halpin, unpub. data). Thus, it is possible that only those males of high competitive ability (as reflected in aggressive interactions) are able to remain and successfully breed in their natal town.

Alternatively, males may leave their natal coteries as a result of female choice (Hoogland 1982). If females recognize their brothers and refuse to mate with them, young males may have no choice but to leave their coterie of origin and seek a mate elsewhere. Such a mechanism would combine avoidance of inbreeding (by the females) with competition for mates (by the males) as the driving forces of prairie dog dispersal. Although sibling recognition has been documented in other terrestrial squirrels (Davis 1982; Holmes and Sherman 1982; Holmes 1984), it has not been tested in prairie dogs. Furthermore, although the available data (Hoogland 1982) suggest that females may avoid mating with their fathers, experiments aimed at testing inbreeding avoidance with brothers are necessary before the role of female choice can be adequately evaluated.

Ultimately, the question of function is one that must address the issue of costs and benefits and cannot be answered with certainty at the present time because of lack of much essential information. We have no information, for example, on the costs of prairie dog dispersal; we do not know what the probability is that a dispersing male will survive, find a mate, and breed successfully elsewhere. Even if a dispersing male arrives at a prairie dog town, he may not be any more successful in acquiring a coterie than he would have been had he stayed in his natal town. Furthermore, since dispersal generally occurs over inhospitable habitat, the mortality rate of dispersing animals is probably high. Thus, long-distance dispersal is likely to be dangerous and the costs high for most animals. We also do not know how new prairie dog towns are formed, nor do we have adequate information on the effects of inbreeding in prairie dogs. Clearly, in order to adequately test the current hypotheses on the functions of sciurid dispersal, much additional information is necessary.

Another confounding factor is that the conditions under which we study prairie dogs at the present time may have little relevance to the conditions under which they evolved and lived before the arrival of European settlers in North America. Descriptions by early European explorers suggest that black-tailed prairie dog colonies stretched over extensive areas of land and consisted of tens of thousands of animals. As a result of human disturbance and the destruction of their habitat, prairie dogs, over most of their range, now live in small, scattered, and ephemeral populations. Patterns of dispersal that may have been adaptive and involved low risks under the earlier conditions may result in higher costs and less certain benefits in modern-day populations.

## ACKNOWLEDGMENTS

This work was supported by grants from the National Geographic Society, the National Academy of Sciences Marsh-Henry Fund, the University of Missouri Weldon Springs Fund, and by special awards from the University of Missouri–St. Louis. I am grateful to Douglas Dalrymple, David A. Chisholm, Katherine C. Noonan, Connie Quinlan, Marsha Hodges, Barbara Rapoza, Paul Strickberger, Vicki Vasileff, and Mary Merello for their assistance in the field. I thank Charles Darling, director of Quivira National Wildlife Refuge, Pat Walker, and John and Donna Nystrom for their assistance and hospitality. Connie Quinlan helped tabulate the data and made helpful suggestions. I also thank David Chisholm for drawing the figures and Arlene Zarembka and Jill Trainer for reading and commenting on the manuscript.

## REFERENCES

Altmann, J. 1974. Observational study of behavior: Sampling methods. *Behaviour* 49:227–67.

Armitage, K. B. 1981. Sociality as a life-history tactic of ground squirrels. *Oecologia* 48:36–49.

———.1984. Recruitment in yellow-bellied marmot populations: Kinship, philopatry, and individual variability. In *The biology of ground-dwelling squirrels*, ed. J. O. Murie and G. R. Michener, 377–403. Lincoln: University of Nebraska Press.

Chesser, R. K. 1983. Genetic variability within and among populations of the black-tailed prairie dog. *Evolution* 37:320–31.

Davis, L. S. 1982. Sibling recognition in Richardson's ground squirrels (*Spermophilus richardsonii*). *Behavioral Ecology and Sociobiology* 11:65–70.

Dobson, S. F. 1982. Competition for mates and predominant juvenile male dispersal in mammals. *Animal Behaviour* 30:1183–92.

Dunford, C. 1977. Behavioral limitations of round-tailed ground squirrel density. *Ecology* 58:1254–68.

Foltz, D. W., and J. L. Hoogland. 1981. Analysis of the mating system of the black-tailed prairie dog (*Cynomys ludovicianus*) by likelihood of paternity. *Journal of Mammalogy* 62:706–12.

———. 1983. Genetic evidence of outbreeding in the black-tailed prairie dog (*Cynomys ludovicianus*). *Evolution* 37:273–81.

Garret, M. G., J. L. Hoogland, and W. L. Franklin. 1982. Demographic differences between an old and a new colony of black-tailed prairie dogs (*Cynomys ludovicianus*). *American Midland Naturalist* 108:51–59.

Haggerty, S. M. 1968. The ecology of the Franklin's ground squirrel (*Citellus franklinii*) at Itasca State Park, Minnesota. M.S. thesis, University of Minnesota, Minneapolis.

Holekamp, K. E. 1984. Dispersal in ground-dwelling sciurids. In *The biology of ground-dwelling squirrels*, ed. J. O. Murie and G. R. Michener, 297–320. Lincoln: University of Nebraska Press.

Holmes, W. G. 1984. Sibling recognition in thirteen-lined ground squirrels: Effects of genetic relatedness, rearing association and olfaction. *Behavioral Ecology and Sociobiology* 14:225–33.

Holmes, W. G., and P. W. Sherman. 1982. The ontogeny of kin recognition in two species of ground squirrels. *American Zoologist* 22:491–517.

Hoogland, J. L. 1981. Nepotism and cooperative breeding in the black-tailed prairie dog (Sciuridae: *Cynomys ludovicianus*). In *Natural selection and social behavior: Recent research and new theory*, ed. R. D. Alexander and D. W. Tinkle, 238–310. New York: Chiron Press.

———. 1982. Prairie dogs avoid extreme inbreeding. *Science* 215:1639–41.

Hoogland, J. L., and D. W. Foltz. 1982. Variance in male and female reproductive success in a harem-polygynous mammal, the black-tailed prairie dog (Sciuridae: *Cynomys ludovicianus*). *Behavioral Ecology and Sociobiology* 11:155–63.

King, J. A. 1955. Social behavior, social organization, and population dynamics in a black-tailed prairie dog town in the Black Hills of South Dakota. *University of Michigan Laboratory of Vertebrate Biology Contibutions* 67:1–123.

Michener, G. R. 1983. Kin identification, matriarchies, and the evolution of sociality in ground-dwelling sciurids. In *Advances in the study of mammalian behavior*, ed. J. F. Eisenberg and D. G. Kleiman, 528–72. Special Publication, American Society of Mammalogists, Shippensburg, Pa.

Michener, G. R., and D. R. Michener. 1973. Spatial distribution of yearlings in a Richardson's ground squirrel population. *Ecology* 54:1138–42.

Rayor, L. S. 1985. Effects of habitat quality on growth, age of first reproduction, and dispersal in Gunnison's prairie dogs (*Cynomys gunnisoni*). *Canadian Journal of Zoology* 63:2835–40.

Tileston, J. V., and R. R. Lechleitner. 1966. Some comparisons of the black-tailed and white-tailed prairie dogs in north-central Colorado. *American Midland Naturalist* 75:292–316.

Vestal, B. M., and H. McCarley. 1984. Spatial and social relations of kin in thirteen-lined and other ground squirrels. In *The biology of ground-dwelling squirrels*, ed. J. O. Murie and G. R. Michener, 404–23. Lincoln: University of Nebraska Press.

# 8. Dispersal Patterns in Kangaroo Rats (*Dipodomys spectabilis*)

*W. Thomas Jones*

Several authors in this volume (e.g., Lidicker and Patton, chap. 10; Chepko-Sade and Shields et al., chap. 19) note that understanding genetic population structure in mammals depends in part on gathering more data on determinants of population structure, including dispersal. However, since Howard's (1949) study of dispersal in *Peromyscus maniculatus*, few other dispersal studies of mammals have equaled his in terms of sample size and quality of data (but see Sullivan 1977; Tamarin 1977; Jannett 1978; Dobson 1979; Clutton-Brock, Guinness, and Albon 1982; Bunnell and Harestead 1983; King 1983; Smith and Ivins 1983 for examples of other good data sets). Given the difficulties of documenting dispersal patterns in wild populations, this situation is understandable. Small mammals can be followed in large numbers and over periods exceeding the maximum life span of individuals, but maternity is difficult to determine. Consequently, the origins of juvenile dispersal movements are often unknown. Natal sites are generally easier to establish for large mammals, but this advantage is offset by the considerable effort required to obtain substantial samples. Studies that have successfully done so often span more than a decade (e.g., Clutton-Brock, Guinness, and Albon 1982).

My aim here is to provide a large and detailed set of dispersal data for a free-living mammal. The banner-tailed kangaroo rat (*Dipodomys spectabilis*), a solitary, nocturnal, desert granivore, is uniquely suited to dispersal studies. It occurs at such sufficiently high densities that many animals can be studied at once, and maternity, and thus natal sites, can be established reliably for most juveniles. Moreover, this species inhabits burrows contained in large conspicuous mounds, individuals can be readily mapped within the study area, and dispersal distances can be accurately measured when movement occurs.

## METHODS

### Natural History of *D. spectabilis*

Although individual *D. spectabilis* live at most 4 years, the mounds they inhabit persist much longer, probably decades, and are passed on from individual to individual within populations. New mounds are rarely constructed (Holdenreid 1957), a process that requires as long as 2 years (Best 1972), so juveniles must usually acquire existing mounds when they become independent of their parents (Jones 1984). As adults, *D. spectabilis* nearly always live in separate mounds, and males' mounds are interspersed among females' mounds.

Most reproduction occurs from February to April when females have one or two litters of one to three offspring. By September virtually all surviving offspring are independently established in separate mounds (Jones 1984). *D. spectabilis* usually reach reproductive maturity in the winter following their birth.

### Data Collection

I studied dispersal in *D. spectabilis* on a square 36 ha site located 6.4 km east northeast of Portal, Cochise County, in southeastern Arizona. The data are from censuses conducted every 2 to 4 months from July 1979 to March 1984. The initial step in each census was to check each mound on the site for such signs of occupancy as well-traveled entrances, runways, and digging patterns characteristic of *D. spectabilis*. In most cases determining occupancy from these signs was straightforward, and all ambiguous cases were clarified by trapping. If I could not trap a rat at a mound with ambiguous signs, I considered the mound unoccupied.

Subsequently I placed at least two Sherman live traps within 1 m of each occupied mound. Each individual I captured received two numbered ear tags, and the tag number, sex, weight, trap location, age class, and reproductive condition were recorded at each capture. Individuals that were reproductively immature, as determined unambiguously from the condition of the nipples or scrota, were considered juveniles. All animals were released where captured immediately after processing.

Repeated trapping within each census allowed me to establish each individual's presence or probable absence in the population, determine each individual's home mound, and determine maternal identity of most juveniles. I assigned ownership of a mound to an individual if I caught it there at least twice and if it was caught there more times than any other adult. I could determine maternity of juveniles because offspring reside in maternal mounds for 1 to 5 months after reaching trappable age (Jones 1984). I considered the maternal identity of a juvenile to be established when I caught it at least twice at an adult female's mound in combination with no captures at any other adult female's mound. To check the accuracy of these methods I radiotracked a subset of the population to their mounds in the daytime using a hand-held antenna. Such radiotelemetry data are accurate to within 1 m.

Repeated censusing provided a means of following dispersal or lack of dispersal from natal mounds, as well as a means of detecting any movements by adults. Natal mounds were those occupied by adult females at the time their litters were born. An individual was considered to have moved to a new mound if its home mound location changed in successive censuses as determined by the criteria of mound ownership stated earlier. I measured dispersal for each juvenile as the distance from the natal mound, or the mound inhabited when first captured as a juvenile if maternity was not definitely known, to the mound occupied during the first breeding season. Thus juvenile dispersal data presented here are for successful dispersal; individuals that died before reaching what would have been their first reproductive season were not included. Adult dispersal movements were measured from the mound occupied at the beginning of their first breeding season to the last mound occupied before disappearance. Adult data were recorded for all individuals that persisted on the site as adults for at least 1 year. To derive distributions of dispersal distances I grouped the data into 50 m intervals, the approximate mean distance between neighboring adults (unpub. data). All distances were measured to the nearest meter from an aerial photograph. The amount of dispersal beyond the study site boundary was estimated by three methods. The first adjusts the distribution of dispersal distances according to the probability that dispersal is contained within the study site (Barrowclough 1978). This probability is a function of radius $R$ of the study site (a circular site is assumed; I used $R = 375$ m to obtain a best-fit circle), the distance $r$ from the site center to the natal site, and the dispersal distance $x$ from the natal site to the breeding site. For $R > r > R - x$ the probability is equal to

$$\{\pi - \cos^{-1} [(R^2 - x^2 - r^2) / 2xr]\} / \pi$$

(Barrowclough 1978, 335). For $0 < r < R - x$ it is 1. For each possible $r$ and $x$ value, I divided the observed number of animals by the probability of detecting dispersal within the site to derive an adjusted distribution. This method can account for movement off the site up to a distance equal to the site's greatest dimension, in this case 600 m or 0.9 km from the site center.

The second method of estimating the amount of movement off the site makes use of an adjacent study area on which J. Brown and others are conducting a study of rodent community structure. On this site kangaroo rats are censused monthly (see Munger and Brown 1981). The adjacent site covers a sector of radius 0.9 km and width 35 degrees centered on my site. By checking how many animals from my site were trapped on the adjacent site I could sample the amount of movement off my site up to nearly a kilometer in 10% of the surrounding area.

The third check for movement off the site was a single long-distance trapping census conducted by P. Waser in autumn 1982. All 735 mounds in a transect 400 m wide and 3 km long were trapped two to three times.

RESULTS

Accuracy of Trapping Data

The methods used to determine home mound location and maternity were reliable. In 62 instances individuals' home mounds were determined by radiotelemetry as well as trapping. These methods yielded conflicting results in only two cases. Twenty-three maternal identities established by repeated trapping were checked with radiotelemetry, and in no instance did these methods disagree.

Juvenile Dispersal

Juvenile dispersal distributions were leptokurtic for both males and females (fig. 8.1). Eighty percent of 96 males and 77% of 99 females remained within 50 m of their natal sites or points of first capture. These data include 49 males and 30 females for which maternity, and thus natal site, was definitely known, and another 47 males and 69 females for which maternity was not definitely known but which were first captured as juveniles and were thus still likely to be in natal vicinities. There were no sex differences in the proportion of individuals moving less than 50 m among either the known-maternity juveniles (0.84 for males and 0.77 for females; $p = .44$, $G$-test) or among all juveniles ($p = .56$, $G$-test). Thirty-seven percent of known-maternity males remained in their natal mounds through reproductive maturity (i.e., moved 0 m), as did 23% of known-maternity females. There was no significant difference between these percentages ($p = .21$, $G$-test), nor between the percentages of all 96 males (42%) and 99 females (41%) that did not move before reproductive maturity ($p = .97$, $G$-test). The median dispersal distances for known-maternity individuals were 17 m for males and 30 m for females. Mean dispersal distances were 29 m for males and 66 m for females among known-maternity juveniles, which were not significantly different ($p > .5$, two-tailed $t$-test with unequal variances). There were no significant sex differences in the distributions of dispersal distances when compared by means of a chi-squared test for either the known-maternity juveniles ($p = .27$) or for all juveniles ($p = .43$). However, there was a female bias in the proportion of long-distance dispersers that remained within the study area. Among known-maternity individuals, none of 49 males moved beyond 200 m while 3 of 30 females (10%) did so ($p = .014$, $G$-test). Similarly, among all juveniles 1 male in 96 (1%) and 6 of 99 females (6%) moved more than 200 m ($p = .047$, $G$-test).

All three methods of estimating the amount of dispersal off the study site indicated that few animals moved beyond the site boundaries. The distribution adjusted for the probability that dispersal is contained within

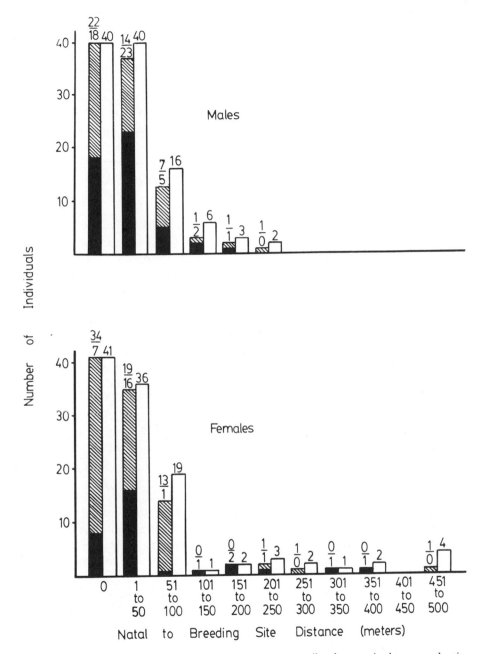

FIGURE 8.1. Distribution of dispersal distances among juveniles that survived to reproductive maturity. Black columns represent known-maternity juveniles, and crosshatched columns represent unknown-maternity juveniles. The open columns correspond to the adjusted distribution. The numbers above each column are the numbers of individuals in each category.

the study site is shown in figure 8.1, along with the observed distribution. The difference between these, 12 males and 12 females, is an estimate of how many individuals disappeared because they left the site. This difference amounted to 12% of the size of the observed sample, and, moreover, most of the difference involved moves of less than 100 m.

The frequency with which animals moved between my site and the adjacent one suggests even more restricted movement off the site than does the previous analysis. In 5 years one male and one female moved from my site to the adjacent one. Similarly, one male moved in the opposite direction.

Of the 458 D. spectabilis trapped in the 3-km-long transect, only 4 had ear tags, and 2 of these were within 50 m of the site boundary.

## Adult Dispersal

Dispersal among adults was more viscous than among juveniles. Seventy percent of 70 males and 61% of 72 females remained in the same mounds throughout their adult lives, and 89% of males and 92% of females remained within 50 m of the mounds they occupied at the beginning of their first breeding season (fig. 8.2). There were no significant sex differences in the proportion remaining in the same mound ($p = .26$, G-test) or in the proportion moving under 50 m ($p = .54$, G-test). The median dispersal distances were 0 m for both males and females. Mean dispersal distances were 15 m for males and 21 m for females, which were not significantly different ($p > .9$, two-tailed $t$-test with unequal variances). A chi-square test revealed no significant sex differences in adult dispersal distributions ($p = .09$).

Few individuals moved off the study site during their adult lives. The difference between the observed and adjusted distributions in figure 8.2 includes one male and three females, or 3% of the size of the observed sample. No adults moved between the two study sites.

## DISCUSSION

D. spectabilis is an exception to the usual pattern of male-biased dispersal in mammals (Greenwood 1980; Dobson 1982). The proportion of juveniles remaining in their natal vicinities was similar in the two sexes, but among the individuals that left their natal vicinities, females tended to move farther than males. However, this trend toward a female bias in longer movements should be viewed as tentative until more data are available, particularly since two of the three individuals that moved between sites were males. Nevertheless, the data allow a high degree of confidence that dispersal is not male-biased in either juvenile or adult D. spectabilis.

The second notable feature of D. spectabilis's dispersal pattern is the predominance of short-distance moves, or even no movement, among the individuals that reached reproductive maturity. Ninety-two percent of ju-

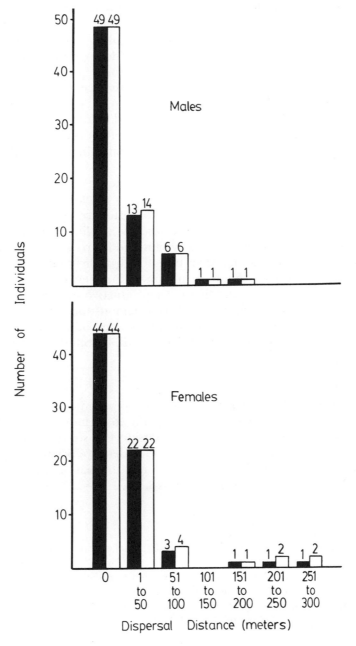

FIGURE 8.2. Distribution of dispersal distances among adults. Black columns represent the observed distribution, and open columns are the adjusted distribution. The numbers above each column are the number of individuals in each category.

veniles and 96% of adults remained within 100 m (about two home range diameters) of their original locations.

Comparison of the high frequency of natal philopatry observed in this species with dispersal distances expected based on demographic considerations (Waser, chap. 16, this volume) indicates that some parents tolerate their offsprings' sharing, as adults, all or part of the maternal home range. Such home range sharing, which may be termed philopatry by parental consent (Waser and Jones 1983), is more generally found in long-lived species (e.g., Mech, chap. 4, Nelson and Mech, chap. 2, and Rogers, chap. 5, this vol.), though it may occur in other short-lived mammals when dispersal risks are extreme (e.g., Smith, chap. 9, this vol.).

From consideration of dispersal patterns alone we might expect that the size of a randomly mating group in *D. spectabilis* would be relatively small. However, the influences of other factors on effective population size are still unknown. *D. spectabilis* is characterized by a solitary dispersed social organization in which both sexes may visit mounds up to about 100 m away from their own, but whether they breed with individuals beyond their immediately adjacent neighbors is unknown. Furthermore, there are no data on variance in annual and lifetime reproductive success of males and females. Nonetheless, these dispersal data demonstrate that the majority of adults of both sexes may breed on or immediately adjacent to their natal sites in some mammalian species (cf. Howard 1949; Smith and Ivins 1983; Smith, chap. 9, this vol.).

## ACKNOWLEDGMENTS

I thank Betsy Bush and Peter Waser for doing much of the fieldwork in this study. Peter Waser kindly allowed me to include his unpublished data from his autumn 1982 trapping outside the study area. Financial support included grants from the Indiana Academy of Sciences, the Theodore Roosevelt Memorial Fund, Sigma-Xi, and the National Science Foundation (DEB 77–6731A2, DEB 80–10848, DEB 81–12773).

## REFERENCES

Barrowclough, G. F. 1978. Sampling bias in dispersal studies based on finite area. *Bird Banding* 49:333–41.

Best, T. L. 1972. Mound development by a pioneer population of the banner-tailed kangaroo rat *Dipodomus spectabilis bailey* Goldman, in eastern New Mexico. *American Midland Naturalist* 87:201–6.

Bunnell, F. L., and A. S. Harestead. 1983. Dispersal and dispersion of black-tailed deer: Models and observation. *Journal of Mammalogy* 64:201–9.

Clutton-Brock, T. H., F. E. Guinness, and S. D. Albon. 1982. *Red deer: Behavior and ecology of two sexes.* Chicago: University of Chicago Press.

Dobson, F. S. 1979. An experimental study of dispersal in the California ground squirrel. *Ecology* 60:1103–9.

————. 1982. Competition for mates and predominant juvenile male dispersal in mammals. *Animal Behavior* 30:1183–92.

Greenwood, P. J. 1980. Mating systems, philopatry, and dispersal in birds and mammals. *Animal Behavior* 28:1140–62.

Holdenreid, R. 1957. Natural history of the banner-tailed kangaroo rat in New Mexico. *Journal of Mammalogy* 38:330–50.

Howard, W. E. 1949. Dispersal, amount of inbreeding, and longevity in a local population of prairie deermice on the George Reserve, Southern Michigan. *University of Michigan Laboratory of Vertebrate Biology Contributions* 43:1–52.

Jannett, F. J., Jr. 1978. The density-dependent formation of extended maternal families of the montane vole, *Microtus montanus nanus. Behavioral Ecology and Sociobiology* 3:245–63.

Jones, W. T. 1984. Natal philopatry in banner-tailed kangaroo rats. *Behavioral Ecology and Sociobiology* 15:151–55.

King, J. A. 1983. Seasonal dispersal in a seminatural population of *Peromyscus maniculatus. Canadian Journal of Zoology* 61:2740–50.

Munger, J. C., and J. H. Brown. 1981. Competition in desert rodents: An experiment with semipermeable exclosures. *Science* 211:510–12.

Smith, A. T., and B. L. Ivins. 1983. Colonization in a pika population: Dispersal vs. philopatry. *Behavioral Ecology and Sociobiology* 13:37–47.

Sullivan, T. 1977. Demography and dispersal in island and mainland populations of the deer mouse, *Peromyscus maniculatus. Ecology* 58:964–78.

Tamarin, R. H. 1977. Dispersal in island and mainland voles. *Ecology* 58:1044–54.

Waser, P. M., and W. T. Jones. 1983. Natal philopatry among solitary mammals. *Quarterly Review of Biology* 58:355–90.

# 9. Population Structure of Pikas: Dispersal versus Philopatry

*Andrew T. Smith*

Determination of the relationship between dispersal/philopatry and population structure is beset with the problem of the spatial scale of movements by individuals in the population. Because animals may move variable distances under a variety of ecological and social situations, it has proven difficult to provide universal definitions for dispersal and philopatry or, more important, to understand the behavioral, ecological, and genetic consequences of these movements. In this chapter I adopt an empirical approach to address this problem of scale. I review a series of investigations of a small alpine lagomorph, the North American pika (*Ochotona princeps*), a species that has proven to be an ideal subject for studies of population structure and dispersal. In many respects parameters of pika ecology conform remarkably closely to those generally modeled by theoreticians of dispersal, and many of these traits also simplify the gathering of field data. In addition, the pika possesses a suite of life-history attributes that differs from most species on which we have substantial data on dispersal, hence increasing our comparative knowledge of this phenomenon. Most important, these investigations incorporate two extreme approaches to the issue of scale of dispersal. Since I began studying pikas in 1969, my approach has varied from examination of the consequences of dispersal at the level of regional populations to examination of those factors that may act to promote or discourage the dispersal and eventual settlement of individuals within populations. Although these two approaches were not studied simultaneously, taken together the data reveal the pattern of dispersal and population structure in pikas.

## BIOLOGY OF *Ochotona princeps*

Studies of pikas are facilitated by the habitat they occupy and their use of space. Pikas live only on talus, which is generally a patchily distributed habitat. Thus immediate comparisons are possible of movement between patches (like metapopulations in dispersal models) and within patches (where the environment is roughly homogeneous). In addition, pikas are individually territorial; average nearest neighbor distances are approximately 20 m (Smith 1981; Smith and Ivins 1984). Territory size (or home range size, depending on the presentation of data) is equivalent between the sexes (Tapper 1973; Smith and Ivins 1984). There is only a small variance in territory size within a patch or throughout the geographic range of *O. princeps* (Smith 1981). Normally it is possible to determine accurately the maximum number of adult pikas that could fit onto a patch of talus. Essentially this means that the carrying capacity and, more important, the degree of saturation of habitat (percentage of potential territories occupied) can be accurately estimated.

A knowledge of demography is essential to understanding the dispersal and population structure of a species because mortality opens up sites available for settlement and fecundity provides the juveniles that are the primary dispersers in most species. For small mammals, pikas are exceptionally long-lived. Some may live to be 6 years old, and survivorship varies between 55% to 63% per year (Millar and Zwickel 1972; Smith 1978). Because of the relatively old age of most adults and the pika's territorial system, few sites become available for colonization in any year, and the appearance of these vacancies is unpredictable in time and space.

Each resident adult female is reproductively equivalent to all others. Females first reproduce as yearlings, and all females initiate two litters each year (Millar 1972, 1973, 1974; Smith 1978). Generally only the first litter is successfully weaned, although if it fails the second breeding attempt may be successful (Smith and Ivins 1983b). Litter size does not vary with age, habitat productivity, or between first and second litters (Millar 1973, 1974; Smith 1978). There is only slight variability in litter size within a locality. Average litter size among sites also varies little ($\bar{x}$ = 2.3–3.7) and may be explained on the basis of spatial and temporal age-specific mortality (Smith 1978). Both gestation and time to weaning is approximately 30 days (Severaid 1950; Millar 1971).

Timing is an important aspect of reproduction in pikas. Apparently selection has favored females that are the first to successfully wean their young each year because these juveniles should have the first opportunity to claim vacant territories left by overwinter or spring mortality. For this reason first litters are conceived, on average, 1 month before snowmelt and the resulting spring flush of alpine vegetation necessary to meet the high metabolic demand of lactation (Millar 1972; Smith 1978; Smith and Ivins 1983b). Breeding this early is possible because pikas do not hibernate and have access to food stored in haypiles (caches of vegetation stored on the talus the preceding summer) and on nearby meadows through the use of snow tunnels. In fact, pikas accumulate fat during the gestation of their first litter for use during lactation (Millar 1973, 1974). One difficulty facing individual pikas, however, is that they apparently cannot determine when snowmelt will occur in any given year. Thus I have determined that variance of initiation of first litters is directly related to the long-term variance of snowmelt at a locality (Smith 1978). Phenology of reproduction is more synchronous where snowmelt is predictable than at comparatively unpredictable sites. In extreme years a female may initiate breeding too early or have insufficient energy reserves coming out of winter and subsequently lose the first litter (Smith and Ivins 1983b). In these cases the second litter becomes important as a backup. Interestingly, the conditions that act to deplete first litters may also act to increase adult mortality; thus in years when second litters are successful there are often even more vacancies available for colonization by these young (Smith and Ivins 1983b).

Body mass has been touted as an important factor in the determination of life-history characteristics (Millar 1977). However, body mass does not appear to be an important variable in pikas. Pikas are sexually monomorphic in size within populations (Smith 1981). There is variation in body mass among populations of *O. princeps* that have been studied ($x = 130–170$ g), but none of this variation appears to be reflected in other life-history attributes (fecundity, survivorship, territory size, etc.; Smith 1978). In addition, pikas demonstrate a remarkable convergence between the sexes in reproductive morphology (Smith 1981).

## STUDY SITES

I have investigated pikas in two extreme situations: (1) high elevations (3,200–3,400 m) in the Sierra Nevada, California, and the Rocky Mountains, Colorado; and (2) low elevations (2,300–2,550 m) near the base of the eastern escarpment of the Sierra Nevada. The high-elevation sites (hereafter referred to as Sierra and Colorado) are each characterized by typical alpine or subalpine vegetation, have long winters with substantial snowpack, and have low daily maximum temperatures in summer. These sites represent the conditions normally found throughout the range of *O. princeps*. The low-elevation sites (Bodie, Masonic, Aurora) are characteristic of many habitats near the distributional boundary of pikas. Here the daily temperatures are high in summer and stressful to pikas. Most of the low-elevation work was conducted at Bodie, an old mining town surrounded by mine tailings that are occupied by pikas. These tailings are essentially habitat islands that vary in size and distance from one another in a sea of sagebrush.

My investigations of the regional biogeography of pikas were conducted at the Sierra and Bodie sites. Data collection was focused on the comparative dispersal abilities of pikas living under contrasting climatic conditions and the resulting effect of dispersal upon regional population structure at the two localities (Smith 1974a, 1974b, 1978, 1980). Most important in this regard was the analysis of occupancy of habitat islands at Bodie. Here many of the habitat islands are small, and pika populations on them are subject to chance extinction. At any one time about 40% of habitable islands (those evidencing prior occupancy by pikas) are unoccupied. This situation results in ideal conditions for the study of dispersal because the factors affecting the success of colonization may be evaluated within a limited geographic area.

Longitudinal investigations on a fully marked population of pikas at the Colorado site focused on the behavioral antecedents of dispersal, the dynamics of settlement in a pika population, and the relationship between dispersal and the pika's mating system (Ivins and Smith 1983; Smith and Ivins 1983a, 1983b, 1984, 1986, 1987). As in all naturalistic investigations, there has been some overlap in these two data sets. Below I highlight the results of these investigations within the framework of a series of progressive topics concerning dispersal.

## ABILITY TO DISPERSE

The ability to disperse independent of complications imposed by intraspecific interactions (vagility) differs between pikas living at high- and low-elevation sites. Pikas are physiologically adapted to cool climatic conditions, such as normally occur at high elevations, and their ability to disperse long distances at these sites does not appear to be impeded by temperature (Smith 1974b). For example, attempts to introduce three animals onto the main study area in Colorado failed because these individuals successfully homed to their original territory 0.5 km to 1.0 km away within 1 to 3 days (Smith and Ivins 1984). Under similar climatic conditions in Alberta, a marked juvenile dispersed 3.0 km to a talus patch that had recently been cleared of pikas (Tapper 1973).

At low-elevation sites it is comparatively warmer at the time of dispersal, and as pikas are not well adapted to these temperatures their vagility is reduced (Smith 1974b). Pikas have relatively high body temperatures for small mammals (over $40°$ C) and low upper lethal temperatures (MacArthur and Wang 1973, 1974; Smith 1974b). When pikas at low elevations are deprived of the opportunity to retreat to favorable microclimates in the talus, they are unable to tolerate the high diurnal temperatures they face. Pikas compensate for their inability to acclimatize to these warm temperatures by curtailing all activity on their territories from about 1000 hr to 1600 hr during summer days. Also, average foraging distances away from the talus edge are less at Bodie than at the Sierra site (Smith 1974b). One important consequence of the stress imposed by warm temperatures on the vagility of pikas is manifested in their relative inability to successfully colonize many of the vacant habitat islands at Bodie. Interisland distances of as short as 300 m appeared to be effective barriers to colonization (Smith 1974a). It is clear from these data that distance as measured in studies of dispersal must be interpreted as more than the linear distance between two points. Instead, dispersal distance must incorporate both the actual distance and some measure of the behavioral/physiological barrier such as might be found near the distributional limit of a species where conditions at the time of dispersal are likely to be more rigorous (Smith 1974a).

Dispersal in pikas may also be inhibited by the threat of predation. Pikas appear to be most susceptible to predation when they are away from talus (Quick 1951; Murie 1961; Ivins and Smith 1983), and their foraging tactics reflect their reticence to venture far from the talus-meadow interface (Huntly, Smith, and Ivins 1986). It follows that the greater the distance of dispersal between patches of talus, the more exposed a pika would be to such predation.

## INTERPATCH DISPERSAL

Pikas may undertake two distinctly different types of dispersal movement: (1) intrapatch dispersal, in which individuals move within the talus on which they are resident; and (2) interpatch dispersal, in which individuals abandon their home talus to venture across inhospitable terrain with the anticipation of encountering a distant patch of talus. During these latter movements a pika has an increased chance of dying as a result of overheating or predation compared with intrapatch movements. In addition there is always the possibility that the pika will never encounter suitable talus for potential colonization; all available evidence indicates that the direction of interpatch dispersal by pikas is random (Smith 1974a).

We have very little direct information about the dynamics of interpatch dispersal by pikas, with the exception of the isolated accounts given earlier. Indirectly it is apparent that pikas live on an unstable habitat, and that populations on talus patches may suffer chance extinction due to local demographic events, predation, or (most likely) catastrophic events (e.g., rockslides, avalanches, extremely severe winters or storms). Yet pikas may be found at close to saturation densities throughout most of their geographic range, even on extremely isolated talus patches. The conclusion from these observations is that pikas can and do disperse successfully between patches of talus. The problem is that we have no idea as to the magnitude of either component—patch extinction or subsequent recolonization—in typical alpine habitat.

Most of our knowledge of interpatch dispersal in pikas comes from my analyses of regional population structure at Bodie (Smith 1974a, 1980). Here, 5 years apart, I censused the entire population of pikas on 78 patches of talus (including three to four large, or "mainland," sites). The results clearly showed that these island populations of pikas represent dynamic equilibria between extinction rate (which was inversely related to island size) and recolonization rate (which was inversely related to interisland distance). The overall results from the two censuses were strikingly similar in spite of the fact that the population size of individual islands changed dramatically (Smith 1980).

When the islands were classified into size and distance categories, two important results surfaced. First, all cases of turnover (extinction and recolonization) occurred on small- or medium-sized islands (those containing five or fewer pikas). Essentially this finding means that extinctions due to chance demographic effects (see MacArthur 1972; Smith 1974a) are probably rare; most pika populations consist of more than five animals. Indeed, the mine tailings at Bodie were probably colonized from a nearby natural population on Sugarloaf Hill that has probably been isolated from the nearby (about 35 km) Sierra population since the Pleistocene–Recent warming trend occurred (see Brown 1971). Second, extinction of populations on islands close to a source of potential colonists was apparently minimized by recurrent colonization (Smith 1980). This result implies that pikas initiating interpatch movements at Bodie either (1) attempt to colonize the first suitable talus that they encounter; or (2) do not survive if they continue to disperse. In either case, there may be profound effects on the resulting genetic structure of the population at Bodie. Even with a habitat that is 40% unsaturated and open for colonization, the observed interpatch dispersal patterns may still yield high levels of inbreeding.

Age and sex of dispersers are important parameters when relating dispersal to population structure. Apparently most interpatch dispersal is by juveniles. In studies using marked animals, most immigrants have been juveniles, and adults have rarely been observed dispersing between patches (Tapper 1973; Smith and Ivins 1983a). In addition, Krear (1965) reported that coyotes, which can only prey on pikas away from talus, take mostly juveniles (which may have been dispersing).

The only available data on interpopulation dispersal by sex are limited by small sample size. Both Tapper (1973) and Smith and Ivins (1983a) determined that males and females may colonize patches separated by nontalus barriers. Such a mixed strategy is likely because settlement of vacant patches by a single sex would not contribute to perpetuation of a population on a patch. At Bodie, sex ratio at weaning favors females (Smith 1974a), thus setting up the possibility that more females disperse between patches than males. This phenomenon has not been confirmed with observations on marked animals.

## INTRAPATCH DISPERSAL

Intrapatch dispersal movements are uncommon in pikas. Instead, both juveniles and adults tend to be philopatric. In the following analysis, based on a 3-year study of marked pikas in Colorado, I define philopatry operationally as the movement of a pika less than 50 m from its center of activity following weaning. Essentially this definition allows a juvenile to occupy either the home range of a parent or a neighboring animal upon its death or disappearance. Animals moving greater than 50 m from the center of their natal home range are classified as dispersers.

Juveniles

Out of 45 juveniles born on the Colorado study area, only 2 (4%) were known to have engaged in intrapatch dispersal: 1 male (170 m during summer) and 1 female (80 m overwinter; Smith and Ivins 1983a). In addition, 2 males remained spatially associated with their natal home range but made occasional long-distance forays. Thirty-three juveniles were born on the study area in 1980 and 1981, years when intensive observational data were collected. As late as mid-September (just prior to the onset of winter in the area), 64% (21) of these juveniles were still alive on the study area, and of these only 1 (5%) had a center of activity greater than 50 m from its natal home range (Smith and Ivins 1983a). Extensive searches on all nearby adjoining talus and talus patches were conducted in an unsuccessful attempt to locate the remaining 12 juveniles. These animals either died on their natal home range, dispersed an intermediate distance and died before the area was censused, or engaged in long-distance interpatch dispersal. As many of these animals were quite small (young and inexperienced) at the time of their disappearance, I believe that in situ death was the most common fate of these juveniles.

Other observational studies also indicate that most juvenile pikas tend to remain close to their site of birth (Krear 1965; Millar 1971; Sharp 1973; Tapper 1973). Tapper (1973) determined that only 16% (8/49) of his juveniles engaged in intrapatch dispersal of greater than 100 m. Both Millar (1971) and Tapper (1973), who worked in Alberta, reported that all juveniles that did disperse greater than 100 m were females, indicating a sex bias in dispersal not found in my Colorado study.

An index of juvenile dispersal is provided by the arrival of unmarked juveniles onto the study area. In 3 years in Colorado 16 (7 male, 9 female) juveniles immigrated onto the study area, and they constituted 26% of all juveniles (Smith and Ivins 1983a). The distance they moved or whether they engaged in inter- or intrapatch dispersal is unknown. Most of these juveniles immigrated in late summer (12 later than mid-August; 9 in September or later). Many were observed for only a short period of time, after which they either died or continued to disperse. One eventually established a territory on the study area. In addition, 3 presumed juveniles (2 male, 1 female) colonized the study area during winter; they were first detected when fieldwork was initiated in May (Smith and Ivins 1983a).

Adults

Once adults have successfully established themselves on a territory, they generally remain there for life (Krear 1965; Tapper 1973; Smith and Ivins 1983a). In Colorado, 85% (35/41) of adults followed through at least one complete season remained on the same home range (Smith and Ivins 1983a). Of the 6 (4 male, 2 female) adults that dispersed, 4 (2 male, 2 female) apparently moved (60, 70, 80, 310 m respectively) to increase their probability of future breeding. In each of these four cases an unpaired adult moved to reside next to an established animal of the opposite sex. The timing of these movements appeared to be facultative, keyed only to a situation in which the disperser increased its fitness as a result of the move. The two remaining instances of adult dispersal are less clear. Each male was associated with a female prior to its move. One male settled next to a different female; the other colonized a vacant patch and became solitary. Both of these movements occurred in early summer before the breeding season was completed.

As expected from the low incidence of dispersal of marked adults, few (2) unmarked adults immigrated onto the study site (Smith and Ivins 1983a). Both of these animals were males, and they appeared in early summer. Tapper (1973) noted that some juvenile males managed to survive winter by occupying the interstices between adult territories and then dispersed in spring or early summer. Thus, there may be a temporal component to juvenile dispersal (if yearling males that have not yet bred are considered juveniles), and Tapper's (1973) observation of female bias in intrapatch dispersal may be negated. In any case, only 2 of the adult dispersers in my Colorado study were spring yearlings, and one was male and the other female. It is unknown whether the two immigrant males mentioned previously were yearlings.

## ANTECEDENTS OF DISPERSAL

The proximate factors responsible for the dispersal movements of an animal are difficult to determine. Dispersal is sufficiently rare in pikas that one is unlikely to observe the process of dispersal directly. Instead we are left with an examination of the social milieu occupied by animals prior to dispersal. Even here analysis is risky because the social factors acting upon an animal that disperses may be similar to those acting upon a philopatric animal. The following specific observations indicate how preliminary our knowledge is on this matter. Neither of the two juveniles that dispersed within the Colorado study area were observed to engage in trial forays of any kind—much less to the areas where they eventually settled. The two males that were observed to make long-distance intrapatch forays remained on their natal home range. One of the juvenile dispersers originally inhabited one of the most crowded localities on the study area (a small meadow centered on the talus which acted like a magnet, drawing in all nearby pikas to forage), and hence could have dispersed due to population pressure. The other juvenile disperser originally inhabited a comparatively low-density area near the edge of a barren shoot of talus and dispersed to a similar site on the opposite side of the talus.

Detailed spatial and behavioral observations help clarify the role of juveniles prior to their incorporation into the population, dispersal, or death.

Juveniles are relatively sedentary throughout the summer and occupy home ranges that are smaller than those of adults (Smith and Ivins 1983a). Juvenile home ranges greatly overlap those of their mother and putative father. In spite of the opportunity for frequent behavioral interaction between parents and young, however, the young demonstrate a strong tendency to temporally avoid adults. Most juvenile activity occurs during periods when parents are not active on their territories (Smith and Ivins 1987). When either two or three young were weaned from the same litter, they avoided one another spatially, each settling in a different sector of the home ranges of their parents (Smith and Ivins 1987).

The young engaged in a relatively even mix of agonistic and affiliative behaviors with their parents, and most of these interactions were initiated by the parents (Smith and Ivins 1983a). In previous investigations only the aggressiveness of adults toward the young was highlighted (Krear 1965; Sharp 1973; Tapper 1973; Svendsen 1979). Adults should be expected to show aggressiveness toward any conspecific, including the young, because adults maintain individual territories. However, the parents tolerate the young on their home range and frequently express affiliative behavior toward them (Smith and Ivins 1983a); these behaviors probably contribute to the high degree of philopatry exhibited by juvenile pikas. It is interesting to note that there was a marked sexual asymmetry of agonistic interactions by adults toward juveniles. In July and August, 69% of adult aggression was directed at juveniles of the same sex (Smith and Ivins 1983a). If aggression leads to dispersal in some instances, then this asymmetry may be explained on the basis that same-sex offspring may eventually compete more directly with parents than will offspring of the opposite sex.

There is no evidence that any of the dispersal movements of adults were preceded in any way by excessive aggressive behavior between neighbors. In fact, as I indicated earlier, adult movement generally occurs when an animal leaves an existing territory in which social encounters are uncommon. Adults normally move to areas of increased likelihood of social interaction with conspecifics.

## SUCCESSFUL SETTLEMENT

Whether an animal disperses or remains philopatric, its ultimate success depends on whether or not it can effectively establish itself into the breeding population. To fully understand this phenomenon in pikas requires some basic understanding of their facultatively monogamous mating system (Smith and Ivins 1984). Although pikas are individually territorial, adjacent territories in Colorado were normally occupied by pikas of the opposite sex, and spatial overlaps were greatest between these heterosexual dyads. Almost all instances of adult affiliative behavior were expressed between spatially contiguous heterosexual dyads, although as expected there was an equal degree of agonistic behavior. Most important, this mating system imposed a permanent structure to the population. Over 3 years, replacement of territories was always by a member of the same sex as the previous occupant ($N = 20$; Smith and Ivins 1984). Thus, although vacancies available for settlement on the talus are few, each vacancy is further restricted, apparently due to the behavior of neighboring resident adults, to animals of a particular sex.

Dispersing and resident (philopatric) pikas are treated differently by resident adults. The behavior of resident adults toward familiar (not yet permanently established) pikas, in almost all cases their progeny, has been outlined earlier. Essentially adults are as tolerant of these animals as they are of their pair mate, resulting in continued philopatry throughout the summer by juveniles.

All three classes of unfamiliar pikas (juveniles on forays, immigrant juveniles, and immigrant adults) are treated with extreme aggression by resident adults (Smith and Ivins 1983a). Rate of aggression directed at the juvenile males while they were on forays was approximately 10 times higher than normal adult-juvenile aggression rates. Most juveniles immigrated onto the study site in late summer when the absolute rate of aggression in the population was low. Nevertheless, in spite of the small number of immigrants and their relative secretiveness, 59% (10/17) of all observations of aggression in September and October were directed at juvenile immigrants. The most extreme cases of directed aggression occurred between residents and immigrant adults. For example, in August 1980, 78% (39/50) of all adult aggressive and interactions were directed at a single immigrant adult (Smith and Ivins 1984).

The asymmetries of behavior directed at potential colonists are reflected in the eventual settlement pattern in a pika population. Neither of the two adult immigrants persisted on the study site. One disappeared shortly after immigrating; the other led a vagrant life before finally settling off the study area. During 3 years, 76% (13/17) of all animals establishing residency were juveniles born on the study area (Smith and Ivins 1983a). Of these only 2 (detailed earlier) were known to have dispersed; the remainder established within 50 m of the center of their natal home range. Four immigrants became established on the study area. Three of these 4 apparently colonized over winter (after we ceased observations in late September/early October); the fourth was a late summer colonist. During late summer/early fall, adults constrict their home ranges to approximately half of their size earlier in the summer (Smith and Ivins 1984). Apparently pika populations are more easily invaded at this time because immigrants are less likely to be detected by adults. At this time adults appear most interested in behaviors that will ensure their overwinter survival, and there is apparently no reason for them to control access to vacant territories. If there had been local sites available for colonization by the offspring of adults, they should already have been claimed by this time. In conclusion, it appears that most vacancies are claimed by neighboring juveniles. Most juveniles remain on their natal home range, where they are familiar with resources and are unlikely to be recipients of extreme aggression. If this tactic is unsuccessful, then they may disperse in late summer. These dispersing animals, although receiving most of the agonistic behavior within the population at this time, are still infrequently detected and may colonize those few sites that remain available.

## CONSEQUENCES OF DISPERSAL/PHILOPATRY

A central issue regarding discussion of the consequences of dispersal and philopatry concerns the degree to which either strategy may be involved in the selection for observed behavioral or demographic traits. In the case of philopatry in pikas it is clear that the resulting population structure is composed of highly related individuals of both sexes, which in most cases mate incestuously. Even in the case of a mammal as asocial as the pika, this inbreeding has led to the expression of nepotistic traits. For example, when predators are sighted, resident (but not new immigrant!) pikas of both sexes utter alarm calls (Ivins and Smith 1983; Smith 1984). Similarly, the high levels of agonistic behavior directed at immigrants by resident adults may ultimately repel the immigrants and give one's offspring an increased probability of successful establishment (Smith and Ivins 1984). Such behavior is likely to increase if all neighboring animals are closely related compared with unfamiliar immigrants.

When animals disperse it is less clear how selection should operate on observed behavioral or demographic traits. There are three widely accepted effects of dispersal: maintenance of regional genetic variability, persistence of local populations (hence stability of a species) over large geographic areas, and efficient use of natural resources through the mechanism of population regulation (Lidicker 1975; Gaines and McClenaghan 1980). I address each of these effects with relation to pikas.

The high level of philopatry and genetic inbreeding in pikas does not appear detrimental. However, it is unknown whether the low rate of dispersal (approximately 25% of all juveniles) is sufficient to counter inbreeding depression (should it occur), whether regional populations are sufficiently similar genetically to prevent immigrants from disrupting highly adaptive gene complexes, or whether animals that disperse have an adverse genetic effect on a local population. Experiments are being planned to test these three alternatives. Regardless of their outcome, however, it is clear that the "threat" of decreased fitness due to the genetic consequences of inbreeding or outbreeding in pikas appears to be minimal compared to other aspects of their population structure. Most juveniles are lucky to encounter an available vacant territory that had been occupied by an animal of the same sex. Any small decrease in long-term genetic fitness should be a secondary consideration in the efforts of the animal in its attempt to claim the site (Smith and Ivins 1983a).

The effect of dispersal upon the maintenance of regional saturation of habitat or extension of the species' range at first appears independent of individual selection arguments. Certainly no animal disperses in this (or any) context for the good of the species. However, the evolution of life-history traits is tightly linked to both dispersal and regional population structure, as evidenced by a comparison of litter size among several populations of *O. princeps* (Smith 1978). As I argued earlier, the fate of juveniles, whether they disperse or not, appears related to their probability of encountering a vacant territory. In saturated populations, such as those found at the Sierra and Colorado sites, vacancies occur as a function of adult mortality. At Bodie, however, 40% of the habitat is unoccupied and continually available for colonization. Thus it is ironic that at Bodie, where the conditions least favor the ability to disperse, the probability of establishment following interpatch dispersal is highest (Smith 1978). Litter size of pikas at Bodie is the highest of any population that has been studied (Smith 1978). Apparently a fugitive reproductive strategy has developed at Bodie due to the spatially unsaturated system; mothers weaning larger litters appear more likely to promote successful dispersers. Further, Bodie has the lowest adult mortality rate of any population of pikas that has been studied, demonstrating there is no direct feedback between adult mortality and fecundity in spatially unsaturated systems (Smith 1978).

Although dispersal may function in the regulation of populations, selection for dispersal with regard to population regulation should occur only if it is to the advantage of individual dispersers or to the advantage of the residents that force their dispersal. There is no direct evidence that adult pikas encourage the dispersal movements of juveniles. During summer there appears to be an abundance of vegetation, and the small number of weaned young (normally two; three at Bodie) and their separate spacing within the home ranges of their parents apparently preclude severe overgrazing of available vegetation. The result is overlap of most young with their parents throughout the summer. In fall, however, alpine vegetation becomes senescent and lacks the nutrition it contained during summer. A juvenile, now fully adult in size, may not be able to compete with its parents for food. Perhaps this decline in carrying capacity is one reason juveniles disperse at this time. Ultimately, due to the individual territorial system in pikas, no young that remained philopatric at the end of the summer of their birth were found alive at the beginning of the next summer unless they had successfully colonized a vacant territory.

## EPILOGUE

I began by indicating that scale was difficult to incorporate in studies of dispersal and philopatry. Data from investigations of pikas indicate that both our definitions of dispersal/philopatry and their consequences may be influenced by scale. Local populations of pikas may show high levels of inbreeding, yet entirely different behavioral and ecological circumstances may promote this inbreeding at Bodie, versus high-altitude sites. Development of nepotistic behaviors has evolved only among related animals living in close proximity, yet the evolution of life-history traits may be influenced by chance effects of long-distance dispersal. Clearly we cannot say what the genetic effects of dispersal or philopatry are for a given species unless we unambiguously indicate the nature of the dependent variable, be it nepotistic social behavior or life-history traits. Also, intraspecific variation in degree of dispersal and philopatry must be taken into account when undertaking interspecific comparisons. Last, we must recognize that there may be effects of dispersal that do not result in selection for the behavior of dispersal (a so-called ecological pleiotropic effect of dispersal). One of the greatest difficulties we face is in ascribing a linked relationship between social behavior, dispersal, and genetics. To do this requires knowledge of the relative rate and success of dispersal on both local and regional scales.

Finally, there is the problem of defining the boundary between dispersal and philopatry. It has become common to define philopatric movements in terms of the number of home ranges traversed by an animal. Shields, (chap. 1, this vol.) defines philopatry as the limited dispersal of an animal no more than 10 home ranges away from its site of origin. Yet in pikas, an animal that successfully disperses this distance should be considered a long-distance disperser. Such an animal is unlikely to interact behaviorally with close relatives at any time in its life; the opportunity for expression of nepotistic behaviors is very low. In addition, because dispersal of this distance is so unlikely in pikas, in both the short- and long-term, this animal should not be surrounded by similar genotypes. One of the greatest reasons for our interest in philopatry is the relationship between degree of relatedness and social behaviors. Thus, we should strive to measure philopatry in terms of the opportunity for social encounter by related conspecifics rather than by a fixed distance such as number of home ranges moved. Such a measure, although more difficult to quantify than home range diameter, is more biologically sound and should increase our ability to compare the consequences of philopatry among taxa.

## ACKNOWLEDGMENTS

I am grateful to Mark Newton, Harriet Smith, and Peter Smouse for reading and commenting on the manuscript. I would like to acknowledge the contribution made by Barbara Ivins, my co-worker on the Colorado study, to the gathering of much of the data referred to in this review. This chapter is dedicated to George A. Bartholomew, upon the occasion of his retirement; he helped instill in me a love of natural history.

# REFERENCES

Brown, J. H. 1971. Mammals on mountaintops: Nonequilibrium insular biogeography. *American Naturalist* 105:467–78.

Gaines, M. S., and L. R. McClenaghan, Jr. 1980. Dispersal in small mammals. *Annual Review of Ecology and Systematics* 11:163–96.

Huntly, N. J., A. T. Smith, and B. L. Ivins. 1986. Foraging behavior of the pika (*Ochotona princeps*), with comparisons of grazing versus haying (caching). *Journal of Mammalogy* 67:139–48.

Ivins, B. L., and A. T. Smith. 1983. Responses of pikas (*Ochotona princeps*, Lagomorpha) to naturally occurring terrestrial predators. *Behavioral Ecology and Sociobiology* 13:277–85.

Krear, H. R. 1965. An ecological and ethological study of the pika (*Ochotona princeps saxitilis* Bangs) in the Front Range of Colorado. Ph.D. diss., University of Colorado, Boulder.

Lidicker, W. Z., Jr. 1975. The role of dispersal in the demography of small mammals. In *Small mammals: Their productivity and population dynamics*, ed. F. B. Golly, K. Petrusewicz, and L. Ryszkowski, 103–28. New York: Cambridge University Press.

MacArthur, R. H. 1972. *Geographical ecology*. New York: Harper and Row.

MacArthur, R. A., and L. C. H. Wang. 1973. Physiology of thermoregulation in the pika, *Ochotona princeps*. *Canadian Journal of Zoology* 51:11–16.

———. 1974. Behavioural thermoregulation in the pika, *Ochotona princeps:* A field study using radio-telemetry. *Canadian Journal of Zoology* 52:353–58.

Millar, J. S. 1971. Breeding of the pika in relationship to the environment. Ph.D. diss., University of Alberta, Edmonton.

———. 1972. Timing of breeding of pikas in southwestern Alberta. *Canadian Journal of Zoology* 50:665–69.

———. 1973. Evolution of litter size in the pika, *Ochotona princeps* (Richardson). *Evolution* 27:134–43.

———. 1974. Success of reproduction in pikas, *Ochotona princeps* (Richardson). *Journal of Mammalogy* 55:527–42.

———. 1977. Adaptive features of mammalian reproduction. *Evolution* 31:370–86.

Millar, J. S., and F. C. Zwickel. 1972. Determination of age, age structure, and mortality of the pika, *Ochotona princeps* (Richardson). *Canadian Journal of Zoology* 50:229–32.

Murie, A. 1961. Some food habits of the marten. *Journal of Mammalogy* 42:516–21.

Quick, H. 1951. Notes on the ecology of weasels in Gunnison County, Colorado. *Journal of Mammalogy* 32:281–90.

Severaid, J. H. 1950. The gestation of the pika. *Journal of Mammalogy* 31:356–57.

Sharp, P. L. 1973. Behaviour of the pika (*Ochotona princeps*) in the Kananaskis region of Alberta. M.Sc. thesis, University of Alberta, Edmonton.

Smith, A. T. 1974a. The distribution and dispersal of pikas: Consequences of insular population structure. *Ecology* 55:1112–19.

———. 1974b. The distribution and dispersal of pikas: Influences of behavior and climate. *Ecology* 55:1368–76.

———. 1978. Comparative demography of pikas (*Ochotona*): Effect of spatial and temporal age-specific mortality. *Ecology* 59:133–39.

————. 1980. Temporal changes in insular populations of the pika (*Ochotona princeps*). *Ecology* 61:8–13.

————. 1981. Territoriality and social behavior of *Ochotona princeps*. In *Proceedings of the World Lagomorph Conference*, ed. K. Myers and C. D. MacInnes, 310–23. Guelph, Ontario: Guelph University Press.

————. 1984. Family bonds and friendly neighbors. In *The encyclopedia of mammals*, ed. D. Macdonald, 728–29. New York: Facts on File.

Smith, A. T., and B. L. Ivins. 1983a. Colonization in a pika population: Dispersal vs philopatry. *Behavioral Ecology and Sociobiology* 13:37–47.

————. 1983b. Reproductive tactics of pikas: Why have two litters? *Canadian Journal of Zoology* 61:1551–59.

————. 1984. Spatial relationships and social organization in adult pikas: A facultatively monogamous mammal. *Zeitschrift für Tierpsychologie* 66:289–308.

————. 1986. Territorial intrusions by pikas (*Ochotona princeps*) as a function of occupant activity. *Animal Behaviour* 34:392–97.

————. 1987. Temporal avoidance by philopatric juvenile pikas limits conflicts with adults. *Animal Behaviour* (in press).

Svendsen, G. E. 1979. Territoriality and behavior in a population of pikas (*Ochotona princeps*). *Journal of Mammalogy* 60:324  30.

Tapper, S. C. 1973. The spatial organisation of pikas (*Ochotona*), and its effect on population recruitment. Ph.D. diss., University of Alberta, Edmonton.

# III. DISPERSAL PATTERNS AND GENETIC STRUCTURE

# 10. Patterns of Dispersal and Genetic Structure in Populations of Small Rodents

*William Z. Lidicker, Jr., and James L. Patton*

A classic problem in population biology concerns the relationship between gene flow and the dispersion patterns of organisms. This relationship is of fundamental evolutionary interest when organisms are not, as is usually the case, distributed evenly across the landscape (see for example Templeton 1980; Wright 1978). Genetic structure refers to the genetic composition of and differentiation among an assemblage of demes (Shields, chap. 1, this vol.) resulting from such a heterogeneous spatial distribution. Unless completely isolated by physical barriers, demes located near each other are variously connected by the movement of individuals and/or their propagules. Some demes become extinct without contributing to other demes. Some surviving demes contribute more than others to successful immigration into other demes and to the establishment of new demes. Genetic structure is thus a dynamic property of metapopulations, and its study lies at the interface of ecology, behavior, and evolution.

This chapter attempts to relate the genetic structure of small-bodied rodent populations to patterns of dispersal. Dispersal, however, is not the same as gene flow, but rather is defined in behavioral terms. It generally is viewed as constituting movements of individuals (or propagules) away from their home areas, excluding short-term exploratory movements (Lidicker 1975; Lidicker and Caldwell 1982). The more limited definition proposed by Shields (chap. 1, this vol.) is inadequate for small mammals because they often disperse for nonreproductive reasons and on multiple occasions, and because the terms *breeding site* and *breeding group* are ambiguous for these species. The rather complex and heterogeneous behaviors subsumed under *dispersal*, which make our definition more useful for small mammals, are reviewed by Lidicker and Stenseth (in press). When dispersal results in movements out of or into populations, such behavior is termed emigration and immigration, respectively. We are here interested in dispersal that

144

occurs across the boundaries of populations defined by genetic discontinuities among them (that is, demes). Such genetic discontinuities occur when imposed by spatial structuring.

Gene flow among demes requires dispersal, but dispersal can occur without gene flow, as for example when a disperser dies before reestablishing itself and reproducing, or when an immigrant fails to breed successfully. Conversely, effects on genetic structure by dispersers can occur in the absence of gene flow. This may happen when dispersers influence group genetic properties by their mere presence or absence, such as by changing economic or behavioral competition among genotypes, by influencing frequency dependent selection on traits, by changing the behavioral or physiological properties of the group, or in the case of emigration by changing the probabilities of genetic drift occuring. When gene flow does occur (Shields's *successful dispersal*; chap. 1, this vol.), it can influence the genetic composition of source populations, recipient populations, or new colonies.

A further complexity is imposed by the fact that dispersal itself is not a simple or monomorphic behavior (Lidicker 1985a; Lidicker and Stenseth in press). The timing of dispersal relative to phase of population growth, season, or developmental stage can have profound effects on the quality and quantity of the dispersing individuals and on their chances of successful reestablishment. The sex of the disperser is also clearly an important variable. Moreover, the motivation to leave home can be intrinsic (genetic, physiological) or extrinsic: economic, social, or some community influence such as interspecific competition, predation, or parasitism. The genetic consequences of dispersal similarly can be complex, depending on numerous factors including the genetic composition of the dispersers relative to the source and recipient populations.

It is therefore increasingly apparent that a simple relationship between dispersal and genetic structure of metapopulations does not necessarily exist (see also Horn 1983). Population genetic models that do not take the necessary complexity into account can be no more than a first step toward a fuller understanding of this relationship. Our approach here is to examine the dispersal–genetic composition link empirically. By examining species with various patterns of spatial groupings, various patterns of temporal stability, and various kinds of social structure and stability, we anticipate that insights will emerge that would not be possible by other approaches. Rodents represent an ideal group for such an empirical approach because within this diverse assemblage of species (about 1,700 species known), a rich diversity of demographic patterns and social behaviors are known. This diversity may well be as great as in all other mammals taken together. Moreover, their relatively high densities and short generation times make them suitable for population studies covering time spans of years or decades rather than centuries or millenia. They are also often tractable subjects for experimental manipulations both in the field and laboratory.

In considering the evolutionary consequences of dispersal, we must keep in mind that effects may be different if dispersal occurs among existing groups, results in the founding of new groups, or functions only to relieve population pressures. Another critical variable that must be incorporated is the stability that groups exhibit over time. Species that show seasonal or multiannual changes in spatial structure provide an entirely different evolutionary scenario than species with stable groupings.

Finally, we wish to emphasize the importance of social systems to both dispersal and genetic structure. Social factors can determine the nature of dispersers and their propensity to leave, as well as when they leave and the likelihood of successful immigration. Moreover, the relative genetic contribution of individuals within the group as well as the size and stability of groups may all have profound social causes. Heuristically, we suggest that social groups in rodents can be viewed as a series of stages of increasing stability and cohesiveness (table 10.1). The absence of socially mediated structure would occur if populations consisted of promiscuous nomads (level 1). We do not know of any examples of this level of sociality among rodents. At the other extreme, we can imagine groups so cohesive socially that dispersal in or out rarely occurs (level 7). Again, such an extreme does not seem to exist among rodents, but such circumstances can be imagined where social groups are strongly isolated by physical or habitat barriers (islands).

In between these two extremes at least five additional levels of social structure can be usefully distinguished.

2. *Individual territoriality*. Adults of both sexes maintain separate and nonoverlapping territories so that social bonding is limited to the primitive triad of mating, nursing, and sibship.

3. *Single-sex territoriality*. At this level, one sex is the primary territory holder. The other sex adopts the preexisting territorial structure either singly or in groups, or wanders among territories. The mating system can be varied, but the general characteristic of this level is that some regular and nonaggressive association occurs between adults of both sexes.

TABLE 10.1 Heuristic Levels of Social Complexity and Stability as Applied to Rodents

| LEVEL | NAME | PRIMARY FEATURES |
|---|---|---|
| 1 | Nonterritorial | Promiscuity; absence of social structure |
| 2 | Individual territoriality | Primitive mammalian social bonds only |
| 3 | Single-sex territoriality | Some regular affiliative associations occur between adults of both sexes |
| 4 | Single-sex philopatry | Social continuity maintained by philopatric sex |
| 5 | Seasonal social groups | Temporary but exclusive social groups; may be nonreproductive |
| 6 | Social group territoriality | Long-term, exclusive social groups |
| 7 | Closed social groups | Social groups isolated behaviorally or physically from other groups |

4. *Single-sex philopatry.* One sex characteristically disperses at or before puberty, leaving home ranges in possession of related individuals of the other sex. This arrangement results in cohesive groups in which members of one sex are closely related.

5. *Seasonal social groups.* In some species, closed social groups consisting of mixed sex and age components occur over some portion of the annual cycle. They may or may not be reproductive units.

6. *Social group territoriality.* At this level, social groups are characterized by long-term continuity. Membership is consistent although some movements in and out of the group are possible. Territorial defense is often cooperative. Reproduction occurs within the group.

Thinking about available data on rodent social systems in the context of these levels may make it possible to recognize patterns that will be revealing of critical relationships between social biology and the dispersal-genetic connection. Questions we pose at this time are: Is there any consistent relationship between social stability and gene flow? Are effective population sizes correlated with or even causally related to social dynamics? Is there any evolutionary significance to the pattern of alternating isolation and panmixia seen in some species? What is the potential for individual, kin, and intergroup selection among the various structuring patterns seen in rodents? One conclusion that is immediately obvious from a survey of the available information is that very few studies simultaneously address the relevant issues of social system, dispersal, and genetic structure. We have chosen four examples among the better-known species of small rodents to illustrate a diversity of social levels and to indicate the status of our understanding of the issues addressed in this chapter.

### Thomomys bottae

Pocket gophers as a group exhibit a suite of characteristics exemplifying a social system with individual territoriality, our level 2. Members of both sexes are sedentary and maintain exclusive territories throughout both the breeding and nonbreeding season; juveniles necessarily disperse from the maternal burrow subsequent to weaning (see, for example, Williams and Cameron 1984). This system of spatially contiguous territories places a maximum limit on adult population density at about 75 individuals per ha, a figure that is low in comparison to most small rodent species. Presumably, this territorial packing, in combination with the energetic costs of constructing an elaborate underground tunnel system, places a premium on juvenile dispersal. Depending upon the degree of habitat saturation, it also affects the distance of that dispersal. Under low density, the adult sex ratio is even, but a significant skew favoring females develops under high densities.

Data on genetic structure of pocket gopher populations from allozyme surveys are available on a regional scale for most taxa in the family (e.g.,

Nevo et al. 1974; Patton and Yang 1977; Zimmerman and Gayden 1981). Data on a local scale, however, are available for only the single species *Thomomys bottae*, and these are based on long-term population studies in central California (Patton and Feder 1981; Daly and Patton 1986, unpub. data). Initial studies focused on breeding patterns, as established by paternity exclusion analyses, and on the potential impact of gene flow on among-population genetic differentiation deduced from extirpation experiments (Patton and Feder 1981). These data clearly showed that (1) the effective sex ratio was biased substantially by a high variance in male reproductive success and that (2) recolonization of vacant fields resulted in a significant lowering of among-field standardized genetic variance (Wright's $F_{st}$; Wright 1965), while nonrandom breeding patterns within fields generated a potential three- to fourfold increase in this variance in but a single generation. The genetically effective population size ($N_e$) was estimated at between 9 and 27 individuals, based on knowledge of the census population numbers and the skewness in male reproductive success compounding the adult sex ratio (Patton and Feder 1981). These data were generally consistent with the traditional view of small effective population size in pocket gophers, resulting from limited vagility.

Such a view of genetic structure, however, is influenced by its horizontal (i.e., spatial) rather than vertical (i.e., temporal) perspective. It is, after all, the product of the momentary effective size ($N_e$) and the migration rate ($m$) that provides the evolutionarily important measure of population size. Migration rates have been estimated from dispersal studies in *T. bottae* based on a 4-year mark and recapture program where genetic markers have been used to establish breeding performance subsequent to dispersal (Daly and Patton, unpub. data). Migration rates calculated by a measure of the variance of distances moved by juveniles from their natal to breeding sites (Endler 1977; Barrowclough 1980) range from 0.2 km to 0.4 km per generation. Virtually all dispersal is accomplished by young of the year, as adults rarely move once a territory is established. Adult longevity is short, with between 58% and 75% of the population within each field turning over each year. The resulting estimated genetic neighborhood size (Wright 1943, 1951) is over 110 individuals, a very sizable increase over the static ($N_e$) estimate of 9 to 27. See Lande (1979) for further discussion of static versus short- and long-term estimates of $N_e$.

Two different views of genetic demography in pocket gopher populations are apparent. These views emphasize the importance of accurate estimates of dispersal in the examination of population genetic structure. In the first case, populations are viewed as being small, due primarily to low census number, and having skewed sex ratios and high variance in male reproductive success. This combination of traits results in significant among-deme differentiation, due largely to genetic drift between generations. However, under conditions where habitat is relatively continuous and saturated, quite a different view is possible. Under these conditions, migration rates are high enough to override the differentiating effects of drift, and effective population sizes are nearly an order of magnitude larger (Patton 1985).

## Microtus californicus

The California vole is assignable to our social level 3 (Ostfeld, Lidicker, and Heske 1985). Males are territorial throughout the year, whereas females are either nonterritorial or facultatively so. Female home ranges overlap extensively, especially in the better microhabitats and during the breeding season. It is not clear at present whether this overlap results from lack of aggression among females or whether aggression is differentially directed, perhaps against unfamiliar or nonrelated individuals. Results from enclosure experiments suggest that adult females are at least sometimes intolerant of other females (Lidicker 1980). There is also some evidence that interfemale aggression is triggered by the presence of an adult male. Monogamy occurs in enclosure populations and almost surely at low densities. Otherwise, polygyny, with a high variance in male success rates, seems to prevail in this species.

While sex ratios among prereproductives tend to be even (Lidicker 1973), adult ratios are characteristically female biased (Lidicker 1980; Ostfeld, Lidicker, and Heske 1985). Strongest bias is found in the better microhabitats and at high densities. Male home ranges (450 m²) are 36% larger than those of females (330 m²), but those portions of the ranges within the 85% probability-of-occurrence contours are the same size (Ford and Krumme 1979; Lidicker 1980). How these home range dimensions may vary with density or microhabitat has not been studied.

The California vole exhibits varied patterns of density change ranging from annual cycles of abundance (Lidicker 1973; Krohne 1982) to characteristic microtine multiannual cycles (Krebs 1966; Bowen 1982; Cockburn and Lidicker 1983; Ostfeld, Lidicker, and Heske 1985). Where multiannual cycles occur, populations persist at extremely low densities for 2 to 3 years at a time. Average life spans are only a few months, with maximum intervals of 32 weeks (Bowen 1982) or 56 weeks (Cockburn and Lidicker 1983) being recorded. Average residency was 1.85 months in the 41-month study of Ostfeld, Lidicker, and Heske (1985).

Adult voles are reluctant to abandon established home ranges, even after fires have destroyed the support capacity of their habitat (Lidicker, unpub. data). Nevertheless, dispersal is a major component of their life history. At least three kinds of dispersal can be distinguished in this species: (1) Dispersal of juveniles occurs shortly after weaning (natal or "ontogenetic dispersal," Lidicker 1985a). Both sexes disperse in this manner, but males are represented in slightly higher percentages. Such juvenile dispersal occurs even at very low densities and results in colonization of empty habitat

(Lidicker and Anderson 1962; Lidicker 1980, 1985a). (2) Saturation dispersal of adults occurs during periods of declining carrying capacity (summer dry periods), especially when densities are relatively high. Adult females may predominate slightly in this phenomenon (Riggs 1979; Lidicker 1980, 1985b). (3) Adult males engage in considerable dispersal just prior to and during the initial stages of the breeding season (Pearson 1960; Lidicker 1973, 1985b); this behavior is an example of the "seasonal dispersal" of Lidicker (1985a).

Dispersal distances are not well known. In a 41-month study involving four grids about 50 m part, only nine instances of intergrid movement were recorded. Bowen (1982) similarly reported four such movements in 20 months of trapping the same population. Although this is an inadequate sample, root mean square dispersal distance calculated on these nine individuals is 75 m (table 10.2). It seems likely that movements of this order or larger are common as Fisler (1962) found that most voles homed from distances of 130 m or less, and some homed from 200 m. Males consistently performed better than females in these tests.

Local genetic differentiation among populations has been studied at the Russell Reservation (Contra Costa Co., California) for 6.5 years utilizing electrophoretically detectable biochemical traits (Bowen 1982; Lidicker 1985c, unpub. data; Bowen and Koford, chap. 12, this vol.). This period encompassed two peaks in population density. Four polymorphic loci available in blood were utilized. For our purposes, two important conclusions can be drawn from this study. First, the genetic composition of refugial groups following population crashes is considerably influenced by stochastic factors. This process was evident in three out of four loci examined. Moreover, Bowen's data (1982) showed that the four grids behaved independently in this regard, that is, closer grids were not more similar than distant ones. The second conclusion is that heterogeneity among grids was very much greater at low densities when the refugial groups were isolated than at high densities when voles were dispersed more continuously. The range of $F_{ST}$ values for comparisons made at the regional level are not greater than for local groups at low densities (Bowen 1982; table 10.2). Mean heterozygosities for biochemical characters are also given in table 10.2.

*Microtus xanthognathus*

The population structure of the taiga vole has been described by Wolff and Lidicker (1980, 1981) and Wolff (1980). During the breeding season it exhibits our social level 3, and for most of the rest of the year it lives at social level 5. Males are strongly territorial during the breeding period, with two or three adult females sharing each home range. Females may overlap with more than one adult male, however, and so it is not clear if individual males have exclusive mating access to particular females. In large indoor arenas, adult females were aggressive to other females only around

TABLE 10.2 Measures of Genetic Differentiation and Dispersal Distance in Four Species of Small-Bodied Rodents

|  | Thomomys bottae | Microtus californicus | Microtus xanthognathus | Mus musculus |
|---|---|---|---|---|
| $F_{ST}$* |  |  |  |  |
|   Statewide | 0.32[a](21) | — | — | 0.18[b](4) |
|   Regional | 0.31[a](21) | 0.05[c](4) | — | 0.10[d](1) |
|  |  |  |  | 0.17[b](3) |
|   Local | 0.07[e](11) | 0.02, 0.10[c](4) | — | 0.02[d](1) |
|  |  |  |  | 0.03[b](3) |
| $Nm$ |  |  |  |  |
|   Regional | 0.53 | 4.85 | — | 2.25 |
|  |  |  |  | 1.22 |
|   Local | 3.32 | 15.36, 2.38 | — | 15.36 |
|  |  |  |  | 9.75 |
| Mean heterozygosity ($H$) | 0.04–0.19[a] | 0.08[f] | 0.06[f] | 0.00–0.11[g] |
|  |  |  |  | 0.11–0.17[h] |
|  |  |  |  | 0.11[i] |
|  |  |  |  | 0.08[j] |
| Root $\bar{x}$ square dispersal distance (m) | 190[k] | 75[l] | 152[m] | 27[n] |
| Maximum recorded distance (m) | > 700 | 141[l] | 800[m] | 1,500[o] |

Sources: (a) Patton and Yang 1977 (California); Patton and Smith, unpub. data (regional); (b) Wright 1978 (Texas); Selander and Kaufman 1975 (regional and local); (c) Bowen 1982 (high and low densities, respectively); (d) Baker 1981; (e) Patton and Feder 1981; (f) Lidicker, Riggs, and Yang, unpub. data; (g) Berry and Peters 1981; (h) Berry et al. 1981; (i) Selander and Yang 1969; (j) Selander, Hung, and Yang 1969; (k) Daly and Patton, unpub. data; (l) Lidicker, unpub. data ($N$ = 9); (m) calculated from data in Wolff and Lidicker 1980 ($N$ = 38); (n) calculated from data in Baker 1981, fig. 1 and table 2; (o) Berry 1968.

*Number of polymorphic loci on which averages are based are given in parentheses.

their own nest boxes. It would be interesting to know if in the field adult females were nonaggressive toward females sharing an adult male's range yet aggressive toward other females. The area encompassing 95% of the home range of adult males averages 650 m², and that of adult females averages 583 m² (males 11.5% larger). Reflecting this difference in home range size, adult sex ratios are female biased during the breeding season. Juvenile sex ratios are even.

For 7 to 8 months of the annual cycle, taiga voles live in an entirely different kind of social group (Wolff and Lidicker 1981). At the end of the summer they reassemble into overwintering (midden) groups consisting of 5 to 10 individuals ($\bar{x}$ = 7.1). Evidence suggests that group members are mostly unrelated individuals, although in two cases a pair of sisters joined the same group. Otherwise the distribution of sexes and ages in these groups appears random. These midden groups collect a huge cache of rhizomes, which comprises 90% of their winter food. There are also marked thermoregulatory advantages accruing to the midden group. Of greatest significance for us here is the fact that once formed, these are apparently closed groups; no exchange of individuals between groups has been observed during the winter period.

Unlike many other species of *Microtus,* taiga voles exhibit relatively stable annual fluctuations in numbers. There are, however, long-term trends in density that are tied to successional changes in the postburn taiga communities they commonly inhabit. Also, when a recently burned area is discovered, there is initially very rapid growth as the new habitat is colonized. Very likely, there are limited habitats where permanent populations of taiga voles persist, but in the main it is a fugitive species depending on the exploitation of postburn successional habitats. The normal life span reported for these interior Alaskan populations is 12 or 13 months, with a few individuals surviving to 18 months. Thus, adults that participate in the formation of midden groups (14% of the population at this time) succumb during the ensuing winter.

Consistent with the fugitive life habits of this species, dispersal is a conspicuous part of its annual cycle. It also manages an average litter size of 8.8, which is large even by microtine standards. There are two periods of the year characterized by extensive dispersal. The first occurs at the end of the breeding season and culminates in the establishment of overwintering groups. All of the juvenile males, most of the juvenile females, and some adults disperse at this time. Then, beginning in mid-April, the midden groups start to break up, with considerable movement occurring under the snow cover that still persists. By the time of melt-off in mid to late May, the breeding social structure has been reestablished. Only one male from a midden group establishes a breeding territory on the same spot; the others, if there are any, disperse. On the other hand, only about 45% of the females in a midden group were observed to disperse. As usual, dispersal distances are difficult to determine. The longest known dispersal movement is 800 m, but this undoubtedly reflects the limitations on trapping effort rather than indicating true dispersal capabilities. The root mean square dispersal distance calculated from 38 known movements is 152 m (table 10.2); this calculation excludes adult late summer dispersals, which are unlikely to have any genetic consequences.

Genetic data suitable for determining differentiation among local populations are not available for taiga voles. A survey of 36 biochemical loci from 20 voles collected over several square miles, however, provides the estimate for mean heterozygosity given in table 10.2. Although these numbers are the lowest for the four species considered here, they are well within the typical range for small mammal species.

## Mus musculus

The ubiquitous house mouse (*Mus musculus;* see Marshall and Sage 1981 for arguments in favor of using *M. domesticus* as the appropriate species name) has been the subject of numerous investigations relevant here. In spite of this attention, conspicuous gaps remain in our knowledge of important aspects of this species' population biology. The species seems incredibly flexible in its social mores commensurate with its deserved reputation as an ecological opportunist. Its social biology at low densities is poorly known, but at high densities it falls into our social level 6. The importance of olfactory cues in maintaining group associations has been shown by Cox (1984). Social group territories are defended by the dominant male(s) and adult females, especially those that are pregnant or lactating (Petrusewicz and Andrzejewski 1962; Anderson and Hill 1965; Reimer and Petras 1967; Anderson 1970; Lidicker 1976). However, some disagreement with this pattern for feral populations has been expressed (Myers 1974; Baker 1981).

Some populations are characterized by low and stable densities. Others have chronically high, but stable, numbers, and still others are eruptive. Although there are few good estimates, adult sex ratios generally appear to be equal (e.g., Berry 1968). Two minor exceptions have been reported by DeLong (1967) for the feral populations he studied in California: (1) males occurred in excess just prior to periods of population growth, and (2) females predominated at very high densities. Home range size varies with habitat and with the presence or absence of other species, but is the same for both sexes (Lidicker 1966; Quadagno 1968). A study of multiple captures in live traps led Myers (1974) to conclude that adult male and adult male–young male combinations were less likely to occur than chance, whereas adult male–adult female and juvenile combinations were more likely to occur.

At least three kinds of dispersal seem to be discernible in this species. In many regions there is a well-established autumn dispersal involving all sex and age groups ("seasonal dispersal" of Lidicker 1985a). This dispersal is associated with the movement of populations into sheltered overwintering sites such as barns, or, as in coastal California, it coincides with the start of the winter rainy season (DeLong 1967). Another kind of seasonal dispersal occurs in the spring prior to the breeding season on Welsh Islands (Berry 1968). A second class of dispersal characterizes recently weaned young (natal or "ontogenetic dispersal" of Lidicker 1985a). Juveniles of both sexes make exploratory movements at this time. If they find a social or habitat vacancy, they disperse. If no suitable place for dispersal is found, they generally return to their home base (Lidicker 1976). While both sexes indulge in this behavior, males are more persistent, and tend to predominate in samples of such dispersers. The third type of dispersal occurs into vacant or improving habitats ("colonization dispersal" of Lidicker 1985a). Myers (1974) reports on 11 known dispersers of this type. In this group, 7 were females and 4 males. The females were more likely to be in reproductive condition than resident females. A single case of colonization described by Lidicker (1976) also involved 5 females (1 adult and 4 juvenile siblings).

Commensurate with the fugitive or "weedy" nature of house mice, dispersal distances are undoubtedly sometimes great. But few relevant records are available. Berry (1968) reports that on Skokholm Island (Wales) 14 individuals were recorded to have moved 1500 m and at least 30 moved 500 m. Myers's (1974) colonizing dispersers moved between two grids 92 m apart. The root mean square dispersal distance calculated for 38 inter-chicken-coop movements reported by Baker (1981)is 27 m (see table 10.2). However, this statistic is obviously biased by the fact that trapping was only done in five adjacent coops (distance between farthest coops was 87.5 m). Of 352 recapture distances reported by Lidicker (1966), only 5% were greater than 21.3 m, and 2.3% were greater than 33 m. Obviously, much more information is needed on the extent and frequency of long-distance movements in this species.

We come now to the primary question of what is known about the genetic structure of house mouse populations. Some authors have assumed incorrectly that stable social groupings imply that these mice characteristically live in genetically closed groups (Ehrlich and Raven 1969; Selander 1970a; Myers 1974; Baker 1981). This is a misconception for several reasons. The first error is to assume that closed social groups are stable for long periods of time. This sometimes happens where mice live in relatively stable habitats such as barns, granaries, or artificial enclosures. Most populations, however, are subject to frequent turnover of socially dominant individuals and to regular habitat changes. These forces constantly disrupt what might otherwise be stable social groupings. The second error is the failure to appreciate that new social groups are almost always formed by unrelated males and females (Lidicker 1976). Moreover, under most circumstances, the formation of new groups is a frequent phenomenon. Thus, there is no a priori conflict between a high level of sociality (level 6 in this case) and extensive gene flow.

Given the wide range of variabilities exhibited by house mouse habitats, it is not surprising that different investigators have found widely varying levels of genetic structuring. When social groupings are relatively stable, successful movement among groups is accomplished mainly by females (Selander 1970a; Lidicker 1976; Pennycuik et al. 1978), and gene flow moves deliberately from one group to adjacent groups (Lidicker 1976; Baker 1981). Significant genetic differentiation has been found in enclosures (Lidicker 1976), barns (Selander 1970a), and granaries (Anderson 1970) under such conditions, differences being stochastically determined. When less stable conditions prevail, significant structuring may still be found, but it is more ephemeral (Myers 1974; Baker 1981). On this scale, selective forces have been implicated in addition to stochastic factors in affecting the genetic makeup of groups (Myers 1974). On a regional scale, selective gradients are the likely determinants of geographic patterns of genetic change (Selander 1970a, 1970b).

A further measure of the potential for social group differentiation is given by effective population sizes ($N_e$). These have been estimated for feral populations by Petras (1967a, 1967b) as between 5 and 80. Based on the performance of captive social groups, DeFries and McClearn (1972) estimated that $(N_e)$ can be less than four individuals. These values are certainly consistent with the possibility of local demic differentiation as well as being permissive of larger breeding units under appropriate conditions.

## DISCUSSION

The degree of genetic structure in a population is expected to result, at least in part, from the product of the genetically effective population size $(N_e)$ and the extent of genetic migration between subpopulations $(m)$. The effective population size, in turn, depends on the complex interactions of the total number of individuals in the population and the degree of non-random breeding, including biases in sex-ratio and differential variance in male or female reproductive success. Moreover, gene flow is intimately tied to dispersal. All of these variables are potentially influenced by the social dynamics of demes. Therefore, a central question is whether the extent of genetic population structure in a given species has any necessary or consistent relationship with the social structure exhibited by that species (see also Chepko-Sade and Shields, chap. 19, this vol.). This very important theoretical question is fundamental to our estimation of the major modes of evolutionary change that may characterize animal populations (e.g., Bush et al. 1977; Templeton 1980). For most species we have some understanding of their social system; for very few do we have a corresponding view of their genetic structure.

Clearly, social organization may have dramatic effects on the distribution of genotypes in space and in time. However, at this point in our understanding, such effects are neither intuitively obvious nor predictive. This is because the magnitude of the dispersal component of social structure is rarely known for any species, and it cannot be predicted from the level of social complexity that a species exhibits, as is often assumed (e.g., Bush et al. 1977; Bush 1981). As a case in point, the subdivision of prairie dog populations (*Cynomys ludovicianus*) into wards and coteries results in significant among-social-group genetic divergence and apparent within-social-group inbreeding (Chesser 1983; but see Hoogland 1982). On the other hand, the harem breeding system of marmots (*Marmota flaviventris*) and greater spearnose bats (*Phyllostomus hastatus*), expected to generate similar genetic consequences, results in little (marmots) or no (bats) among-harem differentiation because of high levels of effective juvenile dispersal (Schwartz and Armitage 1980, 1981; McCracken and Bradbury 1977, 1981). Although data on virtually every aspect of the genetic demography of rodent populations are lacking, those we summarized here, based largely on our own studies, support the general view that social level and genetic structuring are not closely correlated.

The four taxa for which we have presented data span the full range of hierarchical social stability and cohesiveness seen among rodents (table 10.1, levels 2–6), yet their major genetic features are remarkably similar (table 10.2). With equivalent levels of population heterozygosity ($H$), the degree of mean variance in allele frequencies across local populations ($F_{ST}$) varies over a narrow range, 0.02 to 0.10. At the local level, 90% to 98% of the total genetic diversity is contained within each subpopulation, or deme, regardless of the type of social system present. This consistency occurs despite an apparent order-of-magnitude difference in measured dispersal distances (e.g., *Mus* compared with *Thomomys*; table 10.2). Local populations are defined here as those being within range of individual dispersal capabilities (i.e., neighborhoods).

Where these taxa differ substantially is in their regional and statewide levels of differentiation (table 10.2). *Regional* refers here to areas larger than those traversable by individual dispersers, that is, different farms or populations several kilometers apart. In our sample, regional $F_{ST}$'s vary from 0.05 (*Microtus californicus*) to 0.31 (*Thomomys bottae*). The trend in these values does not correspond to the axis of social level, nor is there, as might be expected, a negative relationship between the amount of regional differentiation and the estimated dispersal distances. Only two figures are available for statewide comparisons, and they are remarkably similar to corresponding regional values. The statewide values suggest that differentiation on this level is positively correlated with dispersal distance and negatively correlated with social level, both results being counterintuitive. Any interpretation of these $F_{ST}$ values must, at this point, be done cautiously. Not only are there very few comparable estimates available, but the number of loci on which they are based varies from 1 to 4 for *Mus* and *Microtus* to 11 to 21 for *Thomomys*.

More satisfying is an apparent negative relationship between dispersal distance and social hierarchy (*Thomomys*, level 2, has a higher average dispersal distance than *Microtus*, levels 3 or 5, which in turn is higher than *Mus*, level 6; table 10.2). Here too though there are several potential inadequacies in the dispersal estimates, which must be kept in mind. First, long-distance dispersal parameters are notoriously difficult to estimate. Few studies are designed to determine such values, and what is available tends to be anecdotal or constrained by trapping grid designs. Second, dispersal is at best only measured over the period of a particular study, usually but one or a few years. If it varies stochastically or even in concert with environmental fluctuations, estimates based on the short-term may not be the same as the actual long-term value.

Another way to evaluate the general findings is to consider the overall gene flow parameter ($N_e m$) relative to social structure. Wright (1943) has provided a direct estimate of $Nm$ from the $F_{ST}$ distribution, where $F_{ST} = 1/(4Nm + 1)$, although this estimate is subject to error when the number of subdivided populations examined is small (Nei, Chakravarti, and Tateno 1977), as is the case in virtually all reported studies. These estimates range from 2.38 to 15.36 for local populations under the most structured situations, and from 0.79 to 4.58 for regions (see table 10.2). Slatkin (1981), based on a graphical measure of relative $Nm$ derived from the conditional average allele frequency, considered an $Nm$ product of 1.25 or higher to represent high gene flow. Clearly, each of the species examined here falls into this category at the local level, despite variation in expressed social organization. But, again, it is only at the regional level that differences among the species are seen, although these differences do not follow a consistent pattern with regard to the hierarchy of social complexity exhibited by these species.

We must therefore tentatively conclude that there does not appear to be a consistent relationship between social and genetic structure in small-bodied rodents, at least among local populations. It must be emphasized, however, that both the number of species examined and the quality of the relevant data on both genetic and social structure are inadequate for this conclusion to be taken as more than a suggestion at this time.

With respect to whether or not there is a relationship between $N_e$ and social level, again we must conclude that the evidence for rodents is negative. We note, for example, that similar estimates of $N_e$ have been obtained for both *Thomomys* (social level 2) and *Mus* (social level 6).

Finally, it seems significant that three of the four species in our sample characteristically go through periods when demes are small and relatively isolated. Such episodes are followed by periods in which many small demes mix panmictically. The house mouse does this because of its ability to find and exploit isolated and often temporary favorable habitat patches. Its alternation of phases is more irregular than in the other two species. The taiga vole goes through an annual pattern of isolation and mixing, and the California vole often exhibits multiannual cyclicity in density. Life-history patterns such as these should provide opportunities for kin and intergroup selection to operate. Although we cannot point to any specific evidence that these types of selection do in fact occur, a clear correspondence exists between population structure in these species and the interdemic selection model of Wright (1980), and the trait group model of selection proposed by Wilson (1975, 1977). Moreover, in the California vole, local differentiation exceeds regional differentiation when population densities are low. This pattern is contrary to expectations and translates into $Nm$ values (table 10.2), which average greater at the regional than local level. This observation seems consistent with the suggestion that local differentiation may be an important evolutionary force in this species.

While this review leads to several important, albeit tentative, conclusions, and raises a number of critical questions, its paramount finding is that inadequate data are available to answer the important issues covered. Perhaps we can take refuge in the claim that there is merit in outlining the issues and calling our collective attention to the deficiencies of the data base.

## ACKNOWLEDGMENTS

We express appreciation to Diane Chepko-Sade and Zuleyma Halpin for inviting us to participate in this important and enjoyable symposium. Helpful criticism of our manuscript was given by W. M. Shields, P. E. Smouse, and N. C. Stenseth.

## REFERENCES

Anderson, P. K. 1970. Ecological structure and gene flow in small mammals. In *Variation in mammalian populations*, ed. R. J. Berry and H. N. Southern, 299–325. Zoological Society of London Symposium No. 26. London: Academic Press.

Anderson, P. K., and J. L. Hill. 1965. *Mus musculus:* Experimental induction of territory formation. *Science* 148:1753–55.

Baker, A. E. M. 1981. Gene flow in house mice: Introduction of a new allele into free-living populations. *Evolution* 35:243–58.

Barrowclough, G. F. 1980. Gene flow, effective population sizes, and genetic variance components in birds. *Evolution* 34:789–98.

Berry, R. J. 1968. The ecology of an island population of the house mouse. *Journal of Animal Ecology* 37:445–70.

Berry, R. J., and J. Peters. 1981. Allozymic variation in house mouse populations. In *Mammalian population genetics*, ed. H. M. Smith and J. Joule, 242–53. Athens: University of Georgia Press.

Berry, R. J., R. D. Sage, W. Z. Lidicker, Jr., and W. B. Jackson. 1981. Genetical variation in three Pacific house mouse (*Mus musculus*) populations. *Journal of Zoology* (London) 193:391–404.

Bowen, B. S. 1982. Temporal dynamics of microgeographic structure of genetic variation in *Microtus californicus. Journal of Mammalogy* 63:625–38.

Bush, G. L. 1981. Stasipatric speciation and rapid evolution in animals. In *Evolution and speciation*, ed. W. R. Atchley and D. S. Woodruff, 201–18. Cambridge: Cambridge University Press.

Bush, G. L., S. M. Case, A. C. Wilson, and J. L. Patton. 1977. Rapid speciation and chromosomal evolution in mammals. *Proceedings National Academy of Science* (USA) 74:3942–46.

Chesser, R. K. 1983. Genetic variability within and among populations of the black-tailed prairie dog. *Evolution* 37:320–31.

Cockburn, A., and W. Z. Lidicker, Jr. 1983. Microhabitat heterogeneity and population ecology of an herbivorous rodent, *Microtus californicus. Oecologia* 59:167–77.

Cox, T. P. 1984. Ethological isolation between local populations of house mice (*Mus musculus*) based on olfaction. *Animal Behaviour* 32:1068–77.

Daly, J. C., and J. L. Patton. 1986. Growth, reproduction, and sexual dimorphism in *Thomomys bottae* pocket gophers. *Journal of Mammalogy* 67:256–65.

DeFries, J. C., and G. E. McClearn. 1972. Behavioral genetics and the fine structure of mouse populations: A study in microevolution. In *Evolutionary biology*, ed. T. Dobzhansky, M. K. Hecht, and W. C. Steere, 5:279–91. New York: Appleton-Century-Crofts.

DeLong, K. T. 1967. Population ecology of feral house mice. *Ecology* 48:611–34.

Ehrlich, P. R., and P. H. Raven. 1969. Differentiation of populations. *Science* 165:1228–32.

Endler, J. A. 1977. *Geographic variation, speciation, and clines*. Princeton: Princeton University Press.

Fisler, G. F. 1962. Homing in the California vole, Microtus californicus. *American Midland Naturalist* 68:357–68.

Ford, R. G., and D. W. Krumme. 1979. The analysis of space use patterns. *Journal of Theoretical Biology* 76:125–55.

Hoogland, J. L. 1982. Prairie dogs avoid extreme inbreeding. *Science* 215:1639–41.

Horn, H. F. 1983. Some theories about dispersal. In *The ecology of animal movement*, ed. I. R. Swingland and P. J. Greenwood, 54–62. Oxford: Clarendon Press.

Krebs, C. J. 1966. Demographic changes in fluctuating populations of *Microtus californicus. Ecological Monographs* 36:239–73.

Krohne, D. T. 1982. The demography of low-litter-size populations of *Microtus californicus. Canadian Journal of Zoology* 60:368–74.

Lande, R. 1979. Effective deme sizes during long-term evolution estimated from rates of chromosomal rearrangement. *Evolution* 33:234–51.

Lidicker, W. Z., Jr. 1966. Ecological observations on a feral house mouse population declining to extinction. *Ecological Monographs* 36:27–50.

———. 1973. Regulation of numbers in an island population of the California vole, a problem in community dynamics. *Ecological Monographs* 43:271–302.

———. 1975. The role of dispersal in the demography of small mammals. In *Small mammals: Their productivity and population dynamics*, ed. F. B. Golley, K. Petrusewicz, and L. Ryszkowski, 103–28. London: Cambridge University Press.

———. 1976. Social behaviour and density regulation in house mice living in large enclosures. *Journal of Animal Ecology* 45:677–97.

———. 1980. The social biology of the California vole. *The Biologist* 62:46–55.

———. 1985a. An overview of dispersal in non-volant small mammals. In *Migration: Mechanisms and adaptive significance*, ed. M. A. Rankin, pp. 369–85. Contributions in Marine Science, Supplement, vol. 27.

———. 1985b. Dispersal. In *Biology of New World* Microtus, ed. R. H. Tamarin, 420–54. Special Publication No. 8, American Society of Mammalogists.

———. 1985c. Population structuring as a factor in understanding microtine cycles. *Acta Zoologica Fennica* 173:23–27.

Lidicker, W. Z., Jr., and P. K. Anderson. 1962. Colonization of an island by *Microtus californicus,* analyzed on the basis of runway transects. *Journal of Animal Ecology* 31:503–17.

Lidicker, W. Z., Jr., and R. L. Caldwell. 1982. *Dispersal and migration*. Benchmark Papers in Ecology, vol. 11. Stroudsburg, Pa.: Hutchinson Ross Publishers.

Lidicker, W. Z., Jr., and N. C. Stenseth. In press. To disperse or not to disperse: Who does it and why? In *Dispersal: Small mammals as a model*, ed. N. C. Stenseth and W. Z. Lidicker, Jr. London: Chapman and Hall Publishers.

McCracken, G. F., and J. W. Bradbury. 1977. Paternity and genetic heterogeneity in the polygynous bat, *Phyllostomous hastatus. Science* 198:303–6.

————. 1981. Social organization and kinship in the polygynous bat *Phyllostomus hastatus*. *Behavioural Ecology and Sociobiology* 8:11–34.

Marshall, J. T., and R. D. Sage. 1981. Taxonomy of the house mouse. *Symposium Zoological Society of London* 47:15–25.

Myers, J. H. 1974. Genetic and social structure of feral house mouse populations on Grizzly Island, California. *Ecology* 55:747–59.

Nei, M., A. Chakravarti, and Y. Tateno. 1977. Mean and variance of $F_{st}$ in a finite number of incompletely isolated populations. *Theoretical Population Biology* 11:291–306.

Nevo, E., Y. J. Kim, C. R. Shaw, and C. S. Thaeler, Jr. 1974. Genetic variation, selection and speciation in *Thomomys talpoides* pocket gophers. *Evolution* 28:1–23.

Ostfeld, R. S., W. Z. Lidicker, Jr., and E. J. Heske. 1985. The relationship between habitat heterogeneity, space use, and demography in a population of California voles. *Oikos* 45:433–42.

Patton, J. L. 1985. Population structure and the genetics of speciation in pocket gophers, genus *Thomomys*. *Acta Zoologica Fennica* 170:109–14.

Patton, J. L., and J. H. Feder. 1981. Microspatial genetic heterogeneity in pocket gophers: Nonrandom breeding and drift. *Evolution* 35:912–20.

Patton, J. L., and S. Y. Yang. 1977. Genetic variation in *Thomomys bottae* pocket gophers: Macrogeographic patterns. *Evolution* 31:697–720.

Pearson, O. P. 1960. Habits of *Microtus californicus* revealed by automatic photographic records. *Ecological Monographs* 30:231–49.

Pennycuik, P. R., P. G. Johnston, W. Z. Lidicker, Jr., and N. H. Westwood. 1978. Introduction of a male sterile allele ($t^{w2}$) into a population of house mice housed in a large outdoor enclosure. *Australian Journal of Zoology* 26:69–81.

Petras, M. L. 1967a. Studies of natural populations of *Mus*. I. Biochemical polymorphisms and their bearing on breeding structure. *Evolution* 21:259–74.

————. 1967b. Studies of natural populations of *Mus*. II. Polymorphisms at the *T* locus. *Evolution* 21:466–78.

Petrusewicz, K., and R. Andrzejewski. 1962. Natural history of a free-living population of house mice (*Mus musculus* Linnaeus) with particular reference to groupings within the population. *Ekologica Polska*, ser. A, 10:85–122.

Quadagno, D. M. 1968. Home range size in feral house mice. *Journal of Mammalogy* 49:149–51.

Reimer, J. D., and M. L. Petras. 1967. Breeding structure of the house mouse, *Mus musculus*, in a population cage. *Journal of Mammalogy* 48:88–99.

Riggs, L. A. 1979. Experimental studies of dispersal in the California vole, *Microtus californicus*. Ph.D. diss., University of California, Berkeley.

Schwartz, O. A., and K. B. Armitage. 1980. Genetic variation in social mammals: The marmot model. *Science* 207:665–67.

————. 1981. Social substructure and dispersion of genetic variation in the yellow-bellied marmot (*Marmota flaviventris*). In *Mammalian population genetics*, ed. M. H. Smith and J. Joule, 139–59. Athens: University of Georgia Press.

Selander, R. K. 1970a. Behavior and genetic variation in natural populations. *American Zoologist* 10:53–66.

————. 1970b. Biochemical polymorphism in populations of the house mouse and old-field mouse. *Symposium Zoological Society of London* 26:73–91.

Selander, R. K., W. G. Hunt, and S. Y. Yang. 1969. Protein polymorphism and genic heterozygosity in two European subspecies of the house mouse. *Evolution* 23:379–90.

Selander, R. K., and D. W. Kaufman. 1975. Genetic structure of populations of the brown snail (*Helix aspersa*). I. Microgeographic variation. *Evolution* 29:385–401.

Selander, R. K., and S. Y. Yang. 1969. Protein polymorphism and genic heterozygosity in a wild population of the house mouse (*Mus musculus*). *Genetics* 63:653–67.

Selander, R. K., S. Y. Yang, and W. G. Hunt. 1969. Polymorphism in esterases and hemoglobin in wild populations of the house mouse (*Mus musculus*). *Studies in Genetics* (University of Texas) 5:271–338.

Slatkin, M. 1981. Estimating levels of gene flow in natural populations. *Genetics* 99:323–35.

Templeton, A. R. 1980. Modes of speciation and inferences based on genetic distances. *Evolution* 34:719–29.

Williams, L. R., and G. N. Cameron. 1984. Demography of dispersal in Attwater's pocket gopher (*Geomys attwateri*). *Journal of Mammalogy* 65:67–75.

Wilson, D. S. 1975. A theory of group selection. *Proceedings National Academy of Science* (USA) 72:143–46.

———. 1977. Structured demes and the evolution of group-advantageous traits. *American Naturalist* 111:157–85.

Wolff, J. O. 1980. Social organization of the taiga vole (*Microtus xanthognathus*). *Biologist* 62:34–45.

Wolff, J. O., and W. Z. Lidicker, Jr. 1980. Population ecology of the taiga vole (*Microtus xanthognathus*) in interior Alaska. *Canadian Journal of Zoology* 58:1800–20.

———. 1981. Communal winter nesting and food sharing in taiga voles. *Behavioural Ecology and Sociobiology* 9:237–40.

Wright, S. 1943. Isolation by distance. *Genetics* 28:114–38.

———. 1951. The genetical structure of populations. *Annals of Eugenics* 15:323–54.

———. 1965. The interpretation of population structure by F-statistics with special regard to systems of mating. *Evolution* 19:395–420.

———. 1978. *Evolution and the genetics of populations*, vol. 4: *Variability within and among natural populations*. University of Chicago Press.

———. 1980. Genic and organismic selection. *Evolution* 34:825–43.

Zimmerman, E. G., and N. A. Gayden. 1981. Analysis of genic heterogeneity among local populations of the pocket gopher, *Geomys bursarius*. In *Mammalian population genetics*, ed. M. H. Smith and J. Joule, 272–87. Athens: University of Georgia Press.

# 11. Phenotypic and Genotypic Mechanisms for Dispersal in *Microtus* Populations and the Role of Dispersal in Population Regulation

*Michael S. Gaines and Michael L. Johnson*

Historically, empirical studies of dispersal of microtine rodents have their roots in the observations of Krebs, Keller, and Tamarin (1969) that populations of voles confined in fenced enclosures reach abnormally high densities, severely overgraze their food supply, and undergo a precipitous decline. This aberrant demography was attributed to the lack of emigration from the population. MacArthur (1972) immediately recognized the biological significance of fencing in populations, thereby inhibiting dispersal, and labeled it the Krebs effect.

Since the discovery of the fence effect, investigations of dispersal in populations of microtines have focused on three disparate areas, all of which have converged over the last 15 years. One major area of research has been the role of dispersal in the population regulation of microtine species. Ecologists have searched unsuccessfully for a unifying mechanism for the unusual population fluctuations exhibited by these rodents ever since they were described in detail by Charles Elton (1926). After it became clear that dispersal was an essential component of the normal demographic machinery of microtine populations, it became a leading candidate as the key to understanding population cycles. As we show later, dispersal did not completely fulfill these expectations. A second area of interest has been the proximate mechanism for dispersal in microtine rodents. Aggression appeared to be an obvious factor. Christian (1970) suggested that as density increased, food resources became limited, resulting in intraspecific competition and aggression. Christian assumed that adult males forced subordinate males to leave the population. Krebs (1978a) suggested that an increase in spacing behavior, due to competition among males for mates during periods of increasing density, leads to an increase in agonistic interaction and dominant males force subordinate males to disperse. Although both of these scenarios seem reasonable, critical experiments have not been performed to test these hypotheses. Finally, there has been considerable speculation about the evolution of dispersal in microtine populations. Christian (1970) used an r-α selection argument, proposing that over evolutionary time such species as *Microtus pennsylvanicus,* inhabiting temporary habitats, evolved higher rates of dispersal than such species as *Microtus ochrogaster* inhabiting more permanent habitats. Lidicker (1975, 1985) has discussed the costs and benefits to an individual when it disperses, and he distinguished between two kinds of dispersal in small mammal populations: presaturation and saturation. Presaturation dispersal occurs before carrying capacity is reached and the dispersing individuals are in good condition, whereas saturation dispersal occurs at carrying capacity when resources are limited and the dispersing individuals are the social outcasts. Lidicker proposed that presaturation dispersers play a greater role in the evolution of dispersal behavior because unlike saturation dispersers, they pass their genes on to future generations in newly colonized habitats. As far as we know, there have been no empirical studies examining the selective consequences of dispersal in populations of microtines.

This chapter attempts to summarize some of the work done by both ourselves and others in these three different areas of the study of dispersal in *Microtus* populations. Clearly, all three areas are interrelated. For example, if we could elucidate evolutionary mechanisms for dispersal, it should give us some insight into the role of dispersal in population regulation, which in turn should shed light on proximate mechanisms for dispersal.

## THE ROLE OF DISPERSAL IN POPULATION REGULATION

Much microtine research has been directed toward understanding the role of dispersal in the regulation of microtine populations. The strong emphasis on population regulation is due in part to microtine biologists' compulsive need to explain the causes for population cycles. There are three subsidiary questions related to population regulation. First, what is the magnitude of dispersal in microtine populations? Clearly, if only a few individuals emigrate during density fluctuations then the role of dispersal in population regulation would be minimal. Percentage loss due to dispersal has been used as an index to estimate the amount of dispersal. It is calculated by comparing the number of marked animals disappearing from a control grid who later reappeared on a removal grid from which all animals were continually removed. Percentage loss due to dispersal varies from a high of 54.3% for *M. ochrogaster* in Kansas (Gaines, Vivas, and Baker 1979) to a low of 7% for *M. pennsylvanicus* in Massachusetts (Tamarin 1977). Thus, a substantial amount of dispersal occurs during microtine fluctuations. Second, when does dispersal occur during a density cycle? In a majority of microtine species more losses from control populations are attributable to dispersal during phases of increasing density than during phases of declining density (Gaines and McClenaghan 1980). For example, in *M. ochrogaster* populations in eastern Kansas, percentage loss due to dispersal was 64.4% during phases of increasing density and 42.1% during phases of declining density (Gaines, Vivas, and Baker 1979). We examined the relationship between density and rates of dispersal in fluctuating populations of *M. ochrogaster* in eastern Kansas from data of Gaines, Vivas, and Baker (1979). A dispersal rate for each bimonthly trapping period was calculated by dividing the population density on the control grid into the number of marked animals removed from the removal grid. Product-moment correlation coefficients were calculated for seasonal means of the two variables. There was no obvious association between rates of dispersal and population density (fig. 11.1). We performed the same analysis incorporating 2-, 4-, and 6-week time lags, and again there was no correlation between rate of dispersal and density. We concluded from this analysis that dispersal is density-independent in *M. ochrogaster* populations in eastern Kansas.

The third question is related to qualitative differences between dispersers and residents. Are dispersers a random sample of the resident population? If dispersers are qualitatively different than residents, dispersal during the increase phase of the density cycle could change the demographic characteristics and/or the genetic composition of the population, setting the

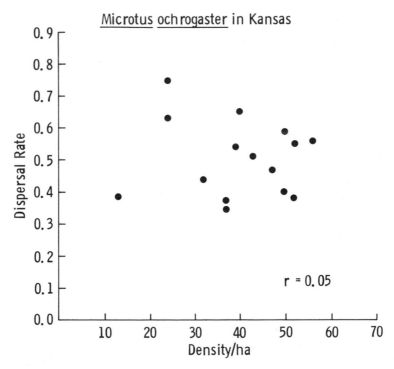

FIGURE 11.1. Dispersal rate versus density for *Microtus ochrogaster* in eastern Kansas.

stage for a decline in density. There is ample documentation that dispersers are not a random subset of the resident population based on demographic parameters (Myers and Krebs 1971; Kozakiewicz 1976; Tamarin 1977; Krebs, Redfield, and Taitt 1978; Gaines, Vivas, and Baker 1979; Gaines, Baker, and Vivas 1979). There are two consistent results from these studies. First, using body mass as an index of age, the age structure of dispersers is biased in favor of younger animals. Second, sex ratios of dispersers are biased in favor of males.

Genetic differences between dispersers and residents have been assessed with electrophoretic markers. For example, Myers and Krebs (1971) reported genetic differences between dispersers and residents at the TF (transferrin) and LAP (leucine aminopeptidase) loci in *Microtus* populations in southern Indiana. There was increased dispersal of $Tf^E/Tf^E$ homozygotes in males during the late peak and decline in *M. pennsylvanicus* populations, while during periods of increasing numbers, $Tf^C/Tf^E$ heterozygous females were more common among dispersing females than in resident populations. In *M. ochrogaster* populations, the rare $Tf^F/Tf^F$ homozygote was found only among dispersing males. LeDuc and Krebs (1976) also found differences between residents and dispersers with respect to LAP genotypes in populations of *M. townsendii*. These studies concluded that the genetic composition of dispersers is not a random sample of the resident population.

The major problem with these studies documenting demographic and genetic differences between dispersers and residents is that they all employ the removal grid method to identify dispersers. All animals appearing on the removal grids were defined as dispersers, and their characteristics were compared to residents on a nearby control grid. Because many of the animals appearing on the removal grid were unmarked, it was only appropriate to compare them with residents on the control grid if their source of origin was a population with identical demographic and genetic structure such as the control population (P. Anderson, n.d.). Unfortunately, they may not always be identical. An additional problem is that animals captured on the removal grids may not be true dispersers (Baird and Birney 1982).

Another method of identifying dispersers is using fenced enclosures with exit tubes. All emigrants leaving the enclosure through exit tubes can be captured, and the characteristics of these dispersers can be compared with the resident population inside the enclosure. In the few studies using enclosures to compare demographic and genetic attributes of dispersers, the results are ambiguous. Riggs (1979) found that *M. californicus* dispersers were younger than residents and that the sex ratio of dispersers was biased in favor of males, but he found no differences in the genetic composition of dispersers and residents based on seven polymorphic loci. Verner and Getz (1985) found low rates of dispersal through exit gates in enclosed populations of *M. pennsylvanicus*. Furthermore, dispersers were a random sample of the resident population with respect to sex ratio, age structure, and genotypes at the TF and LAP loci.

We recently completed a study of dispersal through exit tubes from an enclosed *M. ochrogaster* population (Johnson 1984; Johnson and Gaines, n.d.). The only consistent difference between dispersers and residents with respect to demographic parameters was that dispersers were younger than residents (table 11.1). In another experiment, two fenced enclosures with exit tubes (grid X and grid Z) were each stocked with populations having different $Tf^E$ allele frequencies. The TF allele frequencies for dispersers captured in exit traps were compared with resident populations inside the enclosure. Although there were no statistically significant differences in allele frequencies at the TF locus between dispersers and residents, there was a trend with dispersers having lower $Tf^E$ frequencies than residents

TABLE 11.1 Demographic Parameters for Resident and Dispersing *Microtus ochrogaster* from Two Fenced Enclosures (grids X and Y) Equipped with Exit Tubes

|  | GRID X | | GRID Y | |
|---|---|---|---|---|
| Sex ratio (proportion males) | | | | |
| Residents | .52 | (237) | .51 | (164) |
| Dispersers | .56 | (97) | .54 | ( 62) |
| Males breeding[a] | | | | |
| Subadults | | | | |
| Residents | .27 | (9) | .50 | (10) |
| Dispersers | .28 | (10) | .25 | (5) |
| Adults | | | | |
| Residents | .87 | (152) | .81 | (69) |
| Dispersers | .82 | (51) | .81 | (35) |
| Females breeding[b] | | | | |
| Subadults | | | | |
| Residents | .08 | (3) | .03 | (1) |
| Dispersers | 0 | (0) | 0 | (0) |
| Adults | | | | |
| Residents | .76 | (111) | .68 | (57) |
| Dispersers | .87 | (34) | .66 | (14) |
| Age structure | | | | |
| Juveniles | | | | |
| Residents | .11 | (52) | .16 | (52) |
| Dispersers | .11 | (19) | .08 | (9) |
| Subadults | | | | |
| Residents | .15 | (70) | .23 | (76) |
| Dispersers | .34* | (59) | .34** | (39) |
| Adults | | | | |
| Residents | .73 | (332) | .61 | (200) |
| Dispersers | .55 | (94) | .58 | (67) |

*Notes:* Numbers in parentheses are sample sizes. All other numbers are proportions. All texts of significance are chi-square tests of homogeneity.

a. Based on males with scrotal testes.

b. Based on females with medium and large nipples.

*p < .05.

**p < .001.

(fig. 11.2). In the grid Z population, which was founded with a $Tf^E$ frequency of .25, the allele frequency steadily increased during the study. In the grid X population, which was founded with a $Tf^E$ frequency of .75, the gene frequency remained stable during the study. Because $Tf^E$ frequencies in natural populations vary from .75 to .90, these results suggest that natural selection is operating on the TF locus, and maybe through differential dispersal. In summary, qualitative differences exist between dispersers and residents in a variety of microtine species. The differences between the two

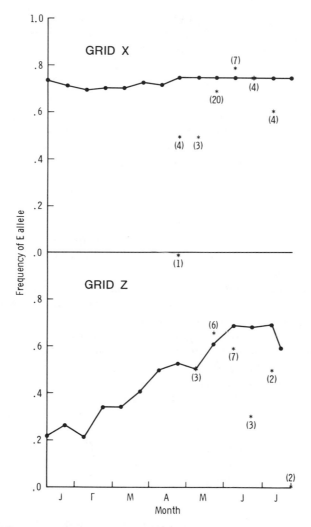

FIGURE 11.2. $Tf^E$ gene frequencies of residents and dispersing *Microtus ochrogaster* in two fenced enclosures. Closed circles are gene frequencies of residents. Asterisks are gene frequencies of dispersers, and numbers of dispersers are in parentheses.

groups are due in part to how a disperser is identified operationally. In the removal grid method, dispersers are successful immigrants, whereas in the fenced enclosure method, dispersers are emigrants that have not experienced the colonization process. What remains to be verified is whether the colonization process selectively removes certain genotypes from the population.

The prospect that dispersal may be linked to population regulation of microtine rodents has produced a number of hypotheses explaining the multiannual fluctuations in density exhibited by these species. Krebs (1978a) suggested that as density increases, increased agonistic interaction leads to greater spacing behavior of individuals, which in turn results in dispersal. Krebs assumed there is a strong genetic component to both spacing behavior and dispersal, which are subject to natural selection. There are several problems with the notion of a genetic-behavioral polymorphism driving a cycle in population numbers. The underlying assumption of the model is that aggression, spacing behavior, and dispersal all have relatively high heritabilities. The only available data on heritability come from a Ph.D. dissertation by Judith Anderson (1975) on behavioral correlations among relatives in *M. townsendii*. The heritability of aggressive behavior was 0, whereas the heritability of tendency to disperse was 0.62. Even with high heritability, it is difficult to imagine that natural selection could increase the frequency of these behaviors to any significant extent in so few generations. Finally, we have no idea how genetic and environmental conditions in the population determine an individual's spacing behavior (Krebs 1985).

Another attempt to relate dispersal to population cycles is Lidicker's dispersal sink concept (1975). Lidicker defined a dispersal sink as an area into which dispersers migrate. Lidicker suggested that with an adequate dispersal sink, a population should exhibit normal demography. Tamarin (1977) proposed that the abnormal demography of fenced populations may be due to the absence of a dispersal sink. Therefore any microtine population without a dispersal sink should exhibit aberrant fluctuations. Tamarin, Reich, and Myer (1984) tested whether a dispersal sink was necessary for meadow vole population cycles by using partially and fully fenced grids that included dispersal sinks in woodland areas. They found that both fenced populations and the control population exhibited identical demographic characteristics, which supports Lidicker's dispersal sink concept. One problem with Tamarin, Reich, and Myer's (1984) experimental design was that a grid without a dispersal sink was not fenced. Also, based on the density figures, we are not convinced the populations underwent multiannual cycles.

Hestbeck (1982) synthesized the views of Krebs (1978a) and Lidicker (1978) to explain how dispersal regulates population density in what he calls the social fence hypothesis. Hestbeck (1982) hypothesized that when density is low in neighboring areas, spacing behavior coupled with the emigration of individuals to dispersal sinks will regulate density. However, when neighboring areas reach high densities, the neighboring groups socially "fence" the control population by inhibiting further emigration. Once emigration is blocked, the population exhausts its food supply.

Hestbeck (1986) tested the social fence hypothesis with *Microtus californicus* in a series of experiments in fenced enclosures. He found that when he created a dispersal sink adjacent to a population by removing animals, dispersal dominated the population regulation process. However, in the absence of dispersal sinks, when habitat occupancy was more uniform, high overall density in surrounding areas resulted in reduced dispersal and resource limitation dominated the regulation process.

Microtine biologists have made great advances over the last 15 years in characterizing dispersing individuals in a variety of vole species. It has been demonstrated convincingly that more dispersal occurs during periods of increasing density than declining density. At this stage, our knowledge about the exact role of dispersal in population cycles is unclear. All we can say is that dispersal seems to be an essential component of the normal demographic machinery in microtine populations. However, experimental approaches, such as those used by Tamarin, Reich, and Myer (1984) and Hestbeck (1986), will go a long way toward advancing our understanding of dispersal as a population regulatory mechanism.

## PROXIMATE MECHANISMS FOR DISPERSAL

The genetic-behavioral polymorphism hypothesis for population cycles in microtine rodents originally proposed by Chitty (1967) and later modified by Krebs (1978b) assumes that aggression is the stimulus for dispersal. Similarly, Christian's (1970) social subordination hypothesis assumes that aggressive males force socially subordinate subadult males to disperse. The two hypotheses differ in that the former relies on a genetic basis for behavioral changes, whereas in the latter, dispersal is a physiological response to the immediate environment. Both hypotheses predict that dispersers will be subordinate individuals. Aggressive behavior of residents and dispersers has been compared in a variety of microtine species. In all cases, behavior was assessed only in male voles by pairwise encounters in a neutral arena in the laboratory. The results are ambiguous. Dispersers from *M. ochrogaster* populations in Indiana (Myers and Krebs 1971) and *M. townsendii* populations (Krebs, Redfield, and Taitt 1978) were subordinate individuals, whereas dispersing *M. pennsylvanicus* (Myers and Krebs 1971) were more aggressive than resident animals. Baker (unpub. data) studied the response of dispersers and residents in *M. ochrogaster* populations in eastern Kansas to naive 30-gram males raised in the laboratory. She found no differences in aggressive behavior between dispersers and residents.

One way to test whether aggression is a proximate stimulus for dispersal is to artificially increase aggression in the population and then monitor dispersal rates. Some progress has been made in manipulating aggressive behavior in microtine populations with testosterone implants. Gipps et al. (1981) manipulated male behavior in field populations of *M. townsendii* on Westham Island near Vancouver, British Columbia, Canada. Subadult males in one population were made precociously aggressive by subcutaneous silastic capsules of testosterone. Adult males in another population were made less aggressive with implants of scopolamine Hbr. There were only two significant demographic effects: (1) the rate of decline in male numbers in spring was lower in the scopolamine population than in the control; and (2) female breeding started earlier, and more females bred in the scopolamine population than a control population that was implanted with shams. There were no significant differences in recruitment or density of the populations. It was concluded that spacing behavior alone is not sufficient to control the demography of *M. townsendii* populations. No attempt was made to monitor dispersal rates.

Taitt and Krebs (1982) manipulated the behavior of female *M. townsendii* using testosterone implants to increase aggressiveness in one study site and force-fed mestranol to reduce aggressiveness in another study site. Testosterone treatment resulted in wounding among females, increased size of female home ranges, a reduction of female survival, and increased female immigration. Males in the testosterone-treated population had the same dynamics as males in a control population. Mestranol treatment had no effect on female dynamics, but males in these populations survived better than control males. Because the testosterone treatment had little effect on the spring decline in numbers, Taitt and Krebs concluded that female spacing behavior may operate in more subtle ways than overt aggression.

Abdellatif (1985) implanted testosterone capsules in prairie vole populations enclosed by fences equipped with exit tubes. There was a statistically significant higher mean level of wounding in males per 2-week trapping period in the testosterone-treated population as compared to a control sham-treated population (sham-treated enclosure: 4.43 ± 1.25 SE wounds per 2 weeks; testosterone enclosure: 9.62 ± 1.66 SE wounds per 2 weeks). This difference in wounding may result in more dispersal because the mean dispersal rate per 2-week trapping period from the testosterone grid was approximately three times higher than from the control grid (sham-treated enclosure: 2.96 ± .92 SE; testosterone enclosure: 8.41 ± 4.98 SE). However, the difference was not statistically significant. Of the 25 animals implanted with testosterone, only 2 dispersed, suggesting that the implanted animals forced others to disperse. Abdellatif compared a number of additional demographic variables such as rates of increase, survival rates, breeding activity based on external reproductive characteristics, body mass, and movement of individuals in the sham and testosterone populations. There were no statistically significant differences in mean values per 2-week trapping period in any of these variables.

Although Abdellatif's results suggest that aggression may be a proximate stimulus for dispersal, the experiment needs to be repeated with a larger number of implanted subadult males in a variety of species. Also, it would be interesting to measure dispersal rates in populations where aggression by females was manipulated with testosterone. Finally, if dispersal is a hormone-mediated response, it is essential to have baseline data on the endogenous level of hormones during population fluctuations. Vivas (1980) monitored plasma corticosterone levels in natural populations of *M. ochrogaster* in eastern Kansas. In individuals, corticosterone was measured using radioimmunoassay techniques. Mean corticosterone levels and density for one population are presented in figure 11.3. There was considerable variation in corticosterone levels within all populations over time, but no apparent relationship was found between corticosterone levels and density. Vivas compared corticosterone levels of male dispersers captured on removal grids with those of male residents on a neighboring control grid. There were no statistically significant differences between the two groups. Furthermore, if the level of corticosterone in the plasma is assumed to be an indicator of stress, this result could be interpreted as indirect evidence against the genetic-behavioral polymorphism and social subordination hypotheses for dispersal. Both assume that dispersers are socially subordinate and are forced out of the population by aggressive residents.

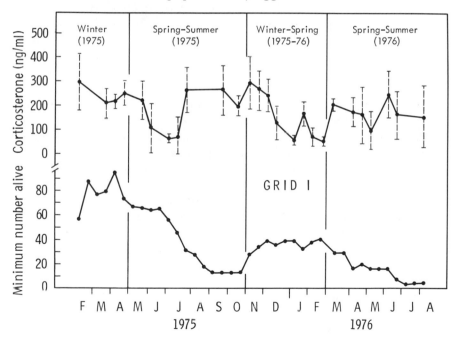

FIGURE 11.3. Minimum number of *Microtus ochrogaster* alive on grid I and mean plasma corticosterone level with SE.

There were some methodological problems with Viva's study. The major one was the response of the voles to traps and handling by the investigator. Although care was taken to minimize handling time of voles to 3 minutes after being released from the traps, some voles might have been stressed as a result of the routine trapping procedure. In spite of these problems, Vivas's study is the first attempt to measure hormone levels over a multiannual density cycle and should be viewed as a trailblazing effort. It would be interesting to measure other hormones, particularly testosterone, that might not respond as sensitively when animals are trapped and handled. Adams, Tamarin, and Calland (1980) were able to relate plasma testosterone levels to breeding activity in male *M. breweri*. However, no attempt was made to relate androgen levels to aggression.

## EVOLUTION OF DISPERSAL

There have been a prodigous number of theoretical models for the evolution of dispersal in populations (e.g. Gadgil 1971; Roff 1975; Hamilton and May 1977; Bengtsson 1978; Comins, Hamilton, and May 1980; Lomnicki 1980), all of which have been reviewed in detail by Stenseth (1983). These models are frustrating to field ecologists because they contain patently unrealistic assumptions and parameters that cannot be measured in natural populations. For example, Hamilton and May (1977) have shown that under conditions of global competition for space, dispersal will be favored even when the dispersing propagules have a very high mortality. Dispersal is maintained in the population as an evolutionarily stable strategy due to the increased fitness of residents that enforce dispersal of their offspring. This result is interesting but not very robust based on the following assumptions of the model: (1) death and replacement of every parent in each generation, (2) absence of vacant patches for colonization, and (3) reproduction by pure parthenogenesis. Comins, Hamilton, and May (1980) extended the Hamilton-May model with a discrete generation island model (Wright 1969) in which reproduction, migration, and competition were stochastic processes. The model allowed for transient patches, and more than one individual could occupy each patch. Although the extended model was more general, it contained parameters that were difficult to measure in the field, such as exogenous extinction rate, chance of surviving migration, and probability of migration.

Charnov and Finerty (1980) proposed a kin selection model that relates dispersal to vole population cycles and may be testable. The model assumes that at low density, there would be a high coefficient of relationship between individuals in the population and hence less aggression. As density increases there is some dispersal and the average coefficient of relationship between individuals in a neighborhood rapidly drops. Concomitant with this drop is an increase in agonistic interactions between individuals. This aggressive behavior leads to more dispersal, which in turn leads to even higher levels of aggression and higher rates of dispersal. Thus, dispersal and aggression are inexorably linked by kin selection. The Charnov-Finerty model makes the following predictions: (1) aggression is positively correlated with population density, (2) average heterozygosity is positively correlated with population density, and (3) dispersal rates are positively correlated with levels of aggression. All three predictions need to be tested in a variety of vole species.

Essential to any evolutionary model of dispersal is information on the costs and benefits associated with philopatry as opposed to those associated with leaving the population. Shields (chap. 1, this vol., tables 1.3 and 1.4) codifies potential costs and benefits of philopatry and dispersal under two major categories—genetic and somatic. Shields considers inbreeding depression associated with philopatry and outbreeding depression caused by the disruption of coadapted gene complexes associated with dispersal to be opposing genetic costs. The potential for increased competition with close kin associated with philopatry and decreased survivorship and fecundity due to greater risks and energy used during dispersal are considered to be opposing somatic costs. This dichotomy may not be useful because somatic effects as defined by Shields will affect the fitness of individuals and should have genetic consequences.

Using Shield's terminology, we recently examined the somatic costs and benefits associated with dispersal in the prairie vole *M. ochrogaster* in eastern Kansas. The experiment was designed to determine the fitness of dispersers placed into optimal habitat as compared to the fitness of nondispersers and "frustrated" dispersers, which were animals that wanted to disperse but were prevented from doing so. While dispersal into optimal habitat may not accurately reflect natural dispersal, it was the logical starting point. If the fitness of dispersers in optimal habitat was lower than the fitness of nondispersers and frustrated dispersers, then the hypotheses that dispersal is advantageous could be clearly rejected.

To begin the experiment, populations were established in enclosures X and Y (fig. 11.4). Exit tubes in these enclosures were opened, and animals caught leaving the enclosures were designated as dispersers. Dispersal was allowed continually on enclosure Y. On enclosure X, one-half of the dispersers were "allowed" to emigrate by moving them to enclosure Z, which was vacant to this point. The other half were returned to enclosure X and were designated frustrated dispersers. The survival of animals on all three enclosures was monitored for approximately 10 weeks, after which time all animals were removed from enclosure Z and the procedure was repeated. Nine replicates were conducted over a 2-year period (see Johnson 1984 for more detail).

During most of the nine replicates, densities on enclosure Z reached or exceeded densities on enclosure X, indicating that enclosure Z was optimal habitat. Survival rates of both adult male and adult female dispersers from

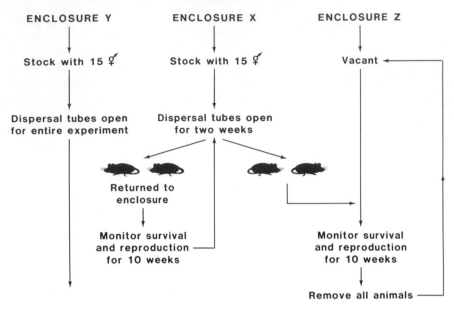

FIGURE 11.4. Experimental design for the fenced enclosure experiment. From Johnson and Gaines (1987).

enclosure Z were higher than the survival rates of nondispersers from enclosures X and Y, who in turn survived better than frustrated dispersers (table 11.2). These results were consistent across replicates. The comparisons with frustrated dispersers were complicated by the fact that these animals were forced to remain within their original population. The typical pattern during each 10-week pulse was for most of the frustrated dispersers to disappear (presumably die) during the first 2 or 4 weeks. However, a few of these animals survived and became established in the population. Their survival during the rest of the 10 weeks was approximately the same as the nondispersers in that enclosure. These preliminary data indicate that there is a cost to being forced to remain in the population rather than emigrate.

While the results from the first experiment are encouraging, some shortcomings in the design need to be tested further. First, moving dispersers to an optimal habitat is probably unrealistic. More often, dispersers en-

TABLE 11.2 Percentage of Adults Surviving per 14 Days

|  | GRID Y RESIDENTS | GRID X RESIDENTS | GRID X FRUSTRATED DISPERSERS | GRID Z DISPERSERS |
|---|---|---|---|---|
| Males | 87.9 | 86.5 | 82.4 | 94.7 |
| Females | 90.3 | 88.7 | 86.8 | 91.7 |

counter established populations or suboptimal habitat. Experiments are currently under way to determine the fitness of dispersers in these situations.

## PROSPECTUS

It has been established beyond question that dispersal is density-independent in fluctuating vole populations, with most dispersal occurring during phases of increasing density. Yet we know nothing about the quantitative distribution of dispersal distance from birth site to breeding site (see Jones, chap. 8, this vol., fig. 8.1, for data on Kangaroo rats) and how it changes with density or phase of the cycle. We are just now developing the technology to mark young in the nest with small radios and to follow them after weaning.

Dispersers are not a random subset of the resident population. The demographic profile of dispersers will vary depending on the methods used to identify them. In the removal grid method, "dispersers" are successful colonizers, whereas "dispersers" from fenced enclosures with exit tubes represent every emigrant. It is unclear whether the differences between dispersers identified by the two techniques are a result of sampling differences or whether the transience phase selectively removes certain phenotypes from the pool of dispersers. Just how these qualitative differences between dispersers and residents translate into a causal mechanism for population cycles is unclear.

Several hypotheses (social subordination hypothesis, genetic-behavioral polymorphism hypothesis, social fence hypothesis, and kin selection hypothesis) have attempted to link dispersal to population cycles. Explicitly stated or implicitly assumed in all these hypotheses is the notion that increasing population density causes changes in spacing behavior, which leads to dispersal. Krebs (1985) postulates that spacing behavior produces an unstable equilibrium at high density due to reproductive inhibition and mortality from aggressive encounters. At this point we do not know what triggers the expression of spacing behavior. The frequency of aggressive individuals and the genetic structure of populations are likely to affect spacing behavior (Krebs 1985). For example, we would expect less aggression if an individual is surrounded by relatives than if an individual is surrounded by genetic strangers. Currently we have little information on the genetic structure of vole populations (Bowen 1982).

The Charnov-Finerty kin selection model generates predictions that can be tested in natural populations of voles. Kin selection can drive population cycles if individuals are more aggressive to strangers than they are to kin. Results from an electrophoretic study of *Peromyscus polionotus* populations are consistent with the kin selection model (Garten 1976). Average heterozygosity over electrophoretic loci was positively correlated with density. Furthermore, individuals with high levels of heterozygosity were more aggressive than homozygous individuals. These results support the kin selection model if heterozygosity can be used as an index of relatedness.

We need more studies directed toward assessing the genetic and somatic costs versus benefits associated with dispersal compared to those associated with philopatry. If dispersal is an evolutionarily stable strategy, we would expect the mean fitness of dispersers to be equal to the mean fitness of residents over evolutionary time. In the studies we performed, dispersers were hand carried to optimal habitat. The transcience phase in which a disperser is searching for a suitable habitat may have a major affect on components of fitness and needs to be investigated in greater detail. Also, it may be possible that our efforts have been focused on the wrong segment of the population. Anderson (1980) argued that more attention needs to be given to the fitness of residents in the population.

An underlying assumption in any evolutionary model of dispersal is that the trait has a high heritability. We only have preliminary information on the heritability of dispersal in *M. townsendii* populations. Krebs (1979) has suggested that cyclic populations would have a high heritability of aggressive behavior whereas noncyclical populations would have a low heritability. This argument could be extended to dispersal behavior.

Finally, we need detailed information on the social structure of different microtine species before we can make any comprehensive statements about population regulation, proximate mechanisms for dispersal, or the evolution of dispersal. It is becoming increasingly apparent that there is a considerable amount of variation in mating systems among microtine species (Wolff 1985). Mating systems range from promiscuity in *M. pennsylvanicus* (Boonstra and Rodd 1983; Rodd and Boonstra 1984; Madison, Fitzgerald, and McShea 1984), polygyny in *M. californicus* (see Lidicker and Patton, chap. 10, this vol.), to monogamy in *M. ochrogaster* (Getz, Carter, and Gavish 1981; Getz and Hofmann 1986). We need to know how the social dynamics of populations change with density during different phases of the cycle. With the development and continued refinement of new technology such as radiotelemetry (Madison 1980) and radionuclides (Tamarin, Sheridan, and Levy 1983), we should make rapid advances in the social dynamics of microtines over the next few years.

## ACKNOWLEDGMENTS

We would like to thank the following people who greatly improved this chapter: P. K. Anderson, R. Boonstra, B. J. Danielson, C. J. Krebs, W. M. Shields, P. E. Smouse, N. C. Stenseth, and R. H. Tamarin. Our own work on vole dispersal was supported by grants from the National Science Foundation (DEB-8020343) and the University of Kansas General Research Fund.

# REFERENCES

Abdellatif, E. M. 1985. Experimental manipulation of aggression and its role as a proximate cause of dispersal in prairie voles, *Microtus ochrogaster*. Ph.D. diss., Department of Systematics and Ecology, University of Kansas.

Adams, M. R., R. H. Tamarin, and I. P. Calland. 1980. Seasonal changes in plasma androgen levels and the gonads of the beach vole, *Microtus breweri*. *General Comparitive Endocrinology* 41:31–40.

Anderson, J. L. 1975. Phenotypic correlations among relatives and variability in reproductive performance in populations of the vole *Microtus townsendii*. Ph.D. diss., University of British Columbia, Vancouver, B.C.

Anderson, P. K. 1980. Evolutionary implications of microtine behavioral systems on the ecological stage. *Biologist* 62:70–88.

———. Evolution of dispersal in rodents: Emigrant or resident fitness. (Unpublished.)

Baird, D. D., and E. C. Birney. 1982. Characteristics of dispersing meadow voles *Microtus pennsylvanicus*. *American Midland Naturalist* 107:262–83.

Bengtsson, B. O. 1978. Avoiding inbreeding: At what cost? *Journal of Theoretical Biology* 73:439–44.

Boonstra, R., and F. H. Rodd. 1983. Regulation of breeding density in *Microtus pennsylvanicus*. *Journal of Animal Ecology* 52:757–80.

Bowen, B. S. 1982. Temporal dynamics of microgeographic structure of genetic variation in *Microtus californicus*. *Journal of Mammalogy* 63:625–38.

Charnov, E. L., and J. Finerty. 1980. Vole population cycles: A case for kin-selection? *Oecologia* 45:1–2.

Chitty, D. 1967. The natural selection of self-regulatory behavior in animal populations. *Proceedings of the Ecological Society of Australia* 2:51–73.

Christian, J. J. 1970. Social subordination, population density, and mammalian evolution. *Science* 168:84–90.

Comins, H. N., W. D. Hamilton, and R. M. May. 1980. Evolutionary stable dispersal strategies. *Journal of Theoretical Biology* 82:205–50.

Elton, C. 1926. Periodic fluctuations in the numbers of animals: Their causes and effects. *British Journal of Experimental Biology* 2:119–63.

Gadgil, M. 1971. Dispersal: Population consequences and evolution. *Ecology* 52:253–61.

Gaines, M. S., C. L. Baker, and A. M. Vivas. 1979. Demographic attributes of dispersing southern bog lemmings (*Synaptomys cooperi*) in eastern Kansas. *Oecologia* 40:91–101.

Gaines, M. S., and L. R. McClenaghan, Jr. 1980. Dispersal in small mammals. *Annual Review of Ecology and Systematics* 11:163–96.

Gaines, M. S., A. M. Vivas, and C. L. Baker. 1979. An experimental analysis of dispersal in fluctuating vole population: Demographic parameters. *Ecology* 60:814–28.

Garten, C. T., Jr. 1976. Relationships between aggressive behavior and genic heterozygosity in the oldfield mouse, *Peromyscus polionotus*. *Evolution* 30:59–72.

Getz, L. L., C. S. Carter, and L. Gavish. 1981. The mating system of the prairie vole, *Microtus ochrogaster:* Field and laboratory evidence for pair-bonding. *Behavioral Ecology and Sociobiology* 8:189–94.

Getz, L. L., and J. E. Hofmann. 1986. Social organization in free-living prairie voles, *Microtus ochrogaster*. *Behavioral Ecology and Sociobiology* 18:275–82.

Gipps, J. H. W., M. Taitt, C. Krebs, and Z. Dundjerski. 1981. Male aggression and the population dynamics of the vole, *Microtus townsendii*. *Canadian Journal of Zoology* 59:147–57.

Hamilton, W. D., and R. M. May. 1977. Dispersal in stable habitats. *Nature* 269:578–81.

Hestbeck, J. 1982. Population regulation of cyclic mammals: The social fence hypothesis. *Oikos* 39:157–63.

———. 1986. Multiple regulation states in populations of the California vole, *Microtus californicus*. *Ecological Monographs* 56:161–81.

Johnson, M. L. 1984. Selective basis for dispersal of the prairie vole, *Microtus ochrogaster*. Ph.D. diss., Department of Systematics and Ecology, University of Kansas.

Johnson, M. L., and M. S. Gaines. 1987. Selective basis for emigration of the prairie vole, *Microtus ochrogaster*. *Ecology* 68:684–94.

Kozakiewicz, M. 1976. Migratory tendencies in a population of bank voles and a description of migrants. *Acta Theriologica* 21:321–38.

Krebs, C. J. 1978a. Aggression, dispersal, and cyclic changes in populations of small mammals. In *Aggression, dominance, and individual spacing*, ed. L. Krames, P. Pliner, and T. Alloway, 49–60. New York: Plenum.

———. 1978b. A review of the Chitty hypothesis of population regulation. *Canadian Journal of Zoology* 56:2463–80.

———. 1979. Dispersal, spacing behaviour, and genetics in relation to population fluctuations in the vole *Microtus townsendii*. *Fortschritte der Zoologie* 25:61–77.

———. 1985. Do changes in spacing behaviour drive population cycles in small mammals? In *British Ecological Society Symposium on Behavioural Ecology, Reading, England*, ed. R. Smith and R. Sibley, 295–312. Oxford: Blackwell Scientific Publication.

Krebs, C. J., B. L. Keller, and R. H. Tamarin. 1969. *Microtus* population biology: Demographic changes in fluctuating populations of *M. ochrogaster* and *M. pennsylvanicus* in southern Indiana. *Ecology* 50:587–607.

Krebs, C. J., J. A. Redfield, and M. J. Taitt. 1978. A pulsed-removal experiment on the vole *Microtus townsendii*. *Canadian Journal of Zoology* 56:2253–62.

LeDuc, J., and C. J. Krebs. 1976. Demographic consequences of artificial selection at the LAP locus in voles (*Microtus townsendii*). *Canadian Journal of Zoology* 53:1825–40.

Lidicker, W. Z., Jr. 1975. The role of dispersal in the demography of small mammal populations. In *Small mammals: Their productivity and population dynamics*, ed. F. B. Golley, K. Petrusewicz, and L. Ryszkowski, 103–28. Cambridge: Cambridge University Press.

———. 1978. Regulation of numbers in small mammal populations: Historical reflections and a synthesis. In *Populations of small mammals under natural conditions*, ed. D. P. Snyder, 122–41. Special Publication Series, Pymatuning Laboratory in Ecology. University of Pittsburgh.

———. 1985. Dispersal. In *Biology of New World* Microtus, ed. R. H. Tamarin, 420–54. Special Publication No. 8, American Society of Mammalogists, Shippensburg, Pa.

Lomnicki, A. 1980. Regulation of population density due to individual differences and patchy environments. *Oikos* 35:185–93.

MacArthur, R. H. 1972. *Geographical ecology.* New York: Harper and Row.

Madison, D. M. 1980. Space use and social structure in meadow voles, *Microtus pennsylvanicus. Behavioral Ecology and Sociobiology* 7:65–71.

Madison, D. M., R. Fitzgerald, and W. McShea. 1984. Dynamics of social nesting in overwintering meadow voles (*Microtus pennsylvanicus*): Possible consequences for population cycles. *Behavioral Ecology and Sociobiology* 15:9–17.

Myers, J. H., and C. J. Krebs. 1971. Genetic, behavioral and reproductive attributes of dispersing field voles *Microtus pennsylvanicus* and *Microtus ochrogaster. Ecological Monographs* 41:53–78.

Riggs, L. A. 1979. Experimental studies of dispersal in the California vole, *Microtus californicus.* Ph.D. diss., Department of Zoology, University of California, Berkeley.

Rodd, F. H., and R. Boonstra. 1984. The spring decline in the meadow vole, *Microtus pennsylvanicus:* The effect of density. *Canadian Journal of Zoology* 62:1464–73.

Roff, D. A. 1975. Population stability and the evolution of dispersal in a heterogeneous environment. *Oecologia* 19:217–37.

Stenseth, N. C. 1983. Causes and consequences of dispersal in small mammals. In *The ecology of animal movements,* ed. I. Swingland and P. Greenwood, 63–101. Oxford: Oxford University Press.

Taitt, M. J., and C. J. Krebs. 1982. Manipulation of female behavior in field populations of *Microtus townsendii. Journal of Animal Ecology* 51:681–90.

Tamarin, R. H. 1977. Dispersal in island and mainland voles. *Ecology* 58:1044–54.

Tamarin, R. H., L. M. Reich, and C. A. Meyer. 1984. Meadow vole cycles within fences. *Canadian Journal of Zoology* 62:1796–1804.

Tamarin, R. H., M. Sheridan, C. K. Levy. 1983. Determining matrilineal kinship in natural populations of rodents using radionuclides. *Canadian Journal of Zoology* 61:271–74.

Verner, L., and L. L. Getz. 1985. Significance of dispersal in fluctuating populations of *Microtus ochrogaster* and *M. pennsylvanicus. Journal of Mammalogy* 66:338–47.

Vivas, A. M. 1980. Adrenocortical activity and population regulation in *Microtus ochrogaster.* Ph.D. diss., Department of Systematics and Ecology, University of Kansas.

Wolff, J. O. 1985. Behavior. In *Biology of New World* Microtus, ed. R. H. Tamarin, 420–54. Special Publication No. 8, American Society of Mammalogists, Shippensburg, Pa.

Wright, S. 1969. *Evolution and the genetics of populations,* vol. 2: *The theory of gene frequencies.* Chicago: University of Chicago Press.

# 12. Dispersal, Population Size, and Genetic Structure of *Microtus californicus*: Empirical Findings and Computer Simulation

*Bonnie S. Bowen and Rolf R. Koford*

Empirical studies of genetic structure are useful (1) for determining the amount of genetic structure in natural populations, thus providing information about the potential role of nonselective factors in evolution (e.g., via Wright's shifting balance model), and (2) for testing theoretically derived predictions concerning the genetic consequences of social and demographic parameters. Comparative studies are especially useful for testing these predictions. One can compare the amount of genetic structure among species with different demographic and social characteristics (Wright 1978; Lidicker and Patton, chap. 10, and O'Brien, chap. 13, this vol.). Alternatively, one can compare the amount of genetic structure within a species of different times, using species that exhibit temporal changes in demographic and social characteristics. In this chapter we use the latter approach, which has two benefits. First, when the demographic and social factors thought to influence patterns of genetic variation vary in time, one can determine whether there are consequent genetic changes and can evaluate the genetic consequences of the various factors. Second, if genetic structure does vary temporally, one can identify the times at which genetic drift is most likely to occur.

Studies of temporal changes in genetic structure are especially appropriate in one group of mammals: microtine rodents. Many of the species in this group undergo dramatic temporal fluctuations in numbers, which are accompanied by changes in other demographic parameters (Krebs and Myers 1974). One therefore expects to observe changes in genetic structure associated with demographic changes in populations of *Microtus*. Such changes have been observed in one species, the California vole, *Microtus californicus* (Bowen 1982).

The temporal changes in genetic structure observed by Bowen (1982) occurred during a 2-year period when population size increased to a peak and then crashed. Bowen (1982) suggested that the changes in genetic structure were due to changing population size and gene flow among subpopulations, although the amount of gene flow, via dispersing individuals, was not measured directly. We have developed a computer simulation model to investigate the relative influences of population size and gene flow on genetic structure. In this chapter we describe the empirical results obtained by Bowen (1978, 1982) and report the results of the simulation model.

The model measures changes in allele frequencies and genetic structure that accompany demographic changes, and it allows us to address quantitatively various questions relating to demography and genetic structure. Specifically, we wanted to evaluate the effects of rapidly changing population size and various migration rates on the amount of differentiation among subpopulations over a period of just a few generations. We use the term *migration* in the sense that it is used in population genetics; the migration rate, $m$, is the fraction of the population composed of immigrants in each generation (Wright 1951; Shields, chap. 1, this vol.). In population genetic models, population size, migration rate, and the breeding system determine the amount of genetic structure expected in the population under equilibrium conditions (Wright 1951, 1965). Such models do not reveal the extent of genetic differentiation expected as a result of demographic changes on a microecological time scale of just a few generations (Spieth 1974; Slatkin 1985). By using a simulation model, we are able to approximate the rapidly changing demographic conditions experienced by *Microtus* populations.

## DYNAMICS OF GENETIC STRUCTURE IN *Microtus*

*Microtus californicus* populations are usually cyclic. During the course of a population cycle, dramatic demographic changes occur, the most obvious of which is a change in density of about two orders of magnitude over a period of 2 to 4 years (Krebs 1966; Pearson 1966; Krebs and Myers 1974). At low densities, animals are distributed in a grassland in discrete patches of a few individuals. After a period of relatively low population density lasting from 1 to 3 years, the population grows rapidly (Krebs 1966). Breeding occurs during the wet season in California, roughly October through May, when green grass, the primary food supply for *M. californicus*, is available. Under favorable field conditions, females can produce a litter of four to six young every 3 to 5 weeks, and young females can be reproductively mature at 3 to 5 weeks of age (Greenwald 1957; Hoffmann 1958). There is thus the potential for very rapid population growth relative to other mammals, and as the population grows, previously unoccupied grassland habitat is colonized as a result of dispersal from the discrete patches. At peak density, the population is distributed continuously throughout the grassland, and discrete patches of animals are no longer discernible. Following peak density, populations typically decline sharply and are again broken up into a series of patches of individuals. Some of the patches persist until the next breeding season, and the cycle continues.

Lidicker and Patton (chap. 10, this vol.) have described three major types of dispersal in *M. californicus* populations: (1) Seasonal dispersal occurs early in the breeding season and is characterized by movements of adult males. (2) Ontogenetic dispersal occurs when young animals disperse shortly after weaning and is especially important when population density is low and population growth is beginning. Young individuals that disperse at that time are important in colonizing unoccupied habitat (Pearson 1960; Lidicker and Anderson 1962; Batzli 1968). (3) Saturation dispersal occurs at the end of the breeding season when the carrying capacity declines due to the summer dry season.

One would theoretically expect that the changing demographic pattern just described would lead to changing genetic structure. When population density is low and the population consists of discrete patches of individuals, one would expect genetic differentiation among the patches. As the population grows and animals disperse between the patches, this differentiation would be reduced as long as immigration into patches is successful (see Shields, chap. 1, this vol., for a definition of successful dispersal). The result would be a genetically homogeneous population throughout the habitat. The rapid decrease in density and the fragmentation of the population that occur after the peak would be expected to result in a bottleneck in population size and multiple founder events. Under these conditions, genetic differentiation should increase.

Bowen (1982) studied a population of *M. californicus* in an 8 ha meadow at Russell Reservation, near Berkeley, California. The study was initiated when animals were distributed in small patches with population density that was moderately low, following one season of breeding after a population crash. Four trapping grids were established in places where voles were present during this period of moderately low density. The grids, each 20 m by 20 m, were arrayed linearly through the meadow, at about 50 m intervals. Animals were trapped approximately every 2 weeks for 20 months. During this time the population density peaked, then crashed to extremely low levels.

Density was measured by censusing the number of animals known to be alive during each sampling period (Krebs 1966). Genetic structure was measured by determining the amount of genetic differentiation in allele frequencies among four grids at four electrophoretically assayed polymorphic loci: *Glucose phosphate isomerase (GPI), Glutamate oxaloacetate transaminase-1 (GOT-1), Leucine aminopeptidase (LAP),* and *Phosphogluconate dehydrogenase (PGD)*. Breeding studies in the laboratory confirmed that all alleles at these loci were inherited in a Mendelian fashion (Bowen and Yang 1978). The measure of genetic structure used was Wright's $F_{ST}$ (Wright 1951, 1965), the standardized variance in allele frequencies, determined for each locus individually and averaged over the four loci.

For analytical purposes, the 20-month study was divided into five time periods representing major demographic periods in the population cycle. The months included in each time period were as follows: period 1, June–October 1975; period 2, November–December 1975; period 3, January–March 1976; period 4, April–May 1976; period 5, June 1976–January 1977. Period 1 was one of moderately low population density. The three central time periods covered those months in which the population was at highest densities, reaching a peak in period 3. Period 5 extended from midway through the crash to a time of very low density following the crash. Little or no breeding occurred during periods 1 and 5 (Bowen 1978). The $F_{ST}$ values reported here are for resident animals, operationally defined as animals that were captured in more than one trapping session. Although some individuals were captured more than once in a time period, they were used only once in each period for calculating $F_{ST}$.

The measure of genetic structure, $F_{ST}$, showed a pattern of change that was the reverse of the pattern of density change (fig. 12.1). Early in the cycle, when density was moderately low, $F_{ST}$ was relatively high. As the density increased, $F_{ST}$ declined to very low levels, indicating essentially no genetic differentiation at that time. As the population density decreased and animals became distributed into discrete patches again, the level of differentiation increased.

Just how high is the maximum level of $F_{ST}$ observed in this study? In absolute terms it is not extremely high (Wright 1978), but considering that these sampling grids were only 50 m apart, the level observed at low density is striking. In fact, the maximum microgeographic level of differentiation is equivalent to that observed across populations sampled over a 10 km range (Bowen 1982).

Three demographic events observed during this study would be expected to affect genetic structure: (1) the low density early in the study, (2) the increase in density during the breeding season, and (3) the rapid decline in density that followed the peak. At the beginning of the cycle, when

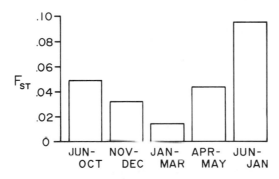

FIGURE 12.1. Amount of genetic differentiation, as measured by $F_{ST}$, among four subpopulations (grids) of *M. californicus* as population density increased then crashed.

population size was low and the population was distributed in discrete patches, founder effects and small size of breeding units would be expected to result in patches that were genetically differentiated from each other. We found differentiation early in the cycle. Similarly, at the end of the study, when density declined rapidly, genetic differences would be expected to appear among the resulting discrete patches of animals. $F_{ST}$ increased as population size decreased between periods 4 and 5. What remains, then, is to understand the decrease in genetic differentiation that occurred during the period of increasing population density.

When reproduction and population growth occur, relatively high population size, emigration (dispersal), and immigration would be expected to cause a shift in the genetic structure of the population. If there is no genetic migration and the subpopulations remain isolated from one another, then increased population size alone would not be expected to cause a decrease in the level of genetic differentiation among them. For *M. californicus*, increasing population size is accompanied by dispersal (Krebs 1966; Riggs 1979). Dispersal itself, however, does not indicate the extent of genetic migration, which depends on the behavior and reproductive success of the dispersers. If the dispersers are able to enter existing patches and breed, the amount of genetic heterogeneity among patches would decrease under the pressure of gene flow. The results obtained by Bowen (1982) suggest that the initiation of breeding at the end of period 1 was accompanied by dispersal and immigration that resulted in gene flow among the grids she studied.

Bowen (1982) did not measure the rate of dispersal or rate of immigration directly, and she could not evaluate the relative contributions of migration and population size in causing the changes in genetic structure observed. We have therefore developed a simulation model to determine what sorts of temporal changes in allele frequencies and genetic structure one might expect with the demographic changes we have described.

## SIMULATION MODEL

We have used the simulation model to explore the genetic changes that would occur over time in neutral alleles. The allele frequencies at one locus were allowed to vary through time in several subpopulations, as demographic events resulted in gains and losses of alleles. The model consisted of repeated time periods (months) during which reproduction, survival, dispersal, and immigration took place.

Males and females were treated separately, although for the runs reported here there were equal numbers of each sex among the offspring produced, the dispersing offspring, and the adults in the initial population. Each sex of a given subpopulation contributed a random sample of alleles to the offspring produced by that subpopulation. The number of offspring produced each month was fixed relative to the number of adult females present in that month. A certain proportion of the offspring stayed in their natal subpopulation to breed in the following month and for the rest of their lives. Genetically, these were a random sample of the offspring produced. The rest of the offspring dispersed, with some going to nearby subpopulations to breed and the remainder disappearing from the population. The sex of immigrating individuals was determined randomly. No adults dispersed between subpopulations. After breeding, mortality (at a fixed rate) occurred among the adults, with the survivors being a random genetic sample of the breeding adults present during that month. Then immigrants (random genetic samples of the offspring produced that month in other subpopulations) and nondispersing offspring were recruited to the breeding (adult) population, and the model was iterated to the next time period.

Changes in allele frequencies were monitored for one locus with two alleles during each simulation run. At the beginning of each run the frequencies of both alleles were 0.5. Allele frequencies were calculated for each subpopulation at the end of each time period by using the alleles present in the surviving adults, the offspring that remained in their natal subpopulations to breed, and the immigrants. The model worked only with allele frequencies, not genotype frequencies. At the end of each time period, the standardized variance, $F_{ST}$, in allele frequencies among the subpopulations was calculated.

The parameter values were chosen to approximate those of a natural population of *Microtus californicus*. There were four subpopulations in the model, all having an equal probability of contributing dispersers to each of the other subpopulations and each receiving a certain proportion of their breeders from immigrants each month. Because each subpopulation was somewhat genetically isolated from the others, we refer to them as demes. Monthly breeding intervals were chosen as a midrange value between the minimum age of females at first parturition (6 weeks) and the minimum interlitter interval for adult females (3 weeks) (Greenwald 1957; Hoffmann 1958). A monthly mortality rate for adults of 25% approximates the disappearance rate of adults in the population studied by Bowen (1978).

The model incorporates several variables that together determine how fast the population grows and how much genetic exchange occurs among demes. The number of offspring produced per female per month (5) approximates the litter size found for *M. californicus* in habitat similar to that found at Russell Reservation (Krohne 1980). The proportion of offspring that disperse is difficult to determine for *M. californicus* because young voles usually do not become trappable until they reach an age at which dispersal can occur. The migration rate (the proportion of the destination deme made up of new immigrants each month) is also unknown for the same reason. We have thus varied dispersal and migration rates for the model in such a way that the combination of values results in a doubling of population size every 5 months; this rate of population growth is within the range of values observed for *M. californicus* in annual grassland habitat (Krebs 1966; Pearson 1966; Heske, Ostfeld, and Lidicker 1984).

Figure 12.2 shows the changes in deme size realized during our simulations, in which we explored the period of population growth (not the decline). In all cases, we started the simulations with 10 adults in each deme, which is approximately the number of adults on each of the grids studied by Bowen (1982) when the population density was moderately low. For migration rates between 0 and 0.26, the population size doubled every 5 months. With these migration rates, the proportion of offspring that remained in their natal deme ranged from 0.03 to 0.15. For a migration rate of 0.35, none of the offspring remained. Even so, the population size more than doubled every 5 months, and is therefore plotted separately.

Figure 12.3 shows the changes in $F_{ST}$, obtained throughout the 10 monthly iterations for the range of migration values. For each migration rate, we have presented the mean value of six runs. We used six runs to approximate the situation encountered in the empirical study when only a few loci were assayed. When there was no immigration into demes ($m = 0$) $F_{ST}$ tended to increase throughout. Thus, increasing population size alone does not produce a decrease in $F_{ST}$, such as that found during the empirical study. For values of $m$ between 0.09 and 0.35, $F_{ST}$ decreased by month 10, after first increasing.

## DISCUSSION

When population size changes through time, as it does in *Microtus* populations, the effective population size, $N_e$, is determined by calculating the harmonic mean of the censused population sizes in each generation (Crow and Kimura 1970; see Chepko-Sade and Shields et al., chap. 19, this vol.). A species that experiences fluctuations in population size has a lower $N_e$ than does a species with a constant population size of the same (arithmetic) average value. On an evolutionary time scale, a species with a low $N_e$ has

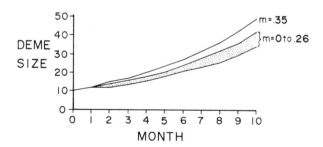

Figure 12.2. Changes in the number of breeding adults (averaged over six runs) during the course of simulations with five levels of interdeme migration.

the potential for genetic drift to play a major role in the evolutionary process (Wright 1982). In addition, short-term genetic consequences of population fluctuations would be expected as a species shifts between periods of low population size, when genetic drift would cause differentiation, and periods of high population size and gene flow, when differentiation due to genetic drift would be less likely.

Finding temporal changes in genetic structure in *Microtus californicus* is important for three reasons. First, it indicates that demographic and social factors are affecting genetic structure, as predicted by population genetic theory. Second, it indicates that this species is more subject to genetic drift when population density is low. There may be evolutionary consequences of this pattern of changes in demographic and genetic structure. If periods of low population size are prolonged, or if differentiated populations become isolated, evolutionary divergence may occur. At such times Wright's shifting balance model of evolution may operate (Wright 1982). Third, the finding of temporal changes in genetic structure indicates that empirical measures of genetic structure at one point in time may not be sufficient for understanding the population genetics of species that experience temporal changes in demographic and social factors.

In the empirical study, the values of $F_{ST}$ measured when population density was low were surprisingly high considering the microgeographic scale of the study. Although the level of differentiation was not extremely high in an absolute sense, there was more differentiation during a period of low population size than at the peak. Further, if the level of structure in the population, and the consequent opportunity for genetic drift, was underestimated using electrophoretic markers, as has been found when genic and pedigree analyses have been combined (see review in Chepko-Sade and Shields et al., chap. 19, this vol.), then the actual level of structure may have been even greater than indicated here.

The results of the simulation model suggest that when the migration rate among demes is 10% to 35% and the population size doubles every 5

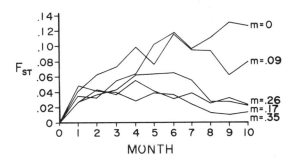

FIGURE 12.3. Changes in the average $F_{ST}$ value during the course of simulations with five levels of interdeme migration.

months, changes in genetic structure can occur over a short period of time. With a low initial deme size, $F_{ST}$ was relatively high in the first few months, even in the face of gene flow. This result is consistent with the empirical result that at moderately low density when the study began, $F_{ST}$ was relatively high. In our simulations, a relatively high level of migration was required to achieve a decrease in $F_{ST}$ similar to that observed by Bowen (1982). This result indicates that if differentiation occurs during periods of low population size, it may persist for several generations when gene flow levels are 10% to 20%.

One use for a model such as the one we have presented is to aid in detecting the operation of selective factors. Our simulation suggests that rates of migration in *M. californicus* populations may be very high, at least as the population approaches peak density. In the model, only the highest migration rate we examined resulted in a level of $F_{ST}$ as low as that found in the empirical study. No higher migration rate was possible without invoking dispersal by adults. We do not regard the existence of such a high migration rate in natural populations as a conclusion of the model, but rather as a prediction. If future findings do not support this prediction, which is based on the action of nonselective factors, then it would be appropriate to look for selective factors that might be acting to bring subpopulations to similar allele frequencies.

Our modeling effort has made clear the need for additional studies and information to help us understand the dynamics of genetic structure in *Microtus* populations. The actual migration rate is unknown for *M. californicus,* but this information, including possible temporal changes, is clearly needed. We have emphasized the migration of young animals, which Lidicker and Patton (chap. 10, this vol.) have called ontogenetic dispersal. The seasonal migration of adult males early in the breeding season may also have had an important effect on genetic structure. One would like to know whether emigration rates and/or immigration rates vary for these two classes of dispersers. Also, we should know the proportions of adult males and females that actually breed. Other important variables in the model are the initial size of the demes and the rate of increase in the size of these demes. We need to know whether the actual population structure of *M. californicus* populations is best characterized by a demic model or whether an isolation-by-distance or stepping-stone model would better describe this species.

Together, the empirical study and the simulation model have demonstrated (1) the temporal pattern of genetic structure on a microgeographic scale and (2) the way this pattern could be produced by changes in the size of subpopulations that have relatively high rates of gene flow among them. It remains to be seen whether similar patterns of genetic structure exist on a macrogeographic scale in microtine populations. If so, simulation models such as the one we used may be helpful for evaluating the relative importance of population size, migration, and selection in determining the amount of genetic structure.

# REFERENCES

Batzli, G. O. 1968. Dispersion patterns of mice in California annual grassland. *Journal of Mammalogy* 49:239–50.

Bowen, B. S. 1978. Spatial and temporal patterns of genetic variation in a local population of the California vole, *Microtus californicus* (Rodentia: Cricetidae). Ph.D. diss., University of California, Berkeley.

———. 1982. Temporal dynamics of microgeographic structure of genetic variation in *Microtus californicus*. *Journal of Mammalogy* 63:625–38.

Bowen, B. S., and S. Y. Yang. 1978. Genetic control of enzyme polymorphisms in the California vole, *Microtus californicus*. *Biochemical Genetics* 16:455–67.

Crow, J. F., and M. Kimura. 1970. *An introduction to population genetics theory.* New York: Harper and Row.

Greenwald, G. S. 1957. Reproduction in a coastal California population of the field mouse, *Microtus californicus*. *University of California Publications in Zoology* 54:421–46.

Heske, E. J., R. S. Ostfeld, and W. Z. Lidicker, Jr. 1984. Competitive interactions between *Microtus californicus* and *Reithrodontomys megalotis* during two peaks of *Microtus* abundance. *Journal of Mammalogy* 65:271–80.

Hoffmann, R. S. 1958. The role of reproduction and mortality in population fluctuations of voles (*Microtus*). *Ecological Monographs* 28:79–109.

Krebs, C. J. 1966. Demographic changes in fluctuating populations of *Microtus californicus*. *Ecological Monographs* 36:239–73.

Krebs, C. J., and J. H. Myers. 1974. Population cycles in small mammals. *Advances in Ecological Research* 8:267–399.

Krohne, D. T. 1980. Intraspecific litter size variation in *Microtus californicus*, II: Variation between populations. *Evolution* 34:1174–82.

Lidicker, W. Z., Jr., and P. K. Anderson. 1962. Colonization of an island by *Microtus californicus*, analyzed on the basis of runway transects. *Journal of Animal Ecology* 31:503–17.

Pearson, O. P. 1960. Habits of *Microtus californicus* revealed by automatic photographic records. *Ecological Monographs* 30:231–49.

———. 1966. The prey of carnivores during one cycle of mouse abundance. *Journal of Animal Ecology* 35:217–33.

Riggs, L. A. 1979. Experimental studies of dispersal in the California vole, *Microtus californicus*. Ph.D. diss., University of California, Berkeley.

Slatkin, M. 1985. Gene flow in natural populations. *Annual Review of Ecology and Systematics* 16:393–430.

Spieth, P. T. 1974. Gene flow and genetic differentiation. *Genetics* 78:961–65.

Wright, S. 1951. The genetical structure of populations. *Annals of Eugenics* 15:323–54.

———. 1965. The interpretation of population structure by F-statistics with special regard to systems of mating. *Evolution* 19:395–420.

———. 1978. *Evolution and the genetics of populations,* vol. 4: *Variability within and among natural populations.* Chicago: University of Chicago Press.

———. 1982. Character change, speciation, and the higher taxa. *Evolution* 36:427–43.

# IV. DEMOGRAPHY, DISPERSAL PATTERNS, AND GENETIC STRUCTURE

# 13. The Correlation between Population Structure and Genetic Structure in the Hutterite Population

*Elizabeth O'Brien*

Population structure—the pattern of social, historical, and physical relationships between individuals and groups in a population—plays a critical role in mediating the flow of genes through time and space. The importance of population structure for understanding processes of evolution in natural populations was pointed out many years ago in the seminal works of Sewall Wright (1921, 1922, 1938, 1949). Over the years a major effort in theoretical population genetics has been to demonstrate the causal nature of the relationship between population structure and genetic variability.

Associations between population characteristics and genetic variation in human populations are usually derived by one of two means. Population characteristics are commonly inferred from observed genetic variation. For example, the phyletic reconstruction method of Cavalli-Sforza and Edwards (1967) is commonly used to infer historic associations between groups within a population based on their genetic similarity. Conversely, genetic structure is often predicted from particular features of population structure. Various theoretical migration models (Wright 1931, 1943, 1951; Malécot 1955; Kimura 1953) generate predicted values for genetic variability between groups, given the systematic effects of subdivision and migration. Empirically derived migration matrices (Bodmer and Cavalli-Sforza 1968; Smouse and Wood, chap. 14, this vol.) are commonly used to generate similar predictions from observed migration networks.

Empirical examination of human populations, however, sometimes fails to demonstrate expected associations between population structure and genetic variability (Ward and Neel 1970; Rothhammer and Spielman 1971; Martin 1973; Workman et al. 1973; Workman and Jorde 1980; Roberts, Jorde, and Mitchell 1981; Relethford and Lees 1983). It is not clear in the general case, therefore, to what extent population structure can be directly inferred from genetic data, or to what extent genetic predictions are reliably generated from observed structural features. In order to adequately test for agreement between population structure and variation, the structural and genetic features of a population must be independently determined. Reliable if not complete historical information about the structural characteristics of interest is required. However, only a few human populations such as the Hutterites (Steinberg et al. 1967), the Mormons (Jorde 1982), and the people of the Åland Islands (Workman and Jorde 1980), the Parma Valley (Skolnick et al. 1976), and Tristan da Cunha (Roberts 1969, 1971) offer this advantage.

This study evaluates the relationship between the genetic structure and population structure of 44 communal Hutterite farms ("colonies") located in the Great Plains region of the United States and Canada. The Hutterite population is ideally suited to this type of investigation because the data used to measure population structure are complete from the founder group to the census population, and provide the opportunity to test agreement between population characteristics and genetic variation as independently derived sets of information.

Whereas a population may be considered "structured" according to many different criteria, in this analysis the term *population structure* pertains to the segmentation and association of individuals and groups according to their social, historical, and physical juxtaposition. Marriage practices, migration, historic separation between groups and individuals, and geographic propinquity belong to this domain. *Genetic structure* pertains to structural characteristics measured on the basis of genetic data. Genetic distances between groups calculated from allele frequencies belong to this domain.

In this study, colony population structure is defined in terms of kinship, history, geography, and migration. Each aspect of population structure, represented in matrix form, is a network of pairwise distance (or similarity) relationships between each pair of colonies. Blood group gene frequencies are used to calculate genetic distance between every pair of colonies; they represent the genetic structure of the population. Two methods are used to determine (1) whether the four measures of population structure define similar or incongruent colony networks, and (2) whether there is agreement between any measure of population structure and the genetic structure of this population.

## MATERIALS AND METHODS
### The Study Population

Approximately 1,200 Hutterites migrated to the United States from Russia in the 1870s. Of the 1,200, approximately 400 settled three communal farms in South Dakota (Bleibtrau 1964; Steinberg et al. 1967; Hostetler 1974). Since 1875 the population has grown rapidly due to religious prohibition against contraception, a naturally high fertility rate, and very little membership attrition (Eaton and Mayer 1953; Tietze 1957; Mange 1964; Sheps 1965). Today there are approximately 30,000 communal-living Hutterites distributed throughout 258 colonies in North and South Dakota, Montana, Alberta, Manitoba, and Saskatchewan.

Each Hutterite colony (or communal farm) is largely self-sufficient and designed to support a communal life-style for approximately 150 members. Because colonies do not expand indefinitely, each must fission periodically. When a split is anticipated a colony purchases a new farm and over a period of a few years builds and puts it into operation. When the new place is able to function as an independent economic enterprise, the colony is divided so that one group becomes the permanent residents of the new farm and one group remains at the old farm. On the average, a Schmiedenleut colony fissions once every 14 years (Olsen 1976).

The Hutterite population contains three subisolates: the Schmiedenleut (S-leut), the Leherleut (L-leut), and the Dariusleut (D-leut). Each leut consists of the colonies descended by fission succession from one of the three communal farms established in 1875. Since World War I inter-leut marriage has occurred only rarely so that the leut constitute subisolates within the population (Steinberg et al. 1967).

Figure 13.1 is a colony tree of the Schmiedenleut. The tree documents the fission history of the colonies as of 1961 and the historic relationship of each contemporary colony to the founding S-leut colony. Figure 13.2 is a geographic map of the same colonies that appear on the tree.

There are two ways in which individuals move between Hutterite colonies: (1) groups migrate when colonies fission, and (2) individuals migrate only on the occasion of intercolony marriage, at which time wives assume the residence of their husbands. Hutterite males, therefore, only move between colonies related by direct fission succession. Females, on the other hand, in theory can marry into any other colony. However, in practice there is an apparent preference for colony endogamous marriage and marriage between individuals of colonies closely related by fission succession (Bleibtrau 1964; Mange 1963).

The genetic ramifications of colony subdivision have not been fully detailed. However, partial analyses of fission processes in the S-leut (Mange 1964) and L-leut (Bleibtrau 1964) demonstrated that average within-colony kinship increased after fission due to nonrandom division of colony members. When a colony splits, married individuals migrate (or stay) together with their unmarried children. In addition, there is a tendency for closely related male heads of households to migrate (or stay) together (Bleibtrau 1964). The resulting phenomenon, known as the *lineal effect,* has been shown in other populations to cause rapid and pronounced genetic divergence between groups (Arends et al. 1967; Neel and Ward 1970; Duggleby 1977; Cheverud, Beuttner-Janusch, and Sade 1978; Fix 1975, 1978; Ober, Olivier, and Beuttner-Janusch 1978; Chepko-Sade and Olivier 1979).

The data used in this study were collected in the late 1950s and early 1960s by Arthur Steinberg, who conducted a medical and genetic survey of the entire Hutterite population. At that time blood samples were obtained from all colony members over 5 years of age, and pedigree records, which the Hutterites have traditionally kept, were collected for each family. The genetic data used in this study were collected from 3,171 individuals in 44 S-leut colonies. These individuals constituted 85.3% of the total S-leut membership over 5 years of age as of 1961. For each colony, seven

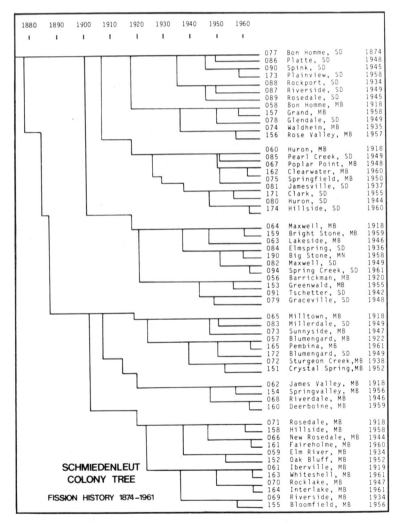

FIGURE 13.1. Schmiedenleut colony tree, depicting the fission history of each colony beginning with Bon Homme (77) in 1874 until 1961.

allele frequencies for four loci—ABO, Rh, MN, Kell—were kindly made available for this study by Alice O. Martin.

## Measures of Population Structure

Four measures of population structure were calculated as intercolony distance matrices. Intercolony genetic distances constitute a fifth matrix to represent the genetic structure of the S-leut colonies.

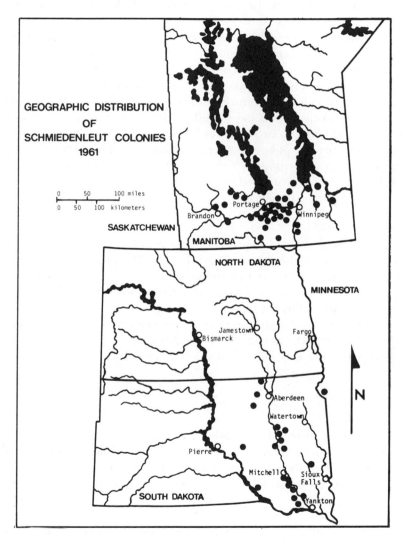

FIGURE 13.2.  Geographic locations of Schmiedenleut colonies in South Dakota and Manitoba, 1961.

1. Genetic distance was calculated as Kurczynski's $D^2_k$ (Kurczynski 1969, 1970), a discrete trait version of Mahalanobis's (1936) measure of generalized distance. The distance values were calculated using seven alleles ($n - 1$ per locus) from four loci.

2. Kinship distance was calculated as average between colony pedigree kinship (Malécot 1948), using pedigree information five generations deep. For each pair of colonies A and B, average kinship distance is the mean of all coefficients generated by pairing each female in A with each female in B. Only females alive at the 1961 census were included in the calculations. These individuals (except those under 5 years) are represented in the genetic data and account for the marriage migration component of kinship relationships as they were distributed between colonies at census.

3. Geographic distance was calculated in arc-transformed mile distances.

4. Historic distance was calculated as the number of fission events that separate a pair of colonies from their closest common ancestral colony.

5. Migration distance was calculated as the total number of links that could be formed between two colonies by connecting each married woman residing in A to any one of her brothers residing in B. This network specifies colony affiliations according to female marriage migration distance from the natal group, here considered to be the male sibs. The 1961 S-leut census included 943 married females.

Statistical Analysis

Two methods were used to assess the effects of population structure on genetic variability in the S-leut. Mantel's (1967) nonparametric test for congruence between two matrices and Ord's (1975) test for autocorrelation on trait measurements due to the effects of distance relationships within a network were applied as follows.

The Mantel method tests congruence between two matrices, $X$ and $Y$, under the null hypothesis that the cell values in $X$ ($X_{ij}$) occur in random relationship to those in $Y$ ($Y_{ij}$). Mantel's test statistic,

$$\Gamma = \sum_{\substack{i=1 \\ i \neq j}}^{n} X_{ij} Y_{ij},$$

is compared to the distribution of all possible values obtained by permuting the rows and columns (keeping pairs intact) of $X$ or $Y$. Each of the $n!$ possible permutations is considered equally likely under the null.

The algorithm used in this analysis to calculate $\Gamma$ and its significance (Costanzo, Hubert, and Gollege 1983) fits the estimated parameters of the permutation distribution (see Mantel 1967, or Baker and Hubert 1981 for derivations) to Pearson's type III distribution. The algorithm also corrects for skewness in the permutation distribution (Mielke 1978). Congruence between each pair of population structure networks, and between each population structure network and genetic distance, is evaluated using Mantel's method.

Autocorrelation pertains to the correlation of a variable with itself, where variates derive from multiple locations usually sampled through time or space (Cliff and Ord 1973). The approach used in this study was particularly influenced by the work of Sokal and colleagues, who have devoted considerable attention to spatial autocorrelation as a means of testing alternative hypotheses about the evolutionary causes of spatial structure in biological traits (Sokal and Oden 1978a, 1978b; Sokal and Wartenburg 1981; Sokal and Friedlaender 1982; Sokal and Menozzi 1982).

The autocorrelation model adopted for this analysis is from Ord (1975); its theoretical specifications are presented in Cliff and Ord (1973). The model requires a series of observations on a variable, $Y$, measured at each of $n$ locations. The effect of spatial interactions between the $n$ locations on the $Y_i$ observation is specified in the following regression form:

$$Y = \rho\, W\, Y + e,$$

where $Y$ is a vector of gene frequency values observed at each of the locations; $W$ is a weight matrix that specifies the level of relationship between each pair of colonies; $e$ is a vector of error terms for the $Y_i$ observations; and $\rho$ is the spatial parameter to be determined.

In this model the estimated values of $\rho$ measure the effect of $W$ on $Y$, that is, the dependence of the $Y_i$ values on each other according to the relationships defined in $W$. The method demonstrates the predictive value of population structure on genetic similarity between colonies. $W$ is scaled so that its row elements sum to 1, each element having been divided by the sum of the elements for the row. Row-normalizing $W$ restricts the range of possible $\rho$ values to lie between the reciprocals of the smallest and largest eigenvalues of $W$, which in the theoretical limit are less than $+1$. Each population structure network was partitioned into an arbitrary number of distance intervals. Newton-Raphson's method (Press et al. 1986) was used to find maximum likelihood solutions for $\rho$ in each distance interval of each network, and for each allele.

## RESULTS

### Mantel Matrix Comparisons

Table 13.1 provides descriptive statistics for the elements of each of the five distance matrices. In table 13.2 the Mantel results for the colony networks demonstrate the level of concordance between each measure of pop-

TABLE 13.1  Descriptive Statistics

|      | KINSHIP | $D^2{}_k$ | HISTORY | MIGRATION | GEOGRAPHY |
|------|---------|-----------|---------|-----------|-----------|
| Mean | .0182   | .6839     | 9.00    | .70       | 214.6     |
| SD   | .0055   | .3835     | 2.51    | .13       | 182.6     |

*Note: N = 44.*

ulation structure, and between each measure of population structure and genetic structure. The matrix correlations, congruence scores, and significance are reported in table 13.2. In that table every distance matrix comparison except one is shown to be significant.

Associations between kinship, historic, and migration distances are highest ranking among the comparisons. This result is expected, given what we know of the life-style factors that determine Hutterite colony organization and the resulting evolution of the population since 1875. Fission (historic) distance and kinship distance are both measures of the pedigree structure of this population, the former pertaining to colony lineages and the latter to individual lineages. Martin (1969, 1970) demonstrated that new colony formation produces increasingly more discrete partitions of a small founder gene pool. More closely related colonies by fission succession should also share a greater proportion of common ancestors. Marriage migration is conservative with respect to fission distance between colonies, and therefore, with respect to kinship as well.

Geographic distance is only a weak correlate of historic association between colonies, this being the lowest ranking significant correlation. The greater association between geographic distance compared to migration and kinship suggests the effects of geographic propinquity on mate choice independent of historic affiliation between colonies. A more detailed examination of the role of geographic distance on marriage migration has shown interesting regional variations in mate choice that implicate differences in colony settlement density (O'Brien 1986).

The associations between measures of population structure and genetic structure rank among the lower significant scores in table 13.2. Geographic distance compared to genetic distance is the only random association between matrices in the series. The comparatively low correlations between genetic distance compared to kinship, migration, and historic distances

TABLE 13.2 Mantel Network Comparisons of S-leut Colonies

| COMPARISON | MANTEL SCORE | CORRELATION |
|---|---|---|
| Migration/kinship | 18.22* | .5885 |
| Kinship/history | 16.53* | .5636 |
| Migration/history | 12.93* | .4178 |
| Migration/geography | 7.51* | .2457 |
| Kinship/geography | 4.99* | .2117 |
| Kinship/$D^2$ | 3.99* | .2011 |
| Migration/$D^2$ | 5.50* | .1854 |
| History/$D^2$ | 3.54* | .1390 |
| History/geography | 3.77* | .1360 |
| Geography/$D^2$ | − .50 | − .0326 |

Note: $N = 44$.

*$p < .002$.

indicate inconsistent associations over all the colonies of the leut. The slightly smaller effect of historic association on genetic distance than of kinship or migration suggests that the entire colony phylogeny is not as relevant to the genetic structure of the population at census as are other measures that document more contemporaneous effects.

## Autocorrelation Effects of Population Structure

The lack of significant geographic association with genetic variation in the S-leut is again demonstrated in table 13.3 and figure 13.3. In table 13.3 one allele, B, shows a consistent and significant negative autocorrelation in every increment of geographic distance. No other allele gives a significant $\rho$ value in any distance interval, and plotted $\rho$ values (fig. 13.3) show no variation with distance. Table 13.4 and figure 13.4 show that historic distance causes small positive autocorrelation effects on gene frequencies. In figure 13.4 six alleles follow a common pattern of slightly increasing $\rho$ values

TABLE 13.3  Geographic Network Autocorrelation

| DISTANCE | | O | $A^1$ | B | M | $R^1$ | $R^2$ | K | N |
|---|---|---|---|---|---|---|---|---|---|
| | | | | ALLELE | | | | | |
| 50.00 | rho | .219 | −.080 | −.332** | −.272 | .088 | −.295 | −.234 | 502 |
| | SD | .224 | .193 | .084 | .242 | .219 | .166 | .142 | |
| 100.00 | rho | .338 | −.017 | −.312** | .168 | .027 | −.346 | −.378* | 396 |
| | SD | .239 | .223 | .010 | .247 | .269 | .200 | .169 | |
| 150.00 | rho | .305 | −.023 | −.510** | −.011 | .049 | −.388 | −.380 | 126 |
| | SD | .267 | .247 | .111 | .304 | .293 | .224 | .192 | |
| 200.00 | rho | .329 | −.009 | −.512** | −.085 | .047 | −.352 | −.389 | 28 |
| | SD | .264 | .250 | .113 | .320 | .299 | .228 | .195 | |
| 250.00 | rho | .329 | −.009 | −.512** | −.085 | .047 | −.352 | −.389 | 0 |
| | SD | .264 | .250 | .113 | .320 | .299 | .229 | .195 | |
| 300.00 | rho | .327 | −.010 | −.514** | −.104 | .044 | −.363 | −.389 | 8 |
| | SD | .265 | .251 | .113 | .322 | .300 | .229 | .195 | |
| 350.00 | rho | .309 | −.029 | −.516** | −.125 | .035 | −.364 | −.391 | 90 |
| | SD | .276 | .256 | .114 | .328 | .305 | .231 | .197 | |
| 400.00 | rho | .319 | −.028 | −.506** | −.150 | .046 | −.355 | −.426 | 296 |
| | SD | .285 | .264 | .118 | .341 | .314 | .238 | .202 | |
| 450.00 | rho | .319 | −.033 | −.521** | −.132 | .039 | −.353 | −.430 | 176 |
| | SD | .293 | .271 | .120 | .347 | .322 | .244 | .207 | |
| 500.00 | rho | .313 | −.037 | −.526** | −.123 | .037 | −.364 | −.429 | 252 |
| | SD | .304 | .278 | .123 | .354 | .331 | .249 | .212 | |
| 550.00 | rho | .313 | −.037 | −.523** | −.123 | .040 | −.368 | −.427 | 18 |
| | SD | .304 | .278 | .124 | .355 | .331 | .250 | .212 | |

*$p \leq .02$.

**$p \leq .002$.

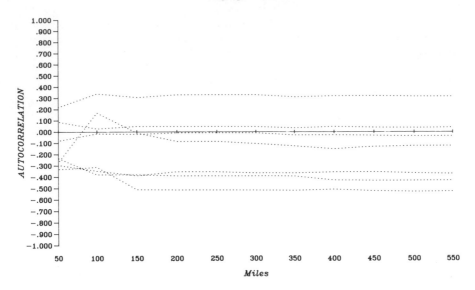

FIGURE 13.3. Rho values plotted for 7 alleles by increments (50 miles) of geographic distance.

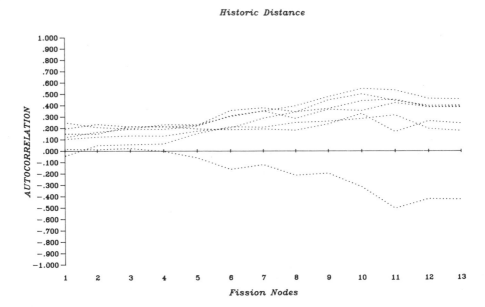

FIGURE 13.4. Rho values for 7 alleles plotted by increments of historic distance measured as the number of fission tree nodes between colonies.

TABLE 13.4  Historic Network Autocorrelation

| DISTANCE | O | $A^1$ | B | M | $R^1$ | $R^2$ | K | N |
|---|---|---|---|---|---|---|---|---|
| 1.00 rho | .103 | .016 | −.045 | .194 | .249* | .150* | .119 | 38 |
| SD | .117 | .086 | .037 | .095 | .089 | .068 | .060 | |
| 2.00 rho | .123 | .012 | .048 | .234* | .207 | .148 | .166* | 14 |
| SD | .120 | .089 | .038 | .094 | .097 | .070 | .059 | |
| 3.00 rho | .133 | .022 | .056 | .212 | .194 | .215* | .191* | 34 |
| SD | .133 | .100 | .043 | .109 | .110 | .074 | .065 | |
| 4.00 rho | .131 | −.003 | .064 | .216 | .232 | .217* | .192* | 40 |
| SD | .140 | .107 | .045 | .114 | .111 | .078 | .068 | |
| 5.00 rho | .174 | −.060 | .153** | .194 | .229 | .224* | .218* | 46 |
| SD | .155 | .124 | .048 | .134 | .129 | .088 | .076 | |
| 6.00 rho | .202 | −.162 | .213** | .186 | .304 | .356** | .310** | 28 |
| SD | .192 | .160 | .058 | .171 | .151 | .096 | .087 | |
| 7.00 rho | .288 | −.120 | .209** | .193 | .351* | .379** | .351** | 122 |
| SD | .201 | .183 | .067 | .194 | .164 | .107 | .095 | |
| 8.00 rho | .342 | −.214 | .250** | .181 | .285 | .345* | .398** | 212 |
| SD | .224 | .220 | .076 | .232 | .210 | .135 | .107 | |
| 9.00 rho | .451 | −.196 | .276** | .237 | .368 | .373* | .480** | 270 |
| SD | .230 | .263 | .088 | .263 | .229 | .156 | .113 | |
| 10.00 rho | .502 | −.315 | .285* | .329 | .355 | .441* | .549** | 380 |
| SD | .256 | .315 | .103 | .285 | .276 | .170 | .121 | |
| 11.00 rho | .442 | −.505 | .316* | .168 | .423 | .452* | .535** | 296 |
| SD | .320 | .365 | .115 | .377 | .289 | .190 | .141 | |
| 12.00 rho | .398 | −.420 | .195 | .262 | .386 | .387 | .462* | 196 |
| SD | .356 | .396 | .136 | .367 | .318 | .218 | .167 | |
| 13.00 rho | .402 | −.424 | .176 | .243 | .393 | .384 | .455* | 16 |
| SD | .356 | .398 | .139 | .375 | .316 | .219 | .169 | |

*$p \le .002$.

**$p \le .02$.

with increasing distance. The highest trend in the graph occurs in the intervals 7–11 where four alleles show significant ρ values. This effect is probably due to the inclusion of more colony pairs as greater intervals are considered. Table 13.4 shows that half the colony pairs in the historic network join at distances of 9 or more fission events. The network effects of long versus short fission distance (within and between colony lineages) associations have also been investigated as a possible explanation for autocorrelation trends in the fission network, but that effect is negligible (O'Brien 1986).

Autocorrelation values within the kinship and migration networks share the characteristic of local variation demonstrated in tables 13.5 and 13.6

TABLE 13.5 Kinship Network Autocorrelation

| Distance | | O | $A^1$ | B | M | $R^1$ | $R^2$ | K | N |
|---|---|---|---|---|---|---|---|---|---|
| .045 | rho | .686** | .510* | .099 | .460 | .037 | .284 | .537** | 2 |
| | SD | .190 | .218 | .151 | .297 | .437 | .259 | .149 | |
| .040 | rho | .549** | .374* | .121 | .282 | .146 | −.045 | .518** | 6 |
| | SD | .152 | .154 | .085 | .214 | .239 | .171 | .083 | |
| .035 | rho | .239 | .046 | .122* | .241 | −.107 | .204 | .188 | 16 |
| | SD | .166 | .133 | .056 | .146 | .160 | .104 | .089 | |
| .030 | rho | −.001 | −.122 | .030 | .264* | .028 | .310** | .124 | 39 |
| | SD | .157 | .110 | .048 | .118 | .138 | .077 | .078 | |
| .025 | rho | .163 | −.049 | .107 | .623** | .457** | .501** | .183 | 134 |
| | SD | .196 | .150 | .062 | .093 | .127 | .080 | .099 | |
| .020 | rho | .485 | .036 | .088 | .379 | .576 | .560 | .612 | 378 |
| | SD | .235 | .254 | .107 | .241 | .178 | .126 | .096 | |
| .015 | rho | .529 | −.876 | .353 | .168 | −.484 | .417 | .516 | 732 |
| | SD | .295 | .447 | .121 | .427 | .581 | .217 | .155 | |
| .010 | rho | .168 | −2.752 | −.052 | −.181 | −.755 | .150 | .401 | 538 |
| | SD | .559 | .272 | .213 | .669 | .757 | .344 | .209 | |
| .005 | rho | .108 | −2.633 | −.221 | −.035 | −1.03 | .195 | .350 | 44 |
| | SD | .602 | .288 | .243 | .607 | .715 | .329 | .227 | |

*$p \leq .02$.

**$p \leq .002$.

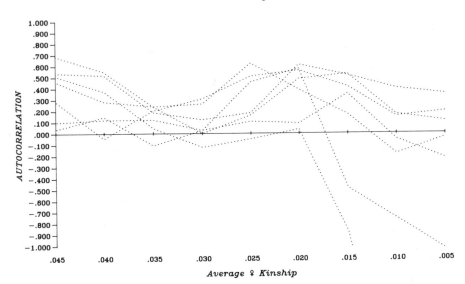

FIGURE 13.5. Rho values for 7 alleles plotted by increments of kinship distance measured as the average between colony value for females.

TABLE 13.6  Migration Network Autocorrelation

| DISTANCE | | O | $A^1$ | B | M | $R^1$ | $R^2$ | K | N |
|---|---|---|---|---|---|---|---|---|---|
| | | | | | ALLELE | | | | |
| 12.00 | rho | .194 | −.035 | .117 | −.026 | .032 | −.029 | .581** | 4 |
| | SD | .329 | .653 | .106 | .310 | .309 | .213 | .090 | |
| 11.00 | rho | .194 | −.035 | .117 | −.026 | .032 | −.029 | .581** | 0 |
| | SD | .329 | .653 | .106 | .310 | .309 | .213 | .090 | |
| 10.00 | rho | .194 | −.035 | .117 | −.026 | .032 | −.029 | .581** | 0 |
| | SD | .329 | .653 | .106 | .310 | .309 | .213 | .090 | |
| 9.00 | rho | .194 | −.035 | .117 | −.026 | .032 | −.029 | .581** | 0 |
| | SD | .329 | .653 | .106 | .310 | .309 | .213 | .090 | |
| 8.00 | rho | .150 | −.213 | .137 | .230 | −.073 | .234 | .481** | 6 |
| | SD | .213 | .381 | .066 | .176 | .194 | .120 | .068 | |
| 7.00 | rho | .176 | −.254 | −.070 | .320* | .001 | .279* | .456** | 2 |
| | SD | .191 | .336 | .062 | .144 | .179 | .104 | .065 | |
| 6.00 | rho | .255 | −.125 | −.027 | .277* | .012 | .289** | .438** | 8 |
| | SD | .143 | .300 | .051 | .122 | .145 | .082 | .052 | |
| 5.00 | rho | .199 | −.200 | .034 | .173 | .064 | .073 | .385** | 14 |
| | SD | .139 | .267 | .047 | .125 | .133 | .091 | .054 | |
| 4.00 | rho | .221 | −.052 | .014 | .180 | .346** | .214* | .296** | 50 |
| | SD | .131 | .271 | .045 | .119 | .098 | .079 | .061 | |
| 3.00 | rho | .310 | −.215 | −.010 | .209 | .461** | .261* | .229* | 84 |
| | SD | .159 | .348 | .059 | .154 | .114 | .100 | .088 | |
| 2.00 | rho | .456** | .000 | −.031 | .310 | .520** | .495** | .379** | 144 |
| | SD | .121 | .000 | .078 | .182 | .017 | .064 | .019 | |
| 1.00 | rho | .604** | −.361 | −.038 | .476* | .549* | .637** | .439** | 364 |
| | SD | .191 | .636 | .104 | .209 | .188 | .109 | .128 | |

*$p \leq .02$.
**$p \leq .002$.

and figures 13.5 and 13.6. The only significant network effect of kinship on genetic autocorrelation is apparent among more closely related colonies (.045–.025). The network effect of kinship distance is not significant for any allele at kinship distances greater than .025. By contrast, less closely related colonies in the migration network give significant ρ values for the greatest number of alleles. One allele, K, is significantly autocorrelated in every distance interval. And ρ values are more unstable by distance and among alleles in the migration network than for other measures of population structure. The less coherent trend is probably due to the skewed distribution of colony pairs among distance intervals (see table 13.6).

## SUMMARY

Kinship, migration, and historic distances bear similarity to one another as measures of association between S-leut colonies, whereas geographic

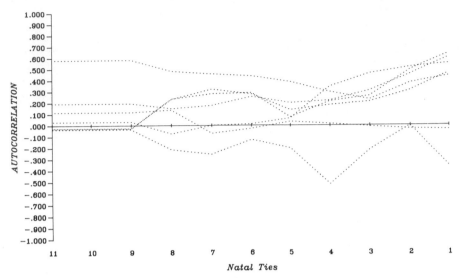

FIGURE 13.6. Rho values for 7 alleles plotted by increments of migration distance measured as the number of ties linking each married woman from one colony to one (or more) brother in another colony.

distance is weakly correlated with them. Though significant, none of the correlations between three measures of population structure and genetic structure are high in the S-leut, and geographic distance is randomly associated with the genetic structure of colonies. It is certain that the genetic information used in this study does not constitute a representative sample of the human genome. The associations between population structure and genetic variation in this population might be improved by sampling more alleles and more loci. In addition, extensions of the Mantel test to include the effects of multiple matrices simultaneously, such as that proposed by Smouse, Long, and Sokal (n.d.), might show that all population structure effects together produce a much higher correlation than any measure alone.

However, neither the allele samples nor the treatment of individual population structure effects distinguish this investigation from others (but see Smouse and Wood, chap. 14, this vol.). The findings of this study, therefore, can be generalized to argue against two very common techniques employed in the study of human microevolution. While in some instances making inferences about the structural features of a population from genetic data might be effective, in this study the practice is shown to have misleading potential. The weak correlation between kinship and genetic distance should be especially regarded, given the frequent use of gene frequency data for estimating kinship. This study suggests that caution should precede any reconstructions of population history from genetic information, and vice versa.

It might be argued that the Hutterite population has unique structural characteristics that do not provide a basis for credible generalizations to other populations. To the contrary, the fission (and migration) requirements of the communal life-style result in a highly regular population structure. This unique characteristic should bias results in favor of finding associations between measures of population structure and genetic variation in the S-leut.

The conclusions presented here do not suggest that clear demonstrations of congruence between population structure and genetic structure do not exist or cannot be found. Rather this example suggests that gene frequency data are not reliable sources of information upon which to make direct inferences about population structure effects. However, in the future we can avoid some of the limitations of traditional electrophoretic and blood group serological analyses by using more recent techniques from molecular genetics which provide access to more highly polymorphic, co-dominant, and selectively neutral sites (Jorde 1985; Lewontin 1985). Molecular analyses of sequence divergence at these sites will allow direct and representative measures of genotypic variation and will produce the data necessary to measure the differential effects of mutation, genetic drift, and gene flow within human populations.

## ACKNOWLEDGMENTS

I am grateful to Arthur Steinberg and Alice Martin for providing the data used in this study. I am also grateful to Alice Martin, Carole Ober, and James Cheverud for offering helpful suggestions during the preparation of this work, and to Richard Kerber, who generously lent his programming expertise for some aspects of data analysis and production of the figures.

## REFERENCES

Arends, T. G., G. Brewer, N. Chagnon, M. L. Gallango, H. Gershowitz, M. Layrisse, J. Neel, D. Sheffler, R. Tashian, and L. Weitkamp. 1967. Intratribal genetic differentiation among the Yanomama Indians of southern Venezuela. *Proceedings of the National Academy of Sciences* 57:1252–59.

Baker, F. B., and L. J. Hubert. 1981. The analysis of social interaction data. *Sociological Methods and Research* 9:339–61.

Bleibtrau, H. K. 1964. Marriage and residence patterns in a genetic isolate. Ph.D. diss., Harvard University.

Bodmer, W., and L. L. Cavalli-Sforza. 1968. A migration matrix model for the study of random genetic drift. *Genetics* 59:565–92.

Cavalli-Sforza, L. L., and A. W. F. Edwards. 1967. Phylogenetic analysis: Models and estimation procedures. *Evolution* 21:550–70.

Chepko-Sade, B. D., and T. J. Olivier. 1979. Coefficient of genetic relationship and the probability of intragenealogical fission in *Macaca mulatta*. *Behavioral Ecology and Sociobiology* 5:263–78.

Cheverud, J. M., J. Beuttner-Janusch, and D. Sade. 1978. Social group fission and the origin of intergroup genetic differentiation among the rhesus monkeys of Cayo Santiago. *American Journal of Physical Anthropology* 49:449–56.

Cliff, A. D., and J. K. Ord. 1973. *Spatial autocorrelation*. London: Pion.

Costanzo, C. M., L. Hubert, and R. G. Golledge. 1983. A higher moment for spatial statistics. *Geographical Analysis* 15:347–51.

Duggleby, C. 1977. Blood group antigens and the population genetics of *Macaca mulatta* on Cayo Santiago. II. Effects of social group division. *Yearbook of Physical Anthropology* 20:263–71.

Eaton, J. W., and A. J. Mayer. 1953. The social biology of very high fertility among the Hutterites: The demography of a unique population. *Human Biology* 25:206–64.

Fix, A. G. 1975. Fission-fusion and lineal effect: Aspects of the population structure of the Semai Senoi of Malaysia. *American Journal of Physical Anthropology* 43:295–302.

———. 1978. The role of kin-structured migration in genetic microdifferentiation. *Annals of Human Genetics* 41:329–39.

Hostetler, J. A. 1974. *Hutterite society*. Baltimore: Johns Hopkins University Press.

Jorde, L. B. 1982. The genetic structure of the Utah Mormons: Migration analysis. *Human Biology* 54:583–97.

Jorde, L. B. 1985. Human genetic distance studies: Present status and future prospects. *Annual Review of Anthropology* 14:343–73.

Kimura, M. 1953. "Stepping stone" model of populations. *Annual Report of the National Institute of Genetics* 3:63–65.

Kurczynski, T. W. 1969. Studies of genetic drift in a human isolate. Ph.D. diss., Case Western Reserve University.

———. 1970. Generalized distance and discrete variables. *Biometrics* 36:525–34.

Lewontin, R. C. 1985. Population genetics. *Annual Review of Genetics* 19:81–102.

Mahalanobis, P. C. 1936. On the generalized distance in statistics. *Proceedings of the National Institute of Science in India* 2:49–55.

Malécot, G. 1948. *Les mathématiques de l'hérédité*. Paris: Masson. Translated in 1969 as *The mathematics of heredity*. San Francisco: Freeman Press.

———. 1955. Decrease of relationship with distance. *Cold Spring Harbor Symposium* 20:52–53.

Mange, A. P. 1963. The population structure of a human isolate. Ph.D. diss., University of Wisconsin.

———. 1964. Growth and inbreeding of a human isolate. *Human Biology* 36(2):104–33.

Mantel, N. 1967. The detection of disease clustering and a generalized regression approach. *Cancer Research* 27(2):209–20.

Martin, A. O. 1969. Recurrent founder effect in a human isolate: History and genetic consequences. Ph.D. diss., Case Western Reserve University.

———. 1970. The founder effect in a human isolate: Evolutionary implications. *American Journal of Physical Anthropology* 32:351–68.

———. 1973. An empirical comparison of some descriptions of population structure in a human isolate. In *Genetic structure of populations*, ed. N. E. Morton, 195–202. Honolulu: University of Hawaii Press.

Mielke, P. W. 1978. Clarification and appropriate inferences for Mantel and Valand's nonparametric multivariate analysis technique. *Biometrics* 34:277–82.

Neel, J. V., and R. H. Ward. 1970. Village and tribal genetic distances among American Indians and the possible implications for human evolution. *Proceedings of the National Academy of Sciences* 65(2):323–30.

Ober, C., T. J. Olivier, and J. Beuttner-Janusch. 1978. Carbonic anhydrase heterozygosity and $F_{st}$ distributions in Kenyan baboon troops. *American Journal of Physical Anthropology* 48:95–100.

O'Brien, E. 1986. Population structure and genetic variability among 44 Hutterite colonies. Ph.D. diss., Northwestern University.

Olsen, C. L. 1976. The demography of new colony formation in a human isolate: Analysis and history. Ph.D. diss., University of Michigan.

Ord, K. 1975. Estimation methods for models of spatial interaction. *Journal of the American Statistical Association* 70(349):120–26.

Press, W. H., B. P. Flannery, S. A. Teukolsky, and W. T. Vetterling. 1986. *Numerical recipes: The art of scientific computing*. New York: Cambridge University Press.

Relethford, J. H., and F. C. Lees. 1983. Correlation analysis of distance measures based on geography, anthropometry, and isonomy. *Human Biology* 55:653–65.

Roberts, D. F. 1969. Consanguineous marriages and calculation of the genetic load. *Annals of Human Genetics* 32:407–10.

———. 1971. The demography of Tristan da Cunha. *Population Studies* 25:465–79.

Roberts, D. F., L. B. Jorde, and R. J. Mitchell. 1981. Genetic structure in Cumbria. *Journal of Biosocial Science* 13:317–36.

Rothhammer, F., and R. S. Spielman. 1971. Anthropometric variation in the Aymara: Genetic, geographic and topographic contributions. *American Journal of Human Genetics* 24:371–80.

Sheps, M. D. 1965. An analysis of reproductive patterns in an American isolate. *Population Studies* 19:65–80.

Skolnick, M., L. L. Cavalli-Sforza, A. Moroni, and E. Siria. 1976. A preliminary analysis of Parma Valley, Italy. In *The demographic evolution of human populations*, ed. R. H. Ward and K. M. Weiss. London: Academic Press.

Smouse, P. E., J. C. Long, and R. R. Sokal. N.d. Multiple regression and correlation extensions of the Mantel test of matrix correspondence. Typescript.

Sokal, R. R., and J. Friedlaender. 1982. Spatial autocorrelation analysis of biological variation on Bougainville Island. In *Current developments in anthropological genetics*, ed. M. H. Crawford and J. H. Mielke, 205–27. New York: Plenum Press.

Sokal, R. R., and P. Menozzi. 1982. Spatial autocorrelations of HLA frequencies in Europe support demic diffusion of early farmers. *American Naturalist* 119:1–17.

Sokal, R. R., and N. L. Oden. 19778a. Spatial autocorrelation in Biology, I: Methodology. *Biological Journal of the Linnean Society* 10:199–228.

———. 1978b. Spatial autocorrelation in biology, II: Some biological applications of evolutionary and ecological interest. *Biological Journal of the Linnean Society* 10:229–49.

Sokal, R. R., and D. E. Wartenburg. 1981. Space and population structure. In *Dynamic spatial models in ecology and human biology,* ed. D. A. Griffith and R. D. MacKinnon, 186–213. New York: Plenum Press.

Steinberg, A. G., H. K. Bleibtrau, T. W. Kurczynski, A. O. Martin, and E. M. Kurczynski. 1967. Genetic studies on an inbred human isolate. In *Proceedings of the Third International Congress of Human Genetics,* ed. J. F. Crow and J. V. Neel, 267–89. Baltimore: Johns Hopkins University Press.

Tietze, C. 1957. Reproductive span and rate of reproduction among Hutterite women. *Fertility and Sterility* 8:89–97.

Ward, R. H., and J. V. Neel. 1970. Gene frequencies and microdifferentiation among the Makiritare Indians. IV. A comparison of a genetic network with ethnohistory and migration matrices: A new index of genetic isolation. *American Journal of Human Genetics* 22:538–61.

Workman, P. L., H. C. Harpending, J. M. Lalouel, C. Lynch, J. D. Niswander, and R. Singleton. 1973. Population studies on southwestern Indian tribes. VI. Papago population structure: A comparison of genetic and migration analyses. In *Genetic structure of populations,* ed. N. E. Morton, 166–94. Honolulu: University of Hawaii Press.

Workman, P. L., and L. B. Jorde. 1980. The genetic structure of the Åland Islands. In *Population structure and genetic disease,* ed. A. W. Eriksson, H. Forsius, H. R. Nevanlinna, and P. L. Workman. New York: Academic Press.

Wright, S. 1921. Systems of mating. *Genetics* 6:1–19.

———. 1922. Coefficients of inbreeding and relationship. *American Naturalist* 56:330–38.

———. 1931. Evolution in Mendelian populations. *Genetics* 16:97–159.

———. 1938. Size of population and breeding structure in relation to evolution. *Science* 87:430–31.

———. 1943. Isolation by distance. *Genetics* 28:114–38.

———. 1949. Population structure in evolution. *Proceedings of the American Philosophical Society* 93:471–78.

———. 1951. The genetical structure of populations. *Annals of Eugenics* 15: 323–54.

# 14. The Genetic Demography of the Gainj of Papua New Guinea: Functional Models of Migration and Their Genetic Implications

*Peter E. Smouse and James W. Wood*

The relationship between gene flow and genetic structure in natural populations has received a great deal of attention. Two broad classes of formal treatment have been used to study this relationship, one based on migration matrix models (Bodmer and Cavalli-Sforza 1968; Smith 1969) and the other on models of isolation by distance (Wright 1943; Malécot 1969). A major attraction of the migration matrix (MM) approach is that any set of pairwise migration rates can be used to predict the elements of a corresponding genetic distance matrix. There are, however, three disadvantages. First, much of the fine detail of an observed migration matrix is ephemeral; while the overall level and geographic pattern of gene flow may be relatively stable over time, the individual migration rates between particular pairs of populations change from generation to generation (Ward and Neel 1970). Second, estimates of migration rates are subject to sampling error, and some of the fine detail may be more apparent than real. Finally, MM models take the observed migration rates as given and thus reveal nothing about the demographic, cultural, ecological, or geographic factors that determine the migration rates themselves. (See, however, Sade et al., chap. 15, this vol.).

The isolation-by-distance (ID) approach, on the other hand, concentrates on broad pattern rather than on fine detail, measuring the rate of decay of genetic affinity with geographic distance. This approach routinely provides a reasonable fit of model to data (e.g., Morton 1969), while avoiding many of the problems associated with overspecification of the MM models. This approach is not without its own limitations, however. First, the treatment is restricted to a single explanatory factor, geographic distance, thus limiting its predictive value. Second, the ID approach sacrifices perhaps too much of the detail. When applied to a set of $n$ populations, for example, the MM approach yields $n^2$ points against which to test the model, whereas

211

the ID approach provides only *n* points of comparison. In other words, ID models are substantially less powerful (in a statistical sense) than the MM models.

We prefer a middle ground, a class of models that explicitly incorporates not only distance, but any number of other explanatory variables as well, while maintaining the statistical power associated with the MM models. Elsewhere, we have described in detail a method for analyzing migration matrices in terms of their underlying explanatory factors (Wood, Smouse, and Long 1985). That method yields model migration matrices that reflect general patterns of migration rather than ephemeral detail or sampling error. In this chapter, we extract genetic predictions from such matrices and show how to test such predictions against genetic data.

We apply this approach to patterns of gene flow in the Gainj, a small population of tribal horticulturalists living on the northern fringes of highland Papua New Guinea. In the course of this application, we pose five questions about the migration rates themselves: How far does the migration pattern depart from random dispersal? How different are the migration patterns for the two sexes? Is geographic distance the best measure of mating isolation? How important are population density effects by comparison? How important are cultural factors? Given the model predicted migration rates, we address two questions about genetic structure: How much of the detailed genetic structure can we predict from these migration rates? How well do these predictions match the corresponding observations on genetic structure?

## THE STUDY POPULATION

We have been studying the genetics and demography of the Takwi Valley Gainj since 1977 (Wood 1980, 1986, 1987; Wood and Smouse 1982; Wood et al. 1982; Wood, Smouse, and Long 1985; Wood, Johnson, and Campbell 1985; Wood et al. 1985; Long 1986; Long et al. 1986). The Takwi Valley population is subdivided into several local groups ranging in size from about 20 to 180 individuals. Each local group occupies a discrete, bounded territory, and members of the group share access to the garden land and forest resources of this territory. In keeping with standard anthropological usage (Hogbin and Wedgwood 1952), we refer to these local residential groups as parishes. There are about 20 Gainj parishes, and the data reported here derive from a subset of 11 that have been sampled intensively for both genetic and demographic data (fig. 14.1).

The westernmost Gainj parishes have received large amounts of gene flow from the Kalam parishes of the neighboring Asai Valley. This is evident both in the genetic marker data (Wood et al. 1982) and in the linguistic composition of individual parishes (Wood, Smouse, and Long 1985). Traditionally, the Gainj practice patrilocal postmarital residence, with the wife joining the husband in his natal parish. Males generally reside within 1 or 2 km of their

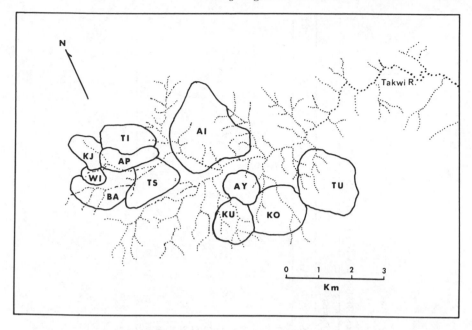

FIGURE 14.1. Map of Gainj parishes in the Takwi Valley, Papua New Guinea. KJ = Kanainj; TI = Tintibun, WI = Winyimbi, AP = Anarup, TS = Tsumbosimbe, AI = Aingdai, AY = Angoinᵞ, KU = Kuak, KO = Komaraga, TU = Tungaga, BA = Baule.

birthplaces throughout their lives, as do females up to about 20 years of age. After age 20, females disperse more widely at the time of marriage (Wood, Smouse, and Long 1985; Wood, Johnson, and Campbell 1985).

The migration data collected from the 11 parishes are presented in the form of backward migration matrices in table 14.1 for males and table 14.2 for females. The tables list the numbers of parents of children born in the $i$th parish who were themselves born in the $j$th parish. The data represent migration occurring over at least the past generation, and we assume that the processes of interest have been constant over that period. It is obvious that the two sexes differ dramatically in migration patterns. Far more females migrate before reproducing than males; over 91% (637/697) of all individuals in the sample were born in their fathers' natal parish, while only about 36% (173/479) were born in their mothers' natal parish. This finding is consistent with the patrilocal rule of postmarital residence.

Travel between parishes is on foot along narrow forest trails. While it is possible to walk from any point in the valley to any other within a single day, such a walk is not taken lightly. Steep, broken terrain, heavy rainfall, frequent mud slides, and dense lower montane rain forest ensure that the level of social interaction between parishes varies inversely with the distance between them. Traditionally, marriage is more common among neighbor-

TABLE 14.1 Observed Numbers of Gainj Children Born in the *i*th Parish, Whose Father was Born in the *j*th Parish

|  | KJ | TI | WI | AP | BA | TS | AI | AY | KU | KO | TU |
|---|---|---|---|---|---|---|---|---|---|---|---|
| | | | | | | **Donor Parish** | | | | | |
| KJ | 42 | | | | | 4 | | | | | 1 |
| TI | | 71 | | | | 6 | 5 | 2 | | 2 | |
| WI | | 1 | 31 | | | | | | | | |
| AP | | | 75 | | 1 | 3 | | | | | |
| BA | | | | | 37 | | | | | | |
| TS | | | | | | 57 | | | 5 | | |
| AI | | 2 | | | 3 | 3 | 94 | 3 | 1 | 3 | |
| AY | | | 3 | | | 2 | | 33 | | | |
| KU | | | | | | 2 | | | 35 | 1 | |
| KO | | | | | | 3 | | | 1 | 91 | |
| TU | | | | | | 3 | | | | | 71 |
| (N) | 67 | 112 | 44 | 121 | 60 | 71 | 167 | 45 | 56 | 117 | 93 |

*Notes:* Parish designations are indicated in fig. 14.1. Estimated population sizes are given in bottom row of table.

TABLE 14.2 Observed Numbers of Gainj Children Born in the *i*th Parish, Whose Mother was Born in the *j*th Parish

| | DONOR PARISH | | | | | | | | | | |
|---|---|---|---|---|---|---|---|---|---|---|---|
| | KJ | TI | WI | AP | BA | TS | AI | AY | KU | KO | TU |
| KJ | 1 | 9 | | | 2 | 1 | | 2 | 2 | 1 | |
| TI | 5 | 19 | | 3 | | 8 | 22 | 3 | 3 | 1 | |
| WI | | | 1 | 1 | | | | | | | |
| AP | 4 | 12 | | 16 | 3 | 6 | 1 | | 3 | 1 | |
| BA | 3 | 3 | | 3 | 9 | 5 | 6 | | 3 | 2 | |
| TS | | 4 | | 4 | | 17 | 6 | 2 | 6 | 4 | |
| AI | | 3 | | 2 | 5 | 2 | 58 | 3 | 7 | 3 | 3 |
| AY | | 2 | | | | 2 | 3 | 3 | 8 | 7 | 4 |
| KU | | 3 | 1 | 4 | 1 | 5 | 2 | 5 | 15 | 2 | 2 |
| KO | | 2 | | 1 | 16 | 18 | 3 | 5 | 5 | 11 | 4 |
| TU | | 3 | | | | 5 | 11 | | 2 | 3 | 24 |
| (N) | 67 | 112 | 44 | 121 | 60 | 71 | 167 | 45 | 56 | 117 | 93 |

*Notes:* Parish designations are indicated in fig. 14.1. Estimated population sizes are given in bottom row of table.

ing parishes than among more distant parishes (Johnson 1982). Melanesian horticulturalists are among the most sedentary of all human populations (Friedlaender 1971), and the Gainj are no exception to this generalization (Wijsman and Cavalli-Sforza 1984). The geographic distances between parish centers, measured with a pedometer along the normal travel routes, are presented in the upper triangular portion of table 14.3.

As mentioned above, the upper Takwi Valley parishes have experienced gene flow from the neighboring Kalam of the Asai Valley. A linguistic survey of the parishes provides an estimate of the amount of this genetic infiltration. (See fig. 14.1 for explanation of codes.) Our best estimates of the proportions of Kalam speakers in these parishes are

| | | | |
|---|---|---|---|
| KJ .922 | AP .661 | KO .969 | TU .010 |
| WI .880 | TI .459 | KU .042 | AI .006 |
| BA .859 | TS .097 | AY .020 | |

Assuming that linguistic and genetic admixture are parallel processes, an estimate of Kalam gene flow is provided by the fraction of Kalam spoken in a particular parish; the difference in admixture between parishes can be expressed in terms of the linguistic distance measure

$$LD_{ij} = \left| \log \frac{\% \text{Kalam}(i)}{\% \text{Gainj}(i)} - \log \frac{\% \text{Kalam}(j)}{\% \text{Gainj}(j)} \right| \qquad (14.1)$$

The linguistic distances are presented in the lower triangular portion of table 14.3. They are correlated but not congruent with the geographic distances.

## MIGRATIONAL ANALYSIS

We now present a method of gauging the impact of demographic, geographic, and cultural factors on the pattern of migration. Consider just

Table 14.3  Distances between Gainj Parishes in the Takwi Valley

| Parish | KJ | TI | WI | AP | BA | TS | AI | AY | KU | KO | TU |
|---|---|---|---|---|---|---|---|---|---|---|---|
| KJ | — | 1.6 | 1.3 | 1.1 | 2.1 | 3.5 | 5.0 | 7.5 | 7.7 | 9.0 | 10.3 |
| TI | 2.3 | — | 2.7 | 1.0 | 2.6 | 2.4 | 4.0 | 5.1 | 6.4 | 7.7 | 9.0 |
| WI | 0.5 | 1.8 | — | 1.3 | 0.6 | 3.0 | 6.1 | 7.4 | 7.2 | 8.5 | 9.8 |
| AP | 1.8 | 0.5 | 1.3 | — | 1.6 | 2.4 | 5.0 | 6.6 | 6.4 | 7.8 | 9.1 |
| BA | 0.7 | 1.6 | 0.2 | 1.1 | — | 2.7 | 5.9 | 6.9 | 6.6 | 8.3 | 9.6 |
| TS | 4.7 | 2.4 | 4.2 | 2.9 | 4.0 | — | 3.7 | 4.3 | 4.2 | 5.6 | 6.9 |
| AI | 7.6 | 5.3 | 7.1 | 5.8 | 6.9 | 2.9 | — | 2.9 | 4.3 | 4.6 | 5.8 |
| AY | 6.4 | 4.1 | 5.9 | 4.6 | 5.7 | 1.7 | 1.2 | — | 1.8 | 1.9 | 3.2 |
| KU | 5.6 | 3.3 | 5.1 | 3.8 | 4.9 | 0.9 | 2.0 | 0.8 | — | 1.8 | 3.1 |
| KO | 5.2 | 2.9 | 4.7 | 3.4 | 4.6 | 0.5 | 2.4 | 1.1 | 0.4 | — | 1.3 |
| TU | 7.1 | 4.8 | 6.6 | 5.3 | 6.4 | 2.4 | 0.5 | 0.7 | 1.5 | 1.8 | — |

Notes: Geographic distances are above the diagonal (in km), and linguistic distances (defined in the text) are below the diagonal. Parish designations are in fig. 14.1.

male migration, since the treatment for female migration is identical. The observed fraction of children born in the $i$th parish whose fathers were born in the $j$th parish is denoted by $m_{ij}$. We have developed a general formulation that can deal with any desired degree of complexity (Wood, Smouse, and Long 1985), but we restrict attention here to a class of very simple models of the form

$$\log (m_{ij}/m_{ii}) = \mu + \alpha(N_i - N_j) + \gamma D_{ij} + \delta LD_{ij}. \tag{14.2}$$

This log-linear model constrains all $m_{ii}$ and $m_{ij}$ to lie between zero and one and ensures that

$$\sum_{j=l}^{n} m_{ij} = 1.$$

The parameter $\mu$ represents the effect of endemicity (or philopatry), the tendency for an individual to reproduce in his or her natal parish. The parameter $\alpha$ represents the effect of a difference in population size between the recipient and donor parishes. The parameters $\gamma$ and $\delta$ measure the respective effects of geographic and linguistic separation on the migration rates between parishes. As is clear from tables 14.1 and 14.2, the endemicity rate ($\mu$) is different for the two sexes and represents a major variable of interest. The density effect ($\alpha$) is included in deference to the traditional "gravity" model of migration (cf. Lees and Relethford 1982) and in deference to the suggestion of a density effect in tables 14.1 and 14.2. The effect of geographic distance ($\gamma$) is visually apparent in tables 14.1 and 14.2, and inclusion of a linguistic divergence ($\delta$) term is motivated by the possibility of reduced exchange between parishes with different Gainj/Kalam mixes.

Parameter estimation is accomplished with a modified version of standard log-linear regression analysis (Wood, Smouse, and Long 1985). A computer program is available on request. Given the estimates from two models, one of which is a submodel of the other, a test of the additional parameters of the more inclusive model can be obtained from a likelihood ratio test, which can be related in turn to a chi-square test with degrees of freedom equal to the difference in the numbers of parameters estimated in the two models (Kendall and Stuart 1979). In addition, any particular model can be compared with the observed migration rates to provide a measure of the residual variation not accounted for by the model.

There are 10 hypotheses of interest, presented here in the form of parameter specifications.

$H_0$: $[0, 0, 0, 0]$          $H_5$: $[\mu, \alpha, \gamma, 0]$

$H_1$: $[\mu, 0, 0, 0]$        $H_6$: $[\mu, \alpha, 0, \delta]$

$H_2$: $[\mu, \alpha, 0, 0]$   $H_7$: $[\mu, 0, \gamma, \delta]$

$H_3$: $[\mu, 0, \gamma, 0]$   $H_8$: $[\mu, \alpha, \gamma, \delta]$

$H_4$: $[\mu, 0, 0, \delta]$   $H_\Omega$: [Observed]

Parameter estimates for these models are provided in table 14.4, along with their residual $\chi^2$-measures. The difference of residuals for a pair of nested hypotheses is the $\chi^2$-test criterion of interest.

As evident in tables 14.1 and 14.2, migration rates depart strongly from uniformity (random dispersal) in both sexes. The overall rate of migration is determined by the level of endemicity; the $\mu$ parameter by itself explains 95% of the variation in male migration rates and 59% of that in female rates. To elucidate the pattern of migration for those who are moving, it is necessary to examine the additional factors of the model ($\alpha$, $\gamma$, $\delta$). After endemicity ($H_1$) is accounted for in males, there is very little residual variation [$\chi^2_{1\Omega} = 183.98$, df $= 109$], and our full model ($H_8$) is only a small improvement [$\chi^2_{18} = 8.55$, df $= 3$]. Endemicity ($H_1$) accounts for a smaller portion of the variation in female migration rates, but our full model ($H_8$) accounts for a substantial fraction of the residual variation [$\chi^2_{18} = 103.14$, df $= 3$], suggesting the importance of language, geography, and density effects.

Considering just the female migration rates, we note that $H_2$ (density), $H_3$ (geography), and $H_4$ (language) are about equally useful as predictors of the pattern of migration. The combination of density with either or both of the distance measures leads to additional improvements in predictability; this can be seen by comparing $H_5$ (density and geography) with $H_3$ (geography), $H_6$ (density and language) with $H_4$ (language), or $H_8$ (full model)

TABLE 14.4 Parameter Estimates and Residual $\chi^2$-values for Several Hypothesized Models of Migration

| | PARAMETER ESTIMATES | | | | RESIDUAL $\chi^2$ VALUE | DEGREES OF FREEDOM |
|---|---|---|---|---|---|---|
| MODEL | $\mu$ | $\alpha$ | $\gamma$ | $\delta$ | | |
| $H_0$ M | — | — | — | — | 3527.88 | 110 |
| F | — | — | — | — | 1248.41 | |
| $H_1$ M | $-4.6650$ | — | — | — | 183.98 | 109 |
| F | $-1.7033$ | — | — | — | 512.62 | |
| $H_2$ M | $-4.7959$ | $+.0056$ | — | — | 178.88 | 108 |
| F | $-1.5861$ | $-.0081$ | — | — | 467.77 | |
| $H_3$ M | $-4.2433$ | — | $-.0918$ | — | 180.60 | 108 |
| F | $-1.0229$ | — | $-.1529$ | — | 468.58 | |
| $H_4$ M | $-4.3802$ | — | — | $-.0912$ | 181.85 | 108 |
| F | $-1.1164$ | — | — | $-.2027$ | 462.88 | |
| $H_5$ M | $-4.3754$ | $+.0054$ | $-.0910$ | — | 175.66 | 107 |
| F | $-0.8738$ | $-.0083$ | $-.1586$ | — | 421.68 | |
| $H_6$ M | $-4.5010$ | $+.0058$ | — | $-.0963$ | 176.44 | 107 |
| F | $-0.9782$ | $-.0083$ | — | $-.2070$ | 417.58 | |
| $H_7$ M | $-4.2270$ | — | $-.0780$ | $-.0254$ | 180.50 | 107 |
| F | $-0.9569$ | — | $-.0791$ | $-.1366$ | 456.46 | |
| $H_8$ M | $-4.3559$ | $+.0056$ | $-.0687$ | $-.0400$ | 175.43 | 106 |
| F | $-0.7930$ | $-.0084$ | $-.0876$ | $-.1357$ | 409.48 | |

Note: Male values are in the upper row and female values in the lower row of each pair.

with $H_7$ (geography and language). On the other hand, the two distance measures are highly redundant, as can be seen by comparing $H_7$ (geography and language) with $H_3$ (geography) or $H_4$ (language). It is also clear that at least for female migration, there remains a substantial amount of residual variation after fitting the full model ($H_8$). Some of this residual can plausibly be attributed to sampling variation, but some of it is probably attributable to inadequacies of the model. We have employed a four-parameter model to describe $11(10) = 110$ migration rates ($m_{ij}$) and have accounted for 67% of the variation in these rates. There are undoubtedly other factors not yet identified that would account for some of the remaining variation. While the full model ($H_8$) is less than a perfect descripter, we have nonetheless identified clear trends in the pattern of gene flow. There is a strong tendency in males and a weaker one in females to reproduce within the natal parish ($\mu < 0$). Female gene flow decreases with increasing geographic distance and/or linguistic divergence ($\gamma, \delta < 0$), and is preferentially from large to small parishes ($\alpha < 0$).

## GENETIC PATTERN ANALYSES

How well do any of these migrational models predict the pattern of genetic affinity for the eleven parishes? We show elsewhere (Wood 1986) how a genetic distance matrix can be predicted from a migration matrix. Given a stochastic migration matrix **M**, describing the pattern of exchange among subpopulations, and a matrix $\mathbf{U} = \text{diag } \{1/2N_{ei}\}$, where $N_{ei}$ is the effective size of the $i$th parish, the matrix of variances and covariances of gene frequencies among parishes in the $t$th generation $\mathbf{V}(t) = \{v_{ij}^{(t)}\}$ is approximately given by:

$$\mathbf{V}^{(t)} = \sum_{r=0}^{t-1} [\mathbf{M}]^r \, \mathbf{U} \, [\mathbf{M}^T]^r . \tag{14.3}$$

We can compute a corresponding distance matrix, $\mathbf{D}^{(t)} = \{d_{1j}^{2\,(t)}\}$, where

$$d_{ij}^{2(t)} = v_{ii}^{(t)} + v_{jj}^{(t)} - 2v_{ij}^{(t)} . \tag{14.4}$$

The genetic distances $d_{ij}^{2(t)}$ converge rapidly.

The population sizes ($N_i$) of these parishes are listed along the bottom rows of tables 14.1 and 14.2. Conventional wisdom suggests that $N_e$ is between 30% and 35% of the nominal head count, so we set $N_{ei} = (1/3)N_i$ for each parish. The matrix **M** varies in what follows, depending on the migrational model under consideration. For a given model, we use an unweighted average of the predicted matrices for the two sexes, since half the genes are drawn from each sex in each generation. The predicted distance matrix for the full model $H_8$ is presented in the lower triangular portion of table 14.5 by way of example. A similar matrix was computed from each of the other models ($H_0, \ldots , H_7, H_\Omega$).

Table 14.5 Genetic Distance between Gainj Parishes in the Tàkwi Valley

| PARISH | KJ | TI | WI | AP | BA | TS | AI | AY | KU | KO | TU |
|---|---|---|---|---|---|---|---|---|---|---|---|
| KJ | — | .154 | .218 | .155 | .225 | .130 | .133 | .103 | .188 | .090 | .168 |
| TI | .132 | — | .469 | .218 | .355 | .183 | .122 | .105 | .030 | .088 | .128 |
| WI | .153 | .146 | — | .249 | .312 | .266 | .364 | .373 | .556 | .323 | .404 |
| AP | .128 | .108 | .142 | — | .293 | .103 | .085 | .151 | .248 | .120 | .122 |
| BA | .145 | .134 | .153 | .130 | — | .221 | .309 | .240 | .346 | .235 | .253 |
| TS | .150 | .128 | .161 | .128 | .150 | — | .134 | .165 | .227 | .151 | .110 |
| AI | .147 | .124 | .160 | .126 | .148 | .127 | — | .093 | .060 | .057 | .062 |
| AY | .175 | .155 | .186 | .156 | .175 | .156 | .140 | — | .096 | .034 | .142 |
| KU | .165 | .145 | .176 | .146 | .165 | .145 | .134 | .156 | — | .072 | .130 |
| KO | .150 | .128 | .162 | .129 | .150 | .128 | .116 | .142 | .130 | — | .121 |
| TU | .160 | .140 | .172 | .140 | .161 | .140 | .121 | .148 | .141 | .121 | — |

*Notes:* Estimated distances from 11 genetic markers are above the diagonal, and predicted distances from our best migrational model ($H_8$) are below the diagonal. Parish designations are in fig. 14.1.

We have reported elsewhere a set of allele and haplotype frequencies for these 11 parishes (Wood et al. 1982). Using codominant markers at the Rh-C, MN, MDH, $PGM_1$, $PGM_2$, ACP, 6PGD, ADA, GPT, ESD, and Tf loci, we have computed genetic distances by the method described in Smouse (1982). These are presented in the upper triangular portion of table 14.5.

A test of matrix correspondence devised by Mantel (1967) can be used to compare the migrationally predicted and genetically measured distance matrices, a test that relies on the correlation of the two sets of distances. This procedure is also employed by O'Brien (chap. 13, this vol.). To assess the significance of the correlation, we permute the rows and columns of the predicted distance matrix, while holding the observed distance matrix constant, and compute the correlation. Under random permutation of one of these matrices, the expected correlation between corresponding elements of the two matrices is zero. The computed correlations from a series of 1,000 permuted matrices provide a null distribution of correlations against which to judge the significance of the observed correlation for each of the 10 migration models. The results of these analyses are presented in table 14.6.

The random dispersal model $H_0$ leads to no genetic divergence and provides no prediction of the pattern of genetic affinity. Allowing for endemicity (nonrandom dispersal) generates a modest but nonsignificant correlation ($H_1$: $r = .343$). Given endemicity, the addition of either geographic ($H_3$: $r = .460$) or linguistic ($H_4$: $r = .456$) distance improves the prediction notably. The addition of population density to either geographic ($H_5$: $r = .469$) or linguistic ($H_6$: $r = .469$) distance improves the fit between prediction and observation slightly. Here again, the effects of geographic and linguistic distance are somewhat redundant, but the combination of both ($H_7$: $r = .472$) leads to a slight improvement over either alone. The full model ($H_8$: $r = .488$) improves the match between predicted and observed

TABLE 14.6  Mantel (1967) Correlation ($\hat{r}$) between Observed and Predicted Genetic Distance Matrices, along with Probabilities of Exceeding These Values by Chance

| MIGRATION MODEL | PARAMETERS FIT | MANTEL TEST | |
|---|---|---|---|
| | | $r$ | $P$-level |
| $H_0$ | [0, 0, 0, 0] | .000 | .500 |
| $H_1$ | [$\mu$, 0, 0, 0] | .343 | .117 |
| $H_2$ | [$\mu$, $\alpha$, 0, 0] | .335 | .123 |
| $H_3$ | [$\mu$, 0, $\gamma$, 0] | .460 | .037 |
| $H_4$ | [$\mu$, 0, 0, $\delta$] | .456 | .037 |
| $H_5$ | [$\mu$, $\alpha$, $\gamma$, 0] | .469 | .031 |
| $H_6$ | [$\mu$, $\alpha$, 0, $\delta$] | .469 | .027 |
| $H_7$ | [$\mu$, 0, $\gamma$, $\delta$] | .472 | .020 |
| $H_8$ | [$\mu$, $\alpha$, $\gamma$, $\delta$] | .488 | .020 |
| $H_\Omega$ | [ Observed ] | .822 | .000 |

genetic distance slightly, but is still considerably less accurate than predictions from the observed migration matrix ($H_\Omega$: $r = .822$). In overview, the panmictic model ($H_0$) generates no genetic divergence; the endemicity model ($H_1$) produces some divergence, but only partially predicts its pattern; the intermediate models predict the pattern better than does $H_1$, with the improvements paralleling those in the migration model itself; the full functional model ($H_8$) is a real improvement on the endemicity model, but there remains considerable room for improvement. We reiterate that we have used only 4 of the available 110 degrees of freedom for the model. Further additions of meaningful ecological, demographic, and/or cultural variables that improve the migrational description would be expected to better predict the genetic distance pattern as well.

## CONCLUSIONS

Among the Gainj, the migration pattern in both sexes departs substantially from random dispersal. Nevertheless, the two sexes differ dramatically in the migration rates and patterns. Males show a very strong endemicity effect but not much else. Females exhibit a less pronounced endemicity effect, but geographic and/or linguistic distances do have an impact on the pattern of female migration, as does population density. Given any particular migration model, it is possible to predict the pattern of genetic distances among parishes. The correlation between predicted and observed distance matrices increases with the degree of improvement in the functional model of migration, with the pattern being the same in both cases: (panmixia) < (endemicity) < (endemicity + distance) < (endemicity + distance + density) < (Observation).

More generally, this chapter has described an approach to the analysis of migration and genetic population structure that capitalizes on some of the more useful features of both the MM and ID models. This approach allows us to model several explanatory variables simultaneously and provides us with genetic predictions of considerable statistical power. The approach should be of fairly general utility for the study of migration and genetic structure in a variety of organisms in a variety of contexts.

## ACKNOWLEDGMENTS

We wish to thank Steve Frank, Frank Livingstone, Jeffrey Long, James Neel, Alan Rogers, William Shields, and Peter Waser for helpful criticisms. We would also like to thank Patricia Johnson and the staff of the Papua New Guinea Institute of Medical Research for help and encouragement during the course of our field studies. PES was supported by NIH-RO1-GM-30135 and DoE-AC02-86ER60089. JWW was supported by NSF-BNS-77-01499, NIH-5-T32-GM-08123, and NIH-F35-GM-08551.

# REFERENCES

Bodmer, W. F., and L. L. Cavalli-Sforza. 1968. A migration matrix model for the study of random genetic drift. *Genetics* 59:565–92.

Friedlaender, J. S. 1971. Isolation by distance in Bougainville. *Proceedings of the National Academy of Sciences* (USA) 58:704–7.

Hogbin, I., and C. Wedgwood. 1952. Local grouping in Melanesia. *Oceania* 23:241–76.

Johnson, P. L. 1982. Gainj kinship and social organization. Ph.D. diss., Department of Anthropology, University of Michigan, Ann Arbor. University Microfilm 82-24977.

Kendall, M., and A. Stuart. 1979. *The advanced theory of statistics*, vol. 2. 4th ed. New York: Macmillan.

Lees, F. C., and J. H. Relethford. 1982. Population structure and anthropometric variation in Ireland during the 1930s. In *Current developments in anthropological genetics*, vol. 2: *Ecology and population structure*, ed. M. H. Crawford and J. H. Mielke, 385–428. New York: Plenum Press.

Long, J. C. 1986. The allelic correlation structure of Gainj- and Kalam-speaking people. I. The estimation and interpretation of Wright's F-statistics. *Genetics* 112:629–47.

Long, J. C., J. M. Naidu, H. W. Mohrenweiser, H. Gershowitz, P. Johnson, J. W. Wood, and P. E. Smouse. 1986. Genetic characterization of Gainj- and Kalam-speaking people of Papua New Guinea. *American Journal of Physical Anthropology* 70:75–96.

Malécot, G. 1969. *The mathematics of heredity.* Trans. P. M. Yermanos. San Francisco: Freeman.

Mantel, N. A. 1967. The detection of disease clustering and a generalized regression approach. *Cancer Research* 27:209–20.

Morton, N. E. 1969. Human population structure. *Annual Review of Genetics* 3:53–74.

Smith, C. A. B. 1969. Local fluctuations in gene frequency. *Annals of Human Genetics* 32:251–60.

Smouse, P. E. 1982. Genetic architecture of swidden agricultural tribes from the lowland rain forests of South America. In *Current developments in anthropological genetics*, vol. 2: *Ecology and population structure*, ed. M. H. Crawford and J. H. Mielke, 139–78. New York: Plenum.

Ward, R. H., and J. V. Neel. 1970. Gene frequencies and microdifferentiation among the Makiritare Indians. IV. A comparison of a genetic network with ethnohistory and migration matrices; A new index of genetic isolation. *American Journal of Human Genetics* 22:538–61.

Wijsman, E. M., and L. L. Cavalli-Sforza. 1984. Migration and genetic population structure, with special reference to man. *Annual Review of Ecology and Systematics* 15:279–301.

Wood, J. W. 1980. Mechanisms of demographic equilibrium in a small human population: The Gainj of Papua New Guinea. Ph.D. diss., Department of Anthropology, University of Michigan, Ann Arbor. University Microfilm 80-13088.

———. 1986. Convergence of genetic distances in a migration matrix model. *American Journal of Physical Anthropology* 71:209–219.

————. 1987. The genetic demography of the Gainj of Papua New Guinea. 2. Determinants of effective population size. *American Naturalist* 129:165–87.

Wood, J. W., P. L. Johnson, and K. L. Campbell. 1985. Demographic and endocrinological aspects of low natural fertility in highland New Guinea. *Journal of Biosocial Sciences* 17:57–79.

Wood, J. W., P. L. Johnson, R. L. Kirk, K. McLoughlin, N. M. Blake, and F. A. Matheson. 1982. The genetic demography of the Gainj of Papua New Guinea. I. Local differentiation of blood group, red cell enzyme, and serum protein allele frequencies. *American Journal of Physical Anthropology* 57:15–25.

Wood, J. W., D. Lai, P. L. Johnson, K. L. Campbell, and I. A. Maslar. 1985. Lactation and birth spacing in highland New Guinea. *Journal of Biosocial Sciences* (supp.) 9:159–73.

Wood, J. W., and P. E. Smouse. 1982. A method of analyzing density-dependent vital rates with an application to the Gainj of Papua New Guinea. *American Journal of Physical Anthropology* 58:403–11.

Wood, J. W., P. E. Smouse, and J. C. Long. 1985. Sex-specific dispersal patterns in a human population. *American Naturalist* 125:747–68.

Wright, S. 1943. Isolation by distance. *Genetics* 28:114–38.

# 15. Definition and Measurement of Migration in Age-Structured Populations

*Donald Stone Sade, B. Diane Chepko-Sade,*
*Malcolm Dow, and James Cheverud*

An individual migrating into a new population can carry such traits as pathogens, learned or cultural characteristics, or alleles that may differ from the usual ones found in the target population. These can be transmitted respectively through contagion, imitation or socialization, and procreation, thereby altering the characteristics of the target population over a period of time. An immigrant can also reduce the fitnesses of individuals in the target population through competition or increase them through cooperation.

One indication of the potential magnitude of the changes that could be induced in a target population is the rate of immigration of the trait of interest. The rate at which individuals migrate into the target population is the basic rate that must be measured, because traits are generally carried by individuals. This *individual immigration rate* is considered in this chapter. If only some immigrants carry the trait of interest, the formulae given here can easily be expanded to include terms for the proportion of migrants and residents that do, and the proportion of migrants and residents that do not, carry the trait.

Now consider a species, such as the rhesus monkey, with long-lived individuals, populations subdivided into many demes or social groups, and individuals making multiple migrations between the groups. The difficulties in applying the usual population genetical models of migration to the complexities of life history observed at a colony of these monkeys stimulated our interest in the methods developed in this chapter.

## CHARACTERISTICS OF THE COLONY

A more detailed history of the Cayo Santiago colony of rhesus monkeys is given in Sade et al. (1985). Following a systematic reduction of the population through the removal of social groups, completed by the end of

1972, the population approximated its theoretical stable age distribution (Sade et al. 1976). On 1 January 1973 the population consisted of 279 monkeys in four heterosexual social groups, designated F, I, J, and L, and an additional 14 *extragroup males* collectively referred to as EGM in the present study. The total population had increased to 458 monkeys by 1 January 1977.

During this period there were no removals (2 males scheduled for removal in 1972 actually were trapped in 1973), and management was designed to facilitate studies on population biology and social behavior. There was a total of 323 males in the population during the 4 years. By 1977 six heterosexual social groups were recognized, two new groups having formed through two successive divisions of Group F (Chepko-Sade and Sade 1979). The data used in this chapter are from 1973 through 1976. For the purposes of this study we consider Groups F and O as the same group, the division between them occurring late in 1976, and we consider Groups F and M distinct groups for the entire period, although the division between them was not considered complete until August 1973. In addition, Group J divided into Groups J and N in late 1973, but the two groups merged into a single group the following year; we consider both a single group for the purposes of this study.

## LIFE CYCLE OF RHESUS MONKEYS

Although rhesus monkeys can live beyond 28 years (van Wagenen 1972), few in the Cayo Santiago colony have been known to be over 20 years old. Females usually do not conceive until 3.5 years and give birth when 4 years old, usually to a single infant, after a gestation of 168 days. Occasionally a female, especially one of a high-ranking genealogy, may conceive and give birth a year early. Females generally remain with their natal group throughout their lives and during the rare episodes of group fission continue to associate with their closest kin in the daughter group (Chepko-Sade and Olivier 1979).

Males reach puberty at 3.5 years, but may not be fertile until the next mating season, when they are 4.5 (Sade 1964; Zamboni, Conaway, and Van Pelt 1974), although Stern and Smith (1984) suggest that an occasional 3.5-year-old male may have fathered offspring in a captive colony. Two-year-old males may migrate from their natal group, especially if they are accompanying an older brother, but most migrations are of males at puberty or older. However, in the data of the present study a few individuals younger than 2 years are tabulated as migrants because they moved with their relatives from Group F to Group M during the formation of the latter group. Males of high-ranking genealogies may delay migration for several to many years. One remained with his natal group until he was over 12 years old. Some males may migrate several times a year while others remain in a new group for several years, but most, if they live, will join several different social groups during their lives (Chepko-Sade 1982; Colvin 1983; Sade 1980b). It was these multiple migrations that stimulated our interest in the present topic.

Under the crowded conditions of the colony it would be incorrect to consider each social group a separate deme, in the sense of the population geneticist. Yet perhaps 95% of fertile matings during any one mating season occur between members of the same social group (Sade, Chepko-Sade, and Schneider 1977; Sade et al. 1982; Sade 1984), and the social groups are more than randomly differentiated genetically (Cheverud 1981; Cheverud, Buettner-Janusch, and Sade 1978; Cheverud 1981; Duggleby 1978). The population may be a poor model of a natural one in many respects (Melnick and Kidd 1983), yet it also shows remarkable similarity in social processes to the few wild populations studied in comparable detail (Pearl 1982; Pearl and Schulman 1983). Especially because of the abundant and detailed longitudinal information on the population dynamics and social organization of the colony it is well suited for developing and testing new methods and models, although the generality of the specific conclusions derived from them may be suspect, except perhaps as they apply to the special case of a rapidly expanding population.

## WAYS MIGRATION RATES ARE MODELED
### The Case with Discrete Generations

The usual models in population biology represent species in which both migration and reproduction are single events in the lives of individual members of a cohort. Each cohort is a discrete generation, and generations succeed one another without overlapping.

Consider that such a species consists of two demes $i$ and $j$. The individual immigration rate for deme $i$, measured after the cohort's single episode of migration but before any selection takes place, is the proportion of migrants among the total membership of deme $i$, and is given by the expression

$$^{I}r_{ij} = \frac{n_j\, m_{ij}}{n_j\, m_{ij} + n_i\, m_{ii}}, \tag{15.1}$$

where $^{I}r_{ij}$ is the individual migration rate from deme $j$ to deme $i$, $n_j$ is the number of individuals in deme $j$, $m_{ij}$ is the proportion of $n_j$ that moved to deme $i$, $n_i$ is the number of individuals in deme $i$ at the start of the migration event, and $m_{ij}$ is the proportion of $n_j$ that remained in deme $i$ following the cohort's episode of migration. The cohort and the generation are the same in this model, and the individual migration rate and the per generation migration rate are the same.

### The Case with Overlapping Generations

If individuals of a species live long enough to reproduce more than once, and if the natality and mortality rates of a population of such individuals vary as a function of age, then a life table may be used to compute the demographic and evolutionary parameters generally of interest in evolutionary theory (Charlesworth 1980). The usual life table describes a single population incremented only through reproduction. Longitudinal observations show that individuals of such long-lived species as rhesus monkeys or humans may migrate several times during the reproductive portion of their life span, thus spending part of their lives and potentially contributing their unique traits, of which their alleles are but one kind, to each of several different demes.

What is the immigration rate for a population of such a species? A possible answer is to include in formula (15.1) additional terms that represent the average or expected time migrant individuals will remain in each deme. Formula (15.1) thus becomes

$$^Er_{ij} = \frac{n_j \, m_{ij} \, e_{ij}}{n_j \, m_{ij} \, e_{ij} + n_i \, m_{ii} \, e_{ii}} \, . \tag{15.2}$$

where $^Er_{ij}$ is the lifetime individual immigration rate from deme $j$ to deme $i$, $e_{ij}$ is the expected time an individual that originates in deme $j$ and migrates to deme $i$ will live in deme $i$, and $e_{ii}$ is the expected time individuals originating in deme $i$ will live in deme $i$.

Formula (15.2) expresses the immigration rate for deme $i$ as the proportion of the total time-units, such as "monkey-years", expected to be lived in deme $i$ by individuals that originated in deme $j$.

If the life expectancy ($E$) and generation time ($T$) for the entire cohort were equal ($T/E = 1$), as in the case of the semelparous species with discrete, nonoverlapping generations, then formula (15.2) also gives the per generation individual immigration rate for deme $i$. If the generation time is less than life expectancy ($T/E < 1$), then the influence of an immigrant as a carrier, altruist, competitor, or potential progenitor might be thought of as extending beyond a single generation. The proration of life expectancy across more than one generation should reduce the per generation rate as given by formula (15.2).

These arguments suggest adjusting formula (15.2) by generation time as a proportion of life expectancy:

$$^Tr_{ij} = \frac{E \, n_j \, m_{ij} \, e_{ij}}{T(n_j \, m_{ij} \, e_{ij} + n_i \, m_{ii} \, e_{ii})} \, , \tag{15.3}$$

where $T$ is the cohort's generation time and $E$ is the average life expectancy for the entire cohort. $^Tr_{ij}$ is the per generation individual migration rate from deme $j$ to deme $i$.

Should generation time exceed life expectancy, as might be the case if many individuals died without reproducing, or should there be a very high variance in reproductive success, as might be the case among male rhesus monkeys, formula (15.3) might result in a per generation migration rate greater than 1. This result does not appear anomolous when one remembers that this immigration rate is not expressed as a probability but rather in units of expected lifetimes per generation ($E/T$). A per generation migration rate greater than 1 can be interpreted as meaning that more than one lifetime impacts upon a single generation.

### The Case with Many Demes

In practice the term $m_{ij}e_{ij}$, which is the expected time an individual that originates in deme $j$ will spend in deme $i$, can be estimated directly from the deme-specific mortality rates and the rates of movements of individuals between demes, using multistate life tables recently developed by human demographers (reviewed in Keyfitz 1980). The matrix techniques introduced by Rogers and Ledent (1976) facilitate the inclusion of any number of demes into a study. In the multistate case the columns of the ordinary life table are replaced by stacks of matrices.

In the remainder of this chapter, boldface capitals represent matrices, boldface lowercase letters represent vectors, plain lowercase letters represent scalars, left superscripts indicate age classes, and right subscripts indicate the element of the matrix or vector.

$^{x}\mathbf{E} = \{^{x}e_{ij}\}$, the life expectancy of individuals of age $x$, is a square matrix in which the columns represent demes of origin and the rows demes into which migrants move. The elements $^{x}e_{ij}$ are the expected number of years that individuals of age $x$ originating in the $j$th deme will live in the $i$th deme. $^{x}\mathbf{E}$ is computed by the formula

$$^{x}\mathbf{E} = {}^{x}\mathbf{T}{}^{x}\mathbf{L}^{-1},\tag{15.4}$$

where the elements of $^{x}\mathbf{T}$ are the total number of years to be lived in the $i$th deme by individuals originating in the $j$th deme from age $x$ through the end of the table.

$^{x}L$ is formed by setting its off-diagonal elements to 0 and its diagonal elements equal to the column sums of the matrix $^{x}L$. $^{x}L$ is the survivorship matrix, analogous to the $l_x$ entry in the ordinary life table. Its elements are the probabilities that an individual originating in the column deme will be alive and in the row deme at age $x$. The column sums of $^{x}L$, entered as diagonal elements in the $^{x}\mathbf{L}$ matrix, are therefore the survivorship probabilities to age $x$ for individuals born into the corresponding column deme.

$^{x}\mathbf{T}$ is found by

$$^{x}\mathbf{T} = \sum_{x}(({}^{x}\mathbf{L} + {}^{(x+1)}\mathbf{L})/2),\tag{15.5}$$

with the summation from age $x$ through the last age class in the series of $^{x}\mathbf{L}$ matrices.

The radix values, entered as the starting values in the survivorship matrices, determine the numerical units in which all the other life table values will be expressed. Consider first the construction of a multistate life table with radix values of unity. Subsequent values are expressed as proportions of the starting population. Radix values of unity are provided by setting $^{0}L = \mathbf{I}$, where $\mathbf{I}$ is the identity matrix. The subsequent $^{x}L$ matrices are generated by a linear approximation to the function that, according to Schoen (1975) and Rogers and Ledent (1976), relates the $^{x}L$ matrices to the empirical deme-specific mortality rates and interdeme rates of movement, recorded for each age class in the matrix $^{x}\mathbf{M}$. In the special case in which the width of the age class interval is unity,

$$^{(x+1)}L = [\,(\mathbf{I} + {}^{x}\mathbf{M}/2)^{-1}\,(\mathbf{I} - {}^{x}\mathbf{M}/2)]\,{}^{x}L\,. \tag{15.6}$$

Each matrix $^{x}\mathbf{M}$ is formed by entering the negatives of the age-specific interdeme transition rates in the appropriate off-diagonal cells. The deme-specific mortality rates are then added to the sum of the absolute values of the corresponding columns and the results entered as the main diagonal elements. The sum of each column of the $^{x}\mathbf{M}$ matrix is thus equal to the deme-specific death rate.

The numerical interpretation of the elements of $^{x}\mathbf{E}$ will depend upon the choice of radix values. If the identity matrix is chosen to provide radix values of 1, then each element of $^{x}\mathbf{E}$ is the average length of time that an individual of age $x$ born into a column deme will spend in a row deme, combining both the probability of migration to each other specific deme and the probability of surviving within each deme. For $x = 0$, $^{x}\mathbf{E}$ is equivalent to the term $m_{ij}e_{ij}$ in formula (15.2).

If the diagonal elements of the matrix $^{0}L$ were set equal to the original size of each $j$th deme's cohort, then the elements of $\mathbf{E}$ would be equivalent to the term $n_{i}m_{ij}e_{ij}$ in equation (15.2), which would simplify to

$$^{E}r_{ij} = \frac{e_{ij}}{e_{ii}}\,. \tag{15.7}$$

If radix values of unity are used, the deme-specific lifetime individual immigration rates are given by

$$^{E}r_{i} = \frac{\sum\limits_{j \neq i} e_{ij}\, n_{j}}{\sum\limits_{j} e_{ij}\, n_{j}}\,, \tag{15.8}$$

where $n_{j}$ is the size of the cohort of the $j$th deme. How that number is to be determined depends on the specific circumstances of each particular study. If the vector of the sizes of deme-specific cohorts, $n = \{n_{j}\}$, is entered on the main diagonal of the $^{0}L$ matrix, formula (15.8) simplifies to

$$E_{r_i} = \frac{\sum\limits_{j \neq i} e_{ij}}{\sum\limits_{j} e_{ij}}$$    (15.9)

in analogy to formula (15.7).

## INTERPRETATION OF LIFE TABLES

The circumstances and kinds of information available for each study will determine how the mortality and transition rates to be entered into the $^xM$ matrices will be computed. As in the construction of the ordinary life table, longitudinal records on a cohort will permit the construction of a table that summarizes the actual experience of the cohort's individuals. However, such a "cohort-based" life table will be greatly influenced by unique events, such as episodes of unusually high or low mortality, or changing opportunities for dispersal, and thus may lack generality.

On the other hand, a table constructed from events occurring during a limited period of time, short in relation to the longevity of individuals of the species, represents a hypothetical life history composed of events occurring during a cross section of the lives of individuals of different age classes from infancy to old age. In such a "time-specific" life table, secular trends will confound age-specific effects; on the other hand, there is no constraint across age classes of earlier events on later characteristics, as would occur in a system with memory, such as in the actual lives of individual monkeys.

Any interpretation of the table can only refer back to the schedule of age-specific rates, as if no other feature that the natural history of the animals might suggest as causal were having an effect. For instance, in the present study there is no indication in the rates alone whether immigrants from a particular social group of monkeys are more likely to remain than are immigrants from another group.

Neither the cohort-based nor time-specific life table is a predictive tool, as there is no information contained within it that indicates how vital rates may distribute in the future. Any projection made from the life table shows only the potential consequence of a particular schedule of vital rates were they to continue unchanged, but makes no statement about the likelihood that they will remain stable in the future.

Although the arguments given in the previous sections are simple and straightforward, unexpected complexities develop in applying them to a set of actual observations. Careful attention to the implication of each term as used in the next section is required.

## APPLICATION TO A SPECIFIC POPULATION: THE LIFE TABLE FOR MALE RHESUS MONKEYS

We now apply the previous arguments by finding the social group-specific immigration rates for males of the free-ranging colony of rhesus monkeys on Cayo Santiago, Puerto Rico. A more extensive methodological treatment, which includes an alternate treatment of survivorship, is found in Sade, Dow, and Chepko-Sade (1984), and a more detailed mathematical treatment of inferential procedures is found in Dow (1985). The data used in this chapter were tabulated from Sade et al. (1985).

The average annual group-specific mortality and intergroup transfer rates for males for the years 1973 through 1976 were calculated for age classes of one year in width, save that the few males 15 years old and older were combined in the single, terminal age class. The resulting 16 $^x$**M** matrices are listed in the appendix. Difficulties that occurred in determining the number of individuals in each group at risk for migrating or dying, because of multiple transfers of some individual males within the same year, are discussed in detail in Sade, Dow, and Chepko-Sade (1984), but can be ignored here.

The six columns and rows represent the five heterosexual social groups, F, I, J, L, and M, and the special category, EGM (extragroup males). The extragroup males are not a single social group but rather a heterogenous category that includes solitary males, males generally seen in pairs, and larger loose groupings of males. These males are those that show no special tendency to move with or interact with the members of any particular heterosexual social group. As a category they deserve more detailed behavioral study. There are also distinct from the so-called peripheral males that are more or less loosely, although definitely, associated with particular heterosexual social groups.

The category EGM is incremented only through migration, not through births, and therefore the first several age classes will have an N of zero. This fact constrains the choice of radix values for the computation of the $^x$**E** matrices. If the vector **n** is chosen to provide radix values, 0 must be entered for EGM. This results in singular $^x$**L** matrices. Therefore radix values of unity are used in this study, and formula (15.9) is used to compute the group-specific immigration rates.

A program was written to convert the $^x$**M** matrices into the $^x$**E** matrix for each of the 16 age classes, of which two are of special interest for the present study. The $^0$**E** matrix is shown in table 15.1.

TABLE 15.1 $0_E$ for Cayo Santiago Males, 1973–1976

|  | SOURCE GROUP | | | | | |
|---|---|---|---|---|---|---|
|  | EGM | F | I | J | L | M |
| EGM | 5.487 | 1.132 | 1.074 | 1.1666 | .942 | .954 |
| F | 1.680 | 4.479 | 1.202 | 1.098 | .939 | 1.028 |
| I | 1.188 | .747 | 5.624 | .624 | .463 | .792 |
| J | 1.122 | .891 | .822 | 4.558 | .525 | .634 |
| L | 1.574 | .894 | 1.093 | 1.412 | 4.875 | .689 |
| M | .891 | 1.653 | .596 | .618 | .491 | 4.402 |

*Note:* $^0L$ = I to provide radix values of 1.

The $^0\mathbf{E}$ matrix shows the way in which males' total life expectancy at birth is partitioned among the other groups in the population, if only the age-specific intergroup transfer rates and the group- and age-specific mortality rates are considered. The sum of a column (not shown) therefore gives the total life expectancy for the average individual born into the *j*th group.

## IMMIGRATION RATES

To apply formula (15.9), the vector **n,** which contains the group-specific cohort sizes, must be determined. For the present study this was done by finding the average number of males born into each group during the four birth seasons from 1973 through 1976. These values are given in table 15.2 (1st column).

The vector of immigration rates for the respective social groups is shown in table 15.2 (2d column). All the monkey-years lived by males in EGM are lived by immigrants. Among the heterosexual groups, nearly half the monkey-years lived by males in Group M are lived by immigrants, and even in Group I, which had the lowest immigration rate, over one-third of all monkey-years are lived by immigrants. These interpretations are, of course, subject to all the simplifying assumptions discussed earlier in the chapter.

The use of the EGM category creates a special problem. The methods used here do not track the migration histories of individual males. Although no males originate in EGM, males that migrate from EGM to the heterosexual groups ultimately originated in one of them (ignoring the fact that in the actual population some may have originated in social groups that no longer existed during the years of the present study). Although we are interested in group of origin, their immigrations do not increment the proportion of the row group contributed by their true natal group. In fact, the contributions of immigrants from EGM are multiplied by zero because

TABLE 15.2  Vectors of Group-Specific Cohort Size (**n**), and Group-Specific Immigration Rates (*r*)

| Group | n | $E_r(1)$ | $E_r(2)$ | $T_r(3)$ | $T_r(4)$ | $T_r(5)$ | $T_r(6)$ |
|-------|------|------|------|------|-------|-------|-------|
| EGM | 0.00 | 1.00 | 1.00 | .899 | 1.107 | 1.060 | 1.300 |
| F | 10.75 | .425 | .752 | .382 | .471 | .694 | .854 |
| I | 7.50 | .364 | .766 | .327 | .403 | .632 | .778 |
| J | 9.50 | .368 | .776 | .331 | .407 | .698 | .859 |
| L | 8.75 | .462 | .786 | .415 | .511 | .704 | .867 |
| M | 7.75 | .487 | .801 | .438 | .539 | .695 | .856 |

*Notes:* (1) lifetime rates for monkeys at birth; (2) lifetime rates for monkeys at age 4 years; (3) per generation rates for monkeys at birth with life expectancy of 9.28 years and generation time of 8.35 years; (4) per generation rates for monkeys at birth with life expectancy of 9.28 years and generation time of 10.28 years; (5) per generation rates for monkeys at 4 years with life expectancy of 7.91 years and generation time of 8.35 years; and (6) per generation rates for monkeys at 4 years with life expectancy of 7.91 years and generation time of 10.28 years.

none are born into that category, and the actual immigration rates given in the tables are therefore slightly low for the heterosexual groups.

A geneticist, perhaps interested in immigrants only as potential progenitors, might wish to discover the vector of immigration rates for males of breeding age, eliminating from consideration immigrants unlikely to be fertile because of youth. Such an interest would be served by substituting the matrix $^4E$, the matrix of life expectation for males at 4 years of age, for the matrix $^0E$ in the computation of immigration rates. $^4E$ is given in table 15.3. The vector of lifetime individual immigration rates for males 4 years old is given in table 15.2 (3d column).

Considering expection of life at 4 years of age of course eliminates all the monkey-years lived by younger males, and since most of these monkey-years will be spent in the natal group prior to the first migration, the denominator of formula (15.9) will be considerably reduced, producing the expected effect of increasing the immigration rates, in this case up to .75–.80, with EGM maintaining its obligatory immigration rate of 1. The mean immigration rate for the heterosexual social groups is .776.

These rates consider only the male portion of the population. If the female portion were to be considered, the denominator of formula (15.9) would need to be incremented by terms representing the size of the group-specific female cohorts and the life expectancy of females within their natal groups. As females rarely migrate, except during episodes of group fission, the female term for the numerator could be usually assumed to be zero. In fact, the data used in this study includes the division of Group F into Groups F and M; a complete analysis, one that included the female portion of the population for this period, would require a nonzero term representing females in the numerator. However, for the purpose of exposition, we assume no migration of females between social groups, a group-specific cohort sex ratio of 1:1, and a mean life expectancy for females equal to the mean life expectancy of natal resident and immigrant males for each heterosexual social group. Under these simplifying but not totally unreasonable assumptions, the group-specific immigration rates of table 15.2

TABLE 15.3 $^4E$ for Cayo Santiago Males, 1973–1976

| | | SOURCE GROUP | | | | | |
| | | EGM | F | I | J | L | M |
|---|---|---|---|---|---|---|---|
| | EGM | 1.807 | 1.416 | 1.305 | 1.432 | 1.302 | 1.412 |
| | F | 1.620 | 2.223 | 1.443 | 1.356 | 1.215 | 1.478 |
| TARGET | I | 1.164 | .899 | 2.726 | .728 | .656 | 1.055 |
| GROUP | J | 1.097 | 1.093 | 1.011 | 1.789 | .744 | .938 |
| | L | 1.505 | 1.112 | 1.273 | 1.622 | 2.556 | 1.014 |
| | M | .859 | 1.355 | .713 | .769 | .675 | 2.221 |

Note: $^4L$ = I to provide radix values of 1.

need only be divided by 2 to yield the immigration rate that includes the female portion of the population. The mean immigration rate so determined is .388. It is this rate that is the analogue of the migration rate usually specified in discrete-generation population genetical models.

## PER GENERATION IMMIGRATION RATES

What is the per generation immigration rate considering first life expectancy at birth and then life expectancy at the age of 4 years? We have not determined generation time for males as yet because this requires an estimation of the reproductive success of males, difficult under free-ranging conditions (Sade 1980a), although under study through the technique of genetical paternity exclusions (Sade et al. 1982). For females in the Cayo Santiago colony generation, length varied from 7.27 to 10.28 years, depending on social group and dominance status, with the average generation equal to 8.35 years, based on natality and mortality records for the years 1973 and 1974 only (Sade et al. 1976). It is unlikely that the generation length for males would be as short as that for females, because of the later maturation of males and the likelihood of considerably greater variance in reproductive success of males in comparison to females, as mentioned earlier.

Life expectancy, on the other hand, requires only observations on age at death. The average life expectancy of males of age $x$ is simply the average of the column sums of the $^x\mathbf{E}$ matrix computed using radix values of unity, and excluding the EGM column, as no males are born into that category. In the present study, males of age $x = 0$ and $x = 4$ are considered.

Table 15.2 includes the group-specific per generation immigration rates for males at birth (4th and 5th columns) and at 4 years of age (6th and 7th columns). Two estimates of generation time are used in each case. Lacking a measure of generation time for males, we use the average and maximum generation time for females as published previously (Sade et al. 1976). The average generation time for females is certainly an underestimate for males. The maximum is perhaps a reasonable one. The vector of per generation immigration rates, $^T\mathbf{r} = {}^Tr_i$, is found simply by multiplying the vector of lifetime individual immigration rates, $^E\mathbf{r}$, by $T/E$, where $T$ is the estimate of generation time and $E$ is the average life expectancy for the population, according to the reasoning that produced formula (15.3).

Note the per generation immigration rate greater than 1.0 for EGM when the longer generation time, which exceeds life expectancy, is used—a possibility discussed earlier. Interpretation of the per generation migration rates does not appear to be straightforward, and it may be that the lifetime individual immigration rates will prove the most useful in subsequent model building. If so, then the lifetime, rather than the generation, would be the useful unit of time in the study of age-structured populations. On the other hand, the present development of a "per generation" demographic treatment of migration may offer a bridge between the analytical simplicity of the discrete generation model and the greater realism of the demographic model.

## INFERENTIAL PROCEDURES

Are the differences between the immigration rates statistically improbable? No easy answer to this question seems likely. The "delta method" (Keyfitz 1968), in which the variance of the original transition and mortality rates would be estimated and followed through each transformation of the matrices, resulting in a direct estimate of the variance of the group-specific life expectancies, appears difficult or impossible for the multistate case, according to Hoem and Hensen (1982), who suggest applying nonparametric procedures, an approach we follow.

A procedure for testing the degree to which two matrices correspond, known now as "quadratic assignment," was first proposed by Mantel (1967) and elaborated by Hubert and Schultz (1976), Hubert and Baker (1978), Baker and Hubert (1981), and Hubert and Golledge (1981). For two square matrices $\mathbf{A} = \{a_{ij}\}$ and $\mathbf{B} = \{b_{ij}\}$, the index of correspondence, $\Gamma$, is the sum of the cross-products of the entries. $\Gamma$ may be computed for the entire matrices or for the main diagonal elements or off-diagonal elements only:

$$\Gamma_1 = \sum_{i,j} (a_{ij})(b_{ij}) . \tag{15.10}$$

$$\Gamma_2 = \sum_{i \neq j} (a_{ij}) (b_{ij}) . \tag{15.11}$$

$$\Gamma_3 = \sum_{i,i} (a_{ij})(b_{ij}) . \tag{15.12}$$

A probability distribution for each $\Gamma$ can be generated by tabulating the frequency of each index produced by the repeated simultaneous random reordering of the rows and columns of one of the matrices of interest, and recalculating the index each time, until all the possible permutations have been completed. The observed index can then be compared to the exact sampling distribution thus produced. However, since this procedure could become tedious, as $n!$ permutations are required, where $n$ is the dimension of the matrix, it is more reasonable to generate an approximate distribution for the $\Gamma$ indices by randomly sampling (with replacement) from the set of $n!$ possible permutations. This latter approach is employed in the analysis reported later. Also, it will generally be preferable to normalize the raw $\Gamma$ cross-product statistics to obtain a more easily interpretable measure of association between the two matrices. In the Pearson product-moment normalization, the variances of the elements in each matrix are the normalizing constants. Significance levels for these latter correlational measures are obtained by tabulating their frequency of occurrence exactly as previously described. Obviously, the significance levels of the $\Gamma$ statistics and the corresponding correlations are identical.

We test the null hypothesis that the life expectancy of males within each group is random in respect to group of origin. If this were so then differences in the immigration rates for different groups would be due only to the differences in the size of the group-specific cohorts. Under this hypothesis the expected values of the entries for any row in the $^x\mathbf{E}$ matrix would be equal to each other and equal to the mean of the entries for that row. This hypothesis provides us with a matrix of expected structure to compare with the observed structure in the $^x\mathbf{E}$ matrix, using the quadratic assignment procedure.

The matrix of expected structure ($^0\mathbf{S}$) for $^0\mathbf{E}$ under the above hypothesis is given in table 15.4. Note that by construction the $^0\mathbf{E}$ and the $^0\mathbf{S}$ matrices are not independent of one another; that is, their row sums are constrained to be equal. In this case, direct application of the Mantel matrix permutation procedure is invalid, since it will be biased toward finding structural similarity in the two matrices. However, Dow (1985) outlines an extension of the Mantel approach that overcomes the dependency problem and permits valid hypothesis testing. In this extension, the elements in each matrix are converted to Z-scores, treating the set of elements in each matrix as a sample. Note that this simple linear transformation leaves intact any underlying patterns in each matrix. Then, the standardized $^0\mathbf{S}$ matrix is subtracted from the standardized $^0\mathbf{E}$ to yield a "residual" matrix $\mathbf{R}$:

$$\mathbf{R} = {}^0\mathbf{E}^* - {}^0\mathbf{S}^*, \qquad (15.13)$$

where the * indicates standardization. Application of the matrix permutation approach to the $^0\mathbf{E}$ and $\mathbf{R}$ matrices provides a test of the extent to which the pattern in the $^0\mathbf{E}$ matrix is removed by the subtraction of the $^0\mathbf{S}$ matrix. That is, the closer the correlation between $^0\mathbf{E}$ and $\mathbf{R}$ is to zero, the closer the correlation between $^0\mathbf{E}$ and $^0\mathbf{S}$ is to one. A significant association between $^0\mathbf{E}$ and $\mathbf{R}$ thus results in rejecting the hypothesis that all of the pattern in $^0\mathbf{E}$ is captured by $^0\mathbf{S}$.

Application of these procedures to the $^0\mathbf{E}$ and $^0\mathbf{S}$ matrices yields a correlation of .671 between $^0\mathbf{E}$ and $\mathbf{R}$, with an associated significance level of

TABLE 15.5  $^4\mathbf{S}$ Structure Matrix

| | | SOURCE GROUP | | | | | |
| --- | --- | --- | --- | --- | --- | --- | --- |
| | | EGM | F | I | J | L | M |
| | EGM | 1.45 | 1.45 | 1.45 | 1.45 | 1.45 | 1.45 |
| | F | 1.56 | 1.56 | 1.56 | 1.56 | 1.56 | 1.56 |
| TARGET | I | 1.21 | 1.21 | 1.21 | 1.21 | 1.21 | 1.21 |
| GROUP | J | 1.11 | 1.11 | 1.11 | 1.11 | 1.11 | 1.11 |
| | L | 1.51 | 1.51 | 1.51 | 1.51 | 1.51 | 1.51 |
| | M | 1.10 | 1.10 | 1.10 | 1.10 | 1.10 | 1.10 |

Note: The null hypothesis is that each row group receives immigrants at random from each column group.

.155 based on 200 permutations. This result suggests that the hypothesis of "no conformity" between $^0\mathbf{E}$ and $^0\mathbf{S}$ be rejected, that the matrices are similar, and that there is thus little tendency for males to migrate preferentially between groups. However, inspection of the original $^0\mathbf{E}$ matrix shows that the diagonal elements are large in relation to the off-diagonal elements and that the diagonal elements are roughly equal because they include the years that premigratory juveniles are expected to spend in their natal groups. Perhaps structure, that is, nonrandom contribution of life expectancy to each $i$th group by the other groups, is masked by the dominance of the diagonal elements.

Accordingly, we also applied the quadratic assignment procedure to the off-diagonal elements only. In the structure matrix for this test the off-diagonal elements are the mean of the off-diagonal elements of the corresponding row of the original matrix. The resulting correlation between the off-diagonals of $^0\mathbf{E}$ and the corresponding $\mathbf{R}$ was .528 with a significance level of .01, again based on 200 permutations. This result implies that the hypothesized structure matrix does not reflect a similar pattern to the $^0\mathbf{E}$ off-diagonal entries and suggests that there is in fact preferential migration or avoidance of migration between certain groups. Note that in this case both matrices were standardized using means and variances calculated for the off-diagonal elements only.

Is the expected distribution of life expectancy random in respect to group of origin for males that have reached breeding age? This question is answered by applying the quadratic assignment procedure to $^4\mathbf{E}$. The structure matrix, $^4\mathbf{S}$, is given in table 15.5. The correlation obtained between $^4\mathbf{E}$ and the corresponding $\mathbf{R}$ matrix was .558, with an associated significance level of .530 based again on 200 permutations. This result rejects the hypothesis of no conformity between $^4\mathbf{E}$ and $^4\mathbf{S}$ and suggests that life expectancy from age 4-years is not preferentially distributed among the social groups. Note that the diagonal elements of the $^4\mathbf{E}$ matrix (table 15.3) are smaller than the corresponding elements in the $^0\mathbf{E}$ matrix (table 15.1), as they should be following the cohort's episodes of early mortality and mi-

TABLE 15.4  $^0\mathbf{S}$ Structure Matrix

|  |  | SOURCE GROUP | | | | | |
|  |  | EGM | F | I | J | L | M |
|---|---|---|---|---|---|---|---|
|  | EGM | 1.78 | 1.78 | 1.78 | 1.78 | 1.78 | 1.78 |
|  | F | 1.74 | 1.74 | 1.74 | 1.74 | 1.74 | 1.74 |
| TARGET | I | 1.57 | 1.57 | 1.57 | 1.57 | 1.57 | 1.57 |
| GROUP | J | 1.43 | 1.43 | 1.43 | 1.43 | 1.43 | 1.43 |
|  | L | 1.76 | 1.76 | 1.76 | 1.76 | 1.76 | 1.76 |
|  | M | 1.44 | 1.44 | 1.44 | 1.44 | 1.44 | 1.44 |

*Note:* The null hypothesis is that each target group receives immigrants at random from each source group.

gration, and also that they are not excessively greater than some of the off-diagonal elements. However, it is still possible that the diagonal elements of $^4\mathbf{E}$ continue to mask some structure created by nonrandom migrations of nonnatal males. This conjecture receives some support from the association between the off-diagonal elements of $^4\mathbf{E}$ and the appropriate structure matrix. The obtained correlation of .351 and associated significance level of .059 between the off-diagonal $^4\mathbf{E}$ and $\mathbf{R}$ lend some support to this notion. This marginally nonsignificant result was based on 1,000 permutations. The diagonal elements of $^4\mathbf{E}$ continue to mask some of the structure created by nonrandom migrations of nonnatal males. Comparison of the results obtained from the $^0\mathbf{E}$ and the $^4\mathbf{E}$ matrices suggests that a male's first migration may be more constrained by which natal group he comes from than are his later migrations, which may be more random in respect to natal group for the migrant population as a whole.

For the present study we can conclude that migrations of males in the Cayo Santiago colony are structured in some way in respect to their group of origin and group of destination, in spite of the crowded conditions and the lack of geographic distance separating the social groups. Although the lineal effect associated with group fissions may be the chief cause of inter-group genetic differentiation (Cheverud and Dow 1985), structure in the migration patterns of males may also mitigate somewhat the homogenizing effect of migration.

## PREFERENTIAL MIGRATIONS SUGGESTED

As there appears to be some structure in the migrations of males among the social groups of the Cayo Santiago colony, a closer look at the particulars is suggested. The matrix $\mathbf{G} = \{g_{ij}\}$ is constructed by multiplying the elements in the $\mathbf{E}$ matrix by the appropriate element of the vector $\mathbf{n}$:

$$g_{ij} = e_{ij}\, n_j \,. \tag{15.14}$$

The elements of $\mathbf{G}$ are the total monkey-years to be lived in the $i$th group by migrants from the $j$th group. The matrix $\mathbf{Y}$ is formed by setting the elements of $\mathbf{G}$ equal to the proportion each contributes to its respective row total:

$$y_{ij} = \frac{g_{ij}}{\sum\limits_j g_{ij}} \,. \tag{15.15}$$

The off-diagonal elements of $\mathbf{Y}$ are, for each row group, the immigration rates from each column group, excepting those individuals that enter a group from the status EGM, as noted earlier.

Tables 15.6 and 15.7 show respectively the matrices $^0\mathbf{G}$ and $^0\mathbf{Y}$. Inspection of these two tables reveals the close relationship between Groups F and M: 27% of Group M's monkey-years are lived by monkeys originating in

TABLE 15.6 Total Monkey-Years Lived in Target Group by Individuals Originating in Source Group ($^0$G)

|  |  | SOURCE GROUP | | | | | |
|---|---|---|---|---|---|---|---|
|  |  | EGM | F | I | J | L | M |
|  | EGM | 0.00 | 12.17 | 8.06 | 11.08 | 8.24 | 7.39 |
|  | F | 0.00 | 48.15 | 9.02 | 10.43 | 8.22 | 7.97 |
| TARGET | I | 0.00 | 8.03 | 42.18 | 5.93 | 4.05 | 6.14 |
| GROUP | J | 0.00 | 9.58 | 6.17 | 43.30 | 4.59 | 4.91 |
|  | L | 0.00 | 9.61 | 8.20 | 13.41 | 42.66 | 5.34 |
|  | M | 0.00 | 17.77 | 4.47 | 5.87 | 4.30 | 34.12 |

Group F, a consequence of the recent formation of Group M by division of Group F. Group J contributes disproportionately to Group L, although no immediate explanation comes to mind. An examination of longitudinal migration histories (Chepko-Sade 1982) suggests that preferential patterns of migration develop and persist for several years.

The $^4$G and $^4$Y matrices are shown in tables 15.8 and 15.9. Inspection of these two tables reveals essentially the same pattern of migration revealed by the previous two, with a decrease in monkey-years contributed by natal animals and a corresponding increase in the proportion contributed by immigrants.

COMMENTS

Recall that no account is taken of the prior history of individuals in this method. The effect of individuals' personal histories might well be to impose more structure on migration patterns than is apparent in either the r vectors of immigration rates (table 15.2) or in the E matrices. Such could come about, for instance, if immigrants from a particular group to a particular row group tended to stay a longer or a shorter time in a particular target group than the expected stay based upon only the intergroup transition rates as used in the present study.

Ober et al. (1984) present a very detailed study of the month by month changes in gene frequencies within the social groups on Cayo Santiago and

TABLE 15.7 Proportion of Monkey-Years Lived in Each Target Group by Monkeys Originating in Each Source Group ($^0$Y)

|  |  | SOURCE GROUP | | | | | |
|---|---|---|---|---|---|---|---|
|  |  | EGM | F | I | J | L | M |
|  | EGM | 0.00 | .26 | .17 | .24 | .18 | .16 |
|  | F | 0.00 | .57 | .11 | .12 | .10 | .10 |
| TARGET | I | 0.00 | .12 | .64 | .09 | .06 | .09 |
| GROUP | J | 0.00 | .14 | .09 | .63 | .07 | .07 |
|  | L | 0.00 | .12 | .10 | .17 | .54 | .07 |
|  | M | 0.00 | .27 | .07 | .09 | .06 | .51 |

TABLE 15.8  Total Monkey-Years Lived in Target Group by Individuals Originating in Source Group ($^4\mathbf{G}$)

|  |  | SOURCE GROUP | | | | | |
|---|---|---|---|---|---|---|---|
|  |  | EGM | F | I | J | L | M |
| TARGET GROUP | EGM | 0.00 | 15.22 | 9.79 | 13.60 | 11.39 | 10.94 |
|  | F | 0.00 | 23.90 | 10.82 | 12.88 | 10.63 | 11.45 |
|  | I | 0.00 | 9.66 | 20.45 | 6.92 | 5.74 | 8.18 |
|  | J | 0.00 | 11.75 | 7.58 | 17.00 | 6.51 | 7.27 |
|  | L | 0.00 | 11.95 | 9.55 | 15.41 | 22.37 | 7.86 |
|  | M | 0.00 | 14.57 | 5.35 | 7.31 | 5.91 | 17.21 |

show that, over the short term, intergroup migration by males has a major effect. In contrast to the microanalysis of that study, the present chapter illustrates a method of compiling the specific observations into the most general summary of life-history events, the life table, and extending it to the multidimensional case of the subdivided population, an essential feature for the development of a mature theory. The application of the method to a relatively complete set of observations reveals unexpected difficulties that offer opportunity for further mathematical and theoretical extensions.

## SUMMARY

Movements between social groups by males among free-ranging rhesus monkeys offer a model of migration in age-structured populations. In contrast to the usual population genetical model, in which migration and reproduction occur once during the individual's life cycle, and in which generations are discrete and do not overlap, rhesus monkey males may reside in several groups during their 15-year-long reproductive span, and during their residence in each group may mate in successive breeding seasons. We partition the total life expectancy of males among the social groups using multistate life tables recently developed by demographers and find the lifetime and per generation individual migration rates. The concept of migration is thus generalized to a distribution of expectation of

TABLE 15.9  Proportion of Monkey-Years Lived in Each Target Group by Monkeys Originating in Each Source Group ($^4\mathbf{Y}$)

|  |  | SOURCE GROUP | | | | | |
|---|---|---|---|---|---|---|---|
|  |  | EGM | F | I | J | L | M |
| TARGET GROUP | EGM | 0.00 | .25 | .16 | .22 | .19 | .18 |
|  | F | 0.00 | .34 | .16 | .18 | .15 | .16 |
|  | I | 0.00 | .19 | .40 | .14 | .11 | .16 |
|  | J | 0.00 | .23 | .15 | .34 | .13 | .15 |
|  | L | 0.00 | .18 | .14 | .23 | .33 | .12 |
|  | M | 0.00 | .29 | .11 | .15 | .12 | .34 |

time of residency among the breeding units. Using a nonparametric, quadratic assignment procedure we show that life expectancy is distributed nonrandomly among these units in respect to group of birth.

## ACKNOWLEDGMENTS

This work was supported in part by grants from the National Science Foundation and the National Institute of Mental Health to D. S. Sade through Northwestern University. The Cayo Santiago colony was supported by a contract from the Division of Research Resources, National Institutes of Health, to the University of Puerto Rico. We thank the participants in the symposium for their comments, and especially Peter Smouse for his extensive and detailed criticisms.

## APPENDIX

$^X\mathbf{M}$ matrices are compiled for the years 1973–76 as discussed in Sade et al.1984. The column and row labels have been omitted to save space but are the same as in the tables in this chapter. In order, from column 1 to 6 and from row 1 to 6, the labels are EGM, F, I, J, L, M. The direction of migration is from the column ($j$th) group to the row ($i$th) group.

Age Class 0 to 1 Year

| | | | | | |
|---|---|---|---|---|---|
| 0.000 | 0.000 | 0.000 | 0.000 | 0.000 | 0.000 |
| 0.000 | .280 | 0.000 | 0.000 | 0.000 | 0.000 |
| 0.000 | 0.000 | .067 | 0.000 | 0.000 | 0.000 |
| 0.000 | 0.000 | 0.000 | .184 | 0.000 | 0.000 |
| 0.000 | 0.000 | 0.000 | 0.000 | .143 | 0.000 |
| 0.000 | −.120 | 0.000 | 0.000 | 0.000 | .300 |

Age Class 1 to 2 Years

| | | | | | |
|---|---|---|---|---|---|
| 0.000 | 0.000 | 0.000 | 0.000 | 0.000 | 0.000 |
| 0.000 | .151 | 0.000 | 0.000 | 0.000 | 0.000 |
| 0.000 | 0.000 | .095 | 0.000 | 0.000 | 0.000 |
| 0.000 | 0.000 | 0.000 | .032 | 0.000 | 0.000 |
| 0.000 | 0.000 | 0.000 | 0.000 | .095 | 0.000 |
| 0.000 | −.121 | 0.000 | 0.000 | 0.000 | .046 |

Age Class 2 to 3 Years

| | | | | | |
|---|---|---|---|---|---|
| 0.000 | 0.000 | 0.000 | 0.000 | 0.000 | 0.000 |
| 0.000 | .083 | 0.000 | 0.000 | −.077 | 0.000 |
| 0.000 | 0.000 | 0.000 | −.040 | 0.000 | −.056 |
| 0.000 | 0.000 | 0.000 | .040 | 0.000 | 0.000 |
| 0.000 | 0.000 | 0.000 | 0.000 | .154 | 0.000 |
| 0.000 | −.083 | 0.000 | 0.000 | 0.000 | 0.000 |

Age Class 3 to 4 Years

| | | | | | |
|---|---|---|---|---|---|
| .875 | −.051 | −.048 | −.083 | −.143 | 0.000 |
| −.250 | .359 | −.095 | −.042 | −.048 | −.111 |
| −.125 | −.051 | .429 | 0.000 | 0.000 | −.111 |
| −.125 | −.128 | 0.000 | .554 | 0.000 | 0.000 |
| −.250 | −.026 | −.190 | −.429 | .287 | 0.000 |
| −.125 | −.077 | −.048 | 0.000 | −.048 | .333 |

Age Class 4 to 5 Years

| | | | | | |
|---|---|---|---|---|---|
| .834 | 0.000 | −.105 | −.048 | −.100 | −.077 |
| −.167 | .531 | −.053 | −.190 | 0.000 | −.154 |
| −.333 | −.088 | .264 | 0.000 | −.050 | −.077 |
| −.167 | −.147 | −.053 | .476 | −.050 | −.077 |
| −.167 | −.118 | −.053 | 0.000 | .355 | −.077 |
| 0.000 | −.147 | 0.000 | −.190 | −.050 | .462 |

Age Class 5 to 6 Years

| | | | | | |
|---|---|---|---|---|---|
| .642 | −.158 | 0.000 | −.300 | −.071 | −.200 |
| −.214 | .474 | −.214 | −.050 | 0.000 | 0.000 |
| −.071 | 0.000 | .428 | −.050 | 0.000 | 0.000 |
| −.143 | −.158 | −.143 | .503 | 0.000 | −.100 |
| −.214 | −.053 | 0.000 | −.050 | .356 | 0.000 |
| 0.000 | −.105 | −.071 | 0.000 | −.071 | .400 |

Age Class 6 to 7 Years

| | | | | | |
|---|---|---|---|---|---|
| .667 | −.286 | 0.000 | −.333 | −.263 | −.182 |
| −.111 | .714 | 0.000 | 0.000 | −.053 | 0.000 |
| −.056 | 0.000 | .167 | 0.000 | 0.000 | 0.000 |
| −.056 | −.071 | −.167 | .555 | −.053 | −.091 |
| −.333 | 0.000 | 0.000 | −.111 | .369 | 0.000 |
| −.111 | −.286 | 0.000 | −.111 | 0.000 | .373 |

Age Class 7 to 8 Years

| | | | | | |
|---|---|---|---|---|---|
| .738 | −.308 | −.222 | −.454 | −.222 | −.429 |
| −.318 | .639 | 0.000 | 0.000 | 0.000 | 0.000 |
| −.136 | −.077 | .444 | 0.000 | 0.000 | 0.000 |
| −.182 | −.077 | −.111 | .565 | 0.000 | 0.000 |
| −.046 | 0.000 | 0.000 | 0.000 | .555 | 0.000 |
| 0.000 | −.077 | −.111 | 0.000 | 0.000 | .429 |

Age Class 8 to 9 Years

| | | | | | |
|---|---|---|---|---|---|
| .942 | −.143 | 0.000 | −.125 | −.250 | −.250 |
| .235 | .500 | 0.000 | −.125 | −.083 | −.125 |
| −.059 | 0.000 | 0.000 | −.125 | 0.000 | 0.000 |
| −.118 | 0.000 | 0.000 | .750 | 0.000 | 0.000 |
| −.412 | −.071 | 0.000 | −.125 | .333 | 0.000 |
| −.059 | −.286 | 0.000 | 0.000 | 0.000 | .661 |

Age Class 9 to 10 Years

| | | | | | |
|---|---|---|---|---|---|
| .752 | −.389 | 0.000 | 0.000 | −.353 | −.286 |
| −.316 | .618 | −.143 | −.333 | 0.000 | 0.000 |
| −.053 | 0.000 | .429 | 0.000 | −.059 | 0.000 |
| 0.000 | 0.000 | 0.000 | .660 | −.059 | 0.000 |
| −.158 | −.111 | −.286 | −.333 | .744 | −.143 |
| −.158 | −.056 | 0.000 | 0.000 | −.059 | .429 |

Age Class 10 to 11 Years

| | | | | | |
|---|---|---|---|---|---|
| .819 | −.385 | −.167 | −.250 | 0.000 | −.200 |
| −.546 | .616 | 0.000 | 0.000 | −.400 | 0.000 |
| −.091 | −.154 | .167 | 0.000 | 0.000 | 0.000 |
| 0.000 | −.077 | 0.000 | .250 | −.200 | 0.000 |
| −.091 | 0.000 | 0.000 | 0.000 | .600 | −.200 |
| −.091 | 0.000 | 0.000 | 0.000 | 0.000 | .600 |

.Age Class 11 to 12 Years

| | | | | | |
|---|---|---|---|---|---|
| .739 | 0.000 | 0.000 | −.125 | −.167 | −.333 |
| 0.000 | .167 | 0.000 | 0.000 | 0.000 | 0.000 |
| −.143 | 0.000 | .500 | 0.000 | 0.000 | 0.000 |
| −.143 | 0.000 | −.250 | .500 | 0.000 | −.167 |
| −.286 | 0.000 | 0.000 | −.125 | .667 | −.167 |
| 0.000 | 0.000 | 0.000 | −.125 | −.333 | .667 |

Age Class 12 to 13 Years

| | | | | | |
|---|---|---|---|---|---|
| .858 | 0.000 | −.167 | −.333 | −.500 | 0.000 |
| −.286 | .333 | −.333 | 0.000 | 0.000 | −.333 |
| 0.000 | −.111 | .834 | 0.000 | 0.000 | 0.000 |
| −.286 | −.222 | −.167 | .666 | 0.000 | 0.000 |
| −.143 | 0.000 | 0.000 | −.333 | .500 | 0.000 |
| −.143 | 0.000 | 0.000 | 0.000 | 0.000 | .333 |

Age Class 13 to 14 Years

| | | | | | |
|---|---|---|---|---|---|
| 1.000 | 0.000 | 0.000 | −.250 | 0.000 | −1.000 |
| 0.000 | 0.000 | 0.000 | 0.000 | 0.000 | 0.000 |
| −.250 | 0.000 | 0.000 | 0.000 | 0.000 | 0.000 |
| −.500 | 0.000 | 0.000 | .250 | 0.000 | 0.000 |
| −.250 | 0.000 | 0.000 | 0.000 | .500 | 0.000 |
| 0.000 | 0.000 | 0.000 | 0.000 | 0.000 | 1.000 |

Age Class 14 to 15 Years

| | | | | | |
|---|---|---|---|---|---|
| .666 | −.143 | −.500 | 0.000 | 0.000 | 0.000 |
| 0.000 | .596 | 0.000 | −.500 | 0.000 | −1.000 |
| 0.000 | 0.000 | .500 | 0.000 | 0.000 | 0.000 |
| −.333 | −.143 | 0.000 | .500 | 0.000 | 0.000 |
| −.333 | 0.000 | 0.000 | 0.000 | .500 | 0.000 |
| 0.000 | −.143 | 0.000 | 0.000 | 0.000 | 1.000 |

Age Class 15 Years and Older

| | | | | | |
|---|---|---|---|---|---|
| .572 | −.087 | 0.000 | 0.000 | −.670 | 0.000 |
| −.286 | .175 | 0.000 | −.250 | 0.000 | 0.000 |
| 0.000 | 0.000 | 0.000 | 0.000 | 0.000 | 0.000 |
| 0.000 | −.044 | 0.000 | .250 | 0.000 | 0.000 |
| −.286 | 0.000 | 0.000 | 0.000 | .667 | 0.000 |
| 0.000 | 0.000 | 0.000 | 0.000 | 0.000 | 0.000 |
| | | 0.000 | | | |

# REFERENCES

Baker, F., and L. Hubert. 1981. The analysis of social interaction data. *Sociological Methods and Research* 9:339–61.

Charlesworth, B. 1980. *Evolution in age-structured populations.* Cambridge: Cambridge University Press.

Chepko-Sade, B. D. 1982. Role of males in group fission in *Macaca mulatta*. Ph.D. diss., Northwestern University.

Chepko-Sade, B. D., and T. J. Olivier. 1979. Coefficient of genetic relationship and the probability of intragenealogical fission in *Macaca mulatta. Behavioral Ecology and Sociobiology* 5:263–78.

Chepko-Sade, B. D., and D. S. Sade. 1979. Patterns of group splitting within matrilineal kinship groups: A study of social group structure in *Macaca mulatta* (Cercopithecidae: Primates). *Behavioral Ecology and Sociobiology* 5:67–86.

Cheverud, J. M. 1981. Variation in highly and lowly heritable morphological traits among social groups of rhesus macaques *(Macaca mulatta)* on Cayo Santiago. *Evolution* 35:75–83.

Cheverud, J. M., J. Buettner-Janusch, and D. S. Sade. 1978. Social group fission and the origin of intergroup genetic differentiation among the rhesus monkeys of Cayo Santiago. *American Journal of Physical Anthropology* 49:449–56.

Cheverud, J., and M. Dow. 1985. An autocorrelation analysis of the effect of lineal fission on genetic variation among social groups. American Journal of Physical Anthropology 49:449–56.

Colvin, J. 1983. Influences of the social situation on male emigration. In *Primate social relationships*, ed. R. A. Hinde. Sunderland, Mass.: Sinauer.

Dow, M. 1985. Non-parametric inference procedures for multi-state life table analysis. *Journal of Mathematical Sociology* 112:245–63.

Duggleby, C. R. 1978. Blood group antigens and the population genetics of *Macaca mulatta* on Cayo Santiago. I. Genetic differentiation of social groups. *American Journal of Physical Anthropology* 48:35–40.

Hoem, J., and U. F. Jensen. 1982. Multi-state life table methodology: A probabilist critique. In *Multidimensional mathematical demography*, ed. K. Land and A. Rogers. New York: Academic Press.

Hubert, L., and F. Baker. 1978. Evaluating the conformity of sociometric measurements. *Psychometrika* 43:31–41.

Hubert, L., and R. G. Golledge. 1981. A heuristic method for the comparison of related structures. *Journal of Mathematical Psychology* 23:214–26.

Hubert, L., and J. Schultz. 1976. Quadratic assignment as a general data analysis strategy. *British Journal of Mathematical and Statistical Psychology* 32:241–54.

Keyfitz, N. 1968. *Introduction to the mathematics of population.* Reading, Mass.: Addison-Wesley.

———. 1980. Multidimensionality in population analysis. In *Sociological methodology,* ed. S. Leinhardt, 191–218. New York: Jossey-Bass.

Mantel, N. 1967. The detection of disease clustering: A generalized regression approach. *Cancer Research* 27:209–20.

Melnick, D., and K. Kidd. 1983. The genetic consequences of social group fission in a wild population of rhesus monkeys *(Macaca mulatta). Behavioral Ecology and Sociobiology* 12:229–36.

Ober, C., T. J. Olivier, D. S. Sade, J. M. Schneider, J. Cheverud, and J.Buettner-Janusch. 1984. Demographic components of gene frequency change in free-ranging rhesus macaques on Cayo Santiago. *American Journal of Physical Anthropology* 64:223–31.

Pearl, M. C. 1982. Networks of social relations among Himalayan rhesus monkeys *(Macaca mulatta).* Ph.D. diss., Yale University.

Pearl, M. C., and S. R. Schulman. 1983. Techniques for the analysis of social structure in animal societies. In *Advances in the study of behavior,* vol. 13. New York: Academic Press.

Rogers, A., and J. Ledent. 1976. Increment-decrement life tables: A comment. *Demography* 13:287–90.

Sade, D. S. 1964. Seasonal cycle in size of testes of free-ranging *Macaca mulatta. Folia Primatalogica* 2:171–80.

———. 1980a. Can "fitness" be measured in primate populations? In *Dahlem workshop report on evolution of social behavior: Hypotheses and empirical tests,* ed. H. Markel. Weinheim: Verlag Chemie.

———. 1980b. Population biology of free-ranging rhesus monkeys on Cayo Santiago, Puerto Rico. In *Biosocial mechanisms of population regulation,* ed. M. N. Cohen, R. S. Malpass, and H. G. Klein. New Haven: Yale University Press.

———. 1984. Rank, tenure, and paternity among free-ranging rhesus monkeys. Paper presented at the Midwest Regional Animal Behavior Meetings, Carbondale, Ill., April 1984.

Sade, D. S., B. D. Chepko-Sade, and J. M. Schneider. 1977. Paternal exclusions among free-ranging rhesus monkeys on Cayo Santiago. Paper presented at the annual meeting of the Animal Behavior Society, University Park, Pa., June 1977.

Sade, D. S., B. D. Chepko-Sade, J. M. Schneider, S. S. Roberts, and J. T. Richtsmeier. 1985. *Basic demographic observations on free-ranging rhesus monkeys.* New Haven: Human Relations Area Files.

Sade, D. S., M. Dow, and B. D. Chepko-Sade. 1984. Multi-state life tables for free-ranging male rhesus monkeys. *Collegium Antropologicum* 8:237–47.

Sade, D. S., D. L. Rhodes, B. D. Chepko-Sade, J. M. Schneider, C. Duggleby, J. Buettner-Janusch, T. J. Olivier, and C. Ober. 1982. High rank of mother favors early reproduction by sons and daughters among free-ranging rhesus monkeys. Paper presented at the Midwest Regional Animal Behavior Meetings, Champaign, Ill., February 1982.

Sade, D. S., J. Schneider, A. Figueroa, J. Kaplan, K. Cushing, P. Cushing, J. Dunaif, T. Morse, D. Rhodes, and M. Stewart. 1976. Population dynamics in relation to social structure on Cayo Santiago. *Yearbook of Physical Anthropology* 20:253–62.

Schoen, R. 1975. Constructing increment-decrement life tables. *Demography* 12: 313–24.

Stern, B. R., and D. G. Smith. 1984. Sexual behaviour and paternity in three captive groups of rhesus monkeys (*Macaca mulatta*). *Animal Behaviour* 32:23–32.

van Wagenen, G. 1972. Vital statistics from a breeding colony: Reproduction and pregnancy outcome in *Macaca mulatta. Journal of Medical Primatology* 1:3–28.

Zamboni, L., C. H. Conaway, and L. Van Pelt. 1974. Seasonal changes in production of semen in free-ranging rhesus monkeys. *Biology of Reproduction* 11:251–67.

# V. MATHEMATICAL MODELS OF POPULATION STRUCTURE

# 16. A Model Predicting Dispersal Distance Distributions

*Peter M. Waser*

What proportion of juveniles should leave their natal site, and how far should those that disperse move before settling? Traditional explanations of dispersal have dealt with these two questions only in relative terms. For instance, the argument that juveniles should be more likely to remain on the natal site in ecologically saturated habitats does not specify how many are expected to disperse under any particular set of conditions. The suggestion that inbreeding avoidance forces members of one sex to disperse farther than members of the other leaves unasked the question How much farther? Our inability to answer these questions more precisely is a serious limitation, since the distribution of dispersal distances that characterizes a population has obvious consequences for its genetics and its social behavior.

In this chapter, I outline a model predicting the expected distribution of dispersal distances as a function of population demography. The model, based on an argument initially formulated by Murray (1967; see also French 1971; Waser 1985), assumes that young animals disperse when it is in their individual interests to do so, that dispersal per se has risks or costs, and that competition for space, particularly breeding space, is the factor that prevents philopatry in most mammals. Comparison of expected dispersal distances with data from the banner-tailed kangaroo rat (Jones, chap. 8, this vol.) clarifies dispersal dynamics in this species and also suggests that the model has general usefulness in predicting dispersal distance distributions.

## THE MODEL

Assume that young animals approaching reproductive maturity move to the first uncontested site they find, but no farther. At that time, there will be some probability $t$ that their own home range will be uncontested—for instance, if their same-sex parent is dead and they have no same-sex sibs. Since dispersal-age animals under those conditions do not disperse, $t$ will also be the probability of nondispersal or philopatry. Assume also that uncontested sites can arise anywhere with equal likelihood. An animal that leaves its natal range will then find each new home range it encounters uncontested with the same probability $t$. The likelihood that it can settle in the $i$th home range it traverses is $p_i = t(1 - t)^i$.

If the dispersing animal leaves its natal range in a random direction and searches in a straight line for an uncontested site, the probability that its straight-line dispersal distance is $n$ is simply

$$p_n = t (1 - t)^n, \tag{16.1}$$

where $n$ is measured in multiples of the distance between adjacent home range centers. At the opposite extreme, if the dispersing animal searches so as to find the nearest home range in any direction from its natal site, and if home ranges are close packed so that each is surrounded by six neighbors, the probability that its dispersal distance is $n$ is

$$p_n = [1 - (1 - t)^{6n}] [1 - t]^{1 + 3n^2 - 3}. \tag{16.2}$$

The derivations and assumptions underlying these equations are discussed more fully in Waser (1985).

## COMPARISON WITH DATA

Jones (chap. 8, this vol.) tabulates male and female natal-to-breeding den distances for banner-tailed kangaroo rats *Dipodomys spectabilis* in 50 m increments. Since the mean distance between neighboring adults' dens is close to 50 m, these data can be compared directly with the dispersal distance distributions predicted by equations (16.1) and (16.2). Any animal moving less than 50 m either has not dispersed or has dispersed to a nearer-than-average den.

To estimate $t$, I assume that the 79% of males and 77% of females that bred less than 50 m from their natal mounds (Jones, chap. 8, this vol.) are all the animals whose natal home ranges were uncontested when they reached reproductive maturity. With $t = .79$ (for males) and $t = .77$ (for females), equation (16.1) appears to predict the overall distributions of dispersal distances rather closely (fig. 16.1). For males, Kolmogorov-Smirnov $D = .03$ ($p > .05$); for females, $D = .07$ ($p > .05$); for both sexes, $D = .06$ ($p > .05$) (one-sample test for an intrinsic hypothesis, Sokal and Rohlf 1981). Equation (16.2) does slightly less well at matching the data, since it does not predict a tail of animals dispersing more than 2 home ranges; $D$ is .07 for males ($p > .05$), .09 for females ($p \approx .05$), and .08 for the sexes combined ($p \approx .01$).

## DISCUSSION

If equation (16.1) is as successful with other species as with these *Dipodomys*, it provides a useful answer to the second question posed in the

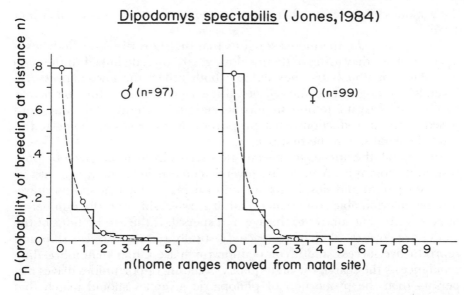

FIGURE 16.1.  Observed (histograms) compared with expected (open circles) dispersal distance distributions for *D. spectabilis*. Expected distributions are calculated from equation (16.1), with *t* equal to the observed probability of dispersing less than 50 m. Data are from Jones 1984, chap. 8.

introduction: How far are individuals expected to disperse? The match of expectations to observations is equally good for at least two other mammalian data sets with sample sizes exceeding 60 for each sex—that of Dice and Howard (1951) for *Peromyscus maniculatus* (Waser 1985) and that of Burt (1940) for *P. leucopus* (B. Keane, pers. comm.).

For these species at least, the assumptions underlying equation (16.1) appear to be valid. Young animals appear to be moving no farther than necessary to find an uncontested site. These distances would be "optimal" if the risk (or the time or energy cost) of moving away from the natal site outweighs all benefits of doing so other than the increase in survival or reproductive success that follows from living on an uncontested rather than a contested site.

However, there are probably other mammals for which this approach will be less successful at predicting precise dispersal distance distributions. Even in the *D. spectabilis* data, a scattering of long-distance dispersers hints that there could be processes beyond competition for breeding space that influence dispersal distances. Inbreeding avoidance could be such a process, as could searching for high-quality mates or territories (other failures of the model's assumptions, such as nonuniform spacing of home ranges, or a bias in favor of resident juveniles when they compete with dispersers, could also lead to deviations). The model could therefore be used to detect the workings of processes other than competition in driving dispersal (Waser 1985).

For *D. spectabilis*, the model suggests that the short distances both sexes disperse when they *do* leave the natal range are directly linked to the high probability of philopatry they exhibit; both reflect the low chances of a juvenile's finding an uncontested site. But what determines, for a particular species, how easy it is to find an uncontested site? In terms of the questions posed in the introduction, what proportion of juveniles are expected to leave the natal site in the first place?

In general, the answer to this question seems likely to be given by population demography. A maturing juvenile contests its home range with its same-sex parent and sibs; its likelihood of being philopatric should therefore be determinable from data on adult survivorship and the number of same-sex sibs that survive to the age of dispersal. If the assumptions of the model are correct—if sharing space drastically reduces survivorship and reproductive success, and if competition for space rather than inbreeding avoidance or the search for high-quality breeding opportunities drives dispersal—then the proportion of philopatric juveniles should match that calculated from survivorship and fecundity.

For *D. spectabilis*, the demographic calculations suggest some interesting complications. Unlike most mammals, and because of its dependence on large, seed-cache-containing mounds (Jones, chap. 8, this vol.), *D. spectabilis* juveniles of both sexes must compete with each other and with the mother (but not the father) to remain in the natal mound. Adult female mortality between one breeding season to the next is .55, independent of age, in the *D. spectabilis* study population (unpub. data). The number of sibs that survive to breeding age (1 year) is approximately 1.7. Assuming that sibs whose mother has died compete on an equal basis for the natal home range, any particular juvenile's chance of remaining there in its first breeding season should be .55/1.7 or .32, independent of sex.

Clearly this estimate is a poor match to the earlier estimates of $t$ (0.77 and 0.79), the likelihood that the natal site is "uncontested." How can we account for the difference?

Some *D. spectabilis* females have secondary mounds within their home ranges, either acquired when the original owner died or constructed as "satellite" mounds by the female and her offspring (Jones 1984). Such mounds are often occupied by juveniles. Jones (chap. 8, this vol.) tabulates the proportion of juveniles that remained not only in their natal home ranges but also in their natal mounds; 40% of males and 41% of females did so. The balance of the philopatric animals, 37 males and 35 females, moved to secondary or other mounds known or suspected to be within the natal home range.

The 41% of juveniles that remained in the natal den is still considerably larger than the 32% expected solely as a consequence of parental death.

Occasionally, a juvenile is tolerated in the natal mound by the parent even past reproductive maturity. Though this option is rare, we have used radio tracking to confirm one case of den sharing by mother and son through the son's first breeding season and until the mother's death when the son was 18 months old (Jones 1984). Moreover, under favorable conditions juvenile females occasionally come into reproductive condition, breed, and raise a litter in the same season they are born, even though they are still sharing the mound with their mother and sibs (unpub. data).

A more significant reason why juveniles find the natal mound uncontested more often than expected from demography alone is evident from Jones's data on adult movements. In the study population, 22 females moved from one mound to another (within 50 m) between breeding seasons; an additional 6 moved more than 50 m. Like juveniles, these females generally moved to secondary mounds, continuing to use at least part of the same home range (Jones 1982). Most, though not all, had offspring that survived to breed in the natal mound. The number of females that moved out of their original mounds, potentially leaving them to one of their offspring (28), is similar to, but somewhat larger than, the number of mounds expected from demography alone (9% of 196, or about 18).

Thus, comparison of the proportion of *D. spectabilis* juveniles breeding less than 50 m from the natal site suggests three ways in which a young animal might find its natal home range "uncontested" even if it is not empty of conspecifics. All three involve what might be termed "philopatry by parental consent" (Waser and Jones 1983). First, a young animal might be forced to leave its natal den but be tolerated by its mother and sibs elsewhere within the natal home range. Second, it might be allowed to share the natal den into adulthood. Third, the mother might abdicate, moving to a nearby area and leaving the natal den to its offspring. Such phenomena seem likely to occur only in populations whose members' chances of successful dispersal are so low that philopatry is in the parent's interest, as well as the juvenile's, conditions that likely characterize these *D. spectabilis* (Jones 1984, 1986).

What relevance, if any, does this model have for the genetics of mammalian populations? Most generally, it suggests a link between population demography and effective population size. To the extent that $N_e$ is influenced by the distances between parent and offspring breeding sites (Wright 1943; Chepko-Sade and Shields et al., chap. 19, this vol.) and that the assumptions of this dispersal model are met, it can be used to predict $N_e$ for a particular population from survivorship and fecundity data. In general, the model suggests that populations with higher adult survival rates, because they have lower rates of home range turnover, will be characterized by greater dispersal distances and thus greater values of $N_e$.

Comparison of *D. spectabilis* dispersal tendencies with those of other species discussed in this volume also suggests two conditions under which mammalian dispersal distances, and $N_e$ values, may differ from those predicted by this model. Animals larger than *Dipodomys* may be at less risk from predation, starvation, and exposure during dispersal, and thus might be able to search beyond the first uncontested site. In general such species are also longer lived than *Dipodomys* and thus more likely to risk close inbreeding if they are philopatric. Both factors might be expected to increase dispersal distances over the expectations of this model and thus to decrease population viscosity. On the other hand, philopatry by parental consent is likely to decrease dispersal distances and $N_e$ not only for bannertailed kangaroo rats but wherever chances of successful dispersal are low, either because of specialized den or habitat requirements (cf. Smith, chap. 9, this vol.) or, in longer-lived animals, because of low home range turnover rates.

## ACKNOWLEDGMENTS

This work was supported by the National Science Foundation (DEB 81-12773).

## REFERENCES

Burt, W. H. 1940. Territorial behavior and populations of some small mammals in Southern Michigan. *Miscellaneous Publications in Zoology, University of Michigan* 45: 1–58.

Dice, L. R., and W. E. Howard. 1951. Distance of dispersal by prairie deermice from birthplaces to breeding sites. *Contributions from the Laboratory of Vertebrate Biology, University of Michigan* 50:1–15.

French, N. R. 1971. Simulation of dispersal in desert rodents. In *Statistical ecology*, ed. G. P. Patil, E. C. Pielou, and W. E. Waters, 3:367–75. University Park, Pa.: Pennsylvania State University Press.

Jones, W. T. 1982. Natal nondispersal in kangaroo rats. Ph.D. diss., Purdue University.

————. 1984. Natal philopatry in bannertailed kangaroo rats. *Behavioral Ecology and Sociobiology* 15:151–55.

————. 1986. Survivorship in philopatric and dispersing kangaroo rats (*Dipodomys spectabilis*). *Ecology* 67:202–7.

Murray, B. G., Jr. 1967. Dispersal in vertebrates. *Ecology* 48:975–78.

Sokal, R. R., and Rohlf, F. O. 1981. *Biometry.* San Francisco: W. H. Freeman.

Waser, P. M. 1985. Does competition drive dispersal? *Ecology* 66:1170–75.

Waser, P. M., and W. T. Jones. 1983. Natal philopatry among solitary mammals. *Quarterly Review of Biology* 50:355–90.

Wright, S. 1943. Isolation by distance. *Genetics* 20:114–38.

# 17. Inferences on Natural Population Structure from Genetic Studies on Captive Mammalian Populations

*Alan R. Templeton*

Wright (1932, 1941) long ago championed the idea that population structure (deme sizes, patterns of gene flow, and system of mating) plays an active role in both microevolution and speciation. For this reason, studies on population structure play a central role in understanding the evolutionary biology of any group. The study of population structure is a demanding one, however, subject to many ambiguities of interpretation. For example, Wilson et al. (1975) argued that such features as harem formation, small social groupings, and so forth imply that mammalian population structure is characterized by extreme population subdivision and inbreeding. Shields (1982) has also argued that natal philopatry, which is common in mammals (Waser and Jones 1983), fosters a highly subdivided, inbred population structure. Other factors that may promote subdivision and inbreeding in mammals are the role of social organization in restricting gene flow (Singleton and Hay 1983), unequal variances of reproductive success between the sexes (Hoogland and Foltz 1982), and the phenomenon of lineal fissioning (Cheney and Seyfarth 1983). In contrast, a large number of behavioral studies imply that most mammalian species are highly outbred rather than inbred (Baker 1981; Hoogland 1982; Lewin 1983; Murray and Smith 1983; Packer 1979). One difficulty with behavioral observations is that although close inbreeding can be excluded, even regular inbreeding between more distantly related animals often cannot be excluded. The cumulative effects of this more remote inbreeding can still result in a highly inbred population, despite avoidance of close inbreeding. In addition, dispersal is far more easily measured by behavioral observations than is gene flow, yet gene flow is the critical determinant of population structure. Because of these difficulties, coupling behavioral studies with surveys of genetic markers can greatly strengthen the inferences about population structure (Daly 1981; Foltz 1981; Foltz and Hoogland 1981; McCracken and Bradbury 1977; Olivier et al. 1981).

Genetic markers have also been used directly to make inferences concerning population structure. The basic idea of these studies is that population structure has certain genetic consequences, and by examining the genetics one can thereby infer the population structure. Typically, these studies utilize Wright's F-statistics (Wright 1965) to examine variability within and between populations. Although many such studies have concluded that mammals are generally outbred rather than inbred (Dracopoli et al. 1983; Melnick, Jolly, and Kidd 1984; Patton and Feder 1981; Schwartz and Armitage 1980), the theoretical work by Prout (1981) indicates that factors such as sex-dependent migration, which is common in mammals, can obscure some of these inferences. For example, when $F_{IS}$ and $F_{IT}$ values are negative, the inference is commonly made that there is a systematic avoidance of consanguineous matings (Schwartz and Armitage 1980). Nevertheless, Prout showed that such negative values could arise directly as a consequence of sex-dependent migration in a model of random mating. Another common aspect of mammalian social organization that has a large impact upon F-statistics is lineal fissioning. Large values of $F_{ST}$ are commonly used to infer a highly subdivided population structure with little gene flow between subdivisions (Wright 1965). However, Rothman, Sing, and Templeton (1974) produced a model of lineal fissioning that yields very high $F_{ST}$ values even when a population is effectively a single breeding unit. Likewise, Fix (1978) performed computer simulations that showed lineal fissioning can yield high $F_{ST}$ values despite extensive gene flow. Hence, it is very important to consider these models when interpreting $F_{ST}$ values, as has been done, for example, by Buettner-Janusch et al. (1983).

This discussion reveals some of the dangers inherent in using genetic surveys to infer population structure. Nevertheless, this technique is potentially powerful. In this chapter I outline an expansion of this approach; it has fewer ambiguities of interpretation than the standard isozyme survey approach because it focuses upon a class of genetic variation that responds directly to a species' system of mating and population structure, namely, the genetically determined response to inbreeding or outbreeding.

## THE GENETIC RESPONSE TO INBREEDING

There is a considerable theoretical and experimental literature relating to how a population responds genetically to outbreeding and inbreeding. Basic population genetic theory indicates that recessive deleterious alleles will accumulate in a large, outbreeding population. Let $u$ be the mutation rate to a recessive, deleterious allele, and let $1 - s$ be the fitness of the affected homozygote relative to the fitness of the unaffected heterozygote and wild-type homozygote. Then, the equilibrium frequency of the deleterious allele is given by $(u/s)^{1/2}$. In contrast, suppose the population is inbred with inbreeding coefficient $f$. Then the equilibrium frequency is $u/(fs)$. To see the impact of inbreeding, let $u = 10^{-6}$ and $s = 1$ ( a recessive lethal). Then, in an outbreeding population, the equilibrium frequency is $10^{-3}$, but in an inbreeding population it would be $10^{-6}/f$. Let $f = .01$, an amount of inbreeding that would be very difficult to detect in most species. Then the equilibrium frequency of the deleterious allele is $10^{-4}$, an order of magnitude less than the outbreeding frequency. As this example illustrates, the accumulation of recessive, deleterious alleles in a population is very sensitive to even small amounts of inbreeding.

Another theoretical expectation of outbreeding is that newly arisen mutants are selected primarily for their effects in heterozygotes (Fisher 1958), whereas under inbreeding both heterozygous and homozygous fitness effects are important (Moran 1962). As a result, heterotic alleles should be more common in outbreeding populations than in inbreeding ones, a prediction empirically confirmed by Moll and Stuber (1971) in maize.

Outbred populations should therefore accumulate both recessive deleterious alleles and heterotic loci. As a consequence, when inbreeding does occur among individuals derived from an outbred population, there is a decrease in fitness due to an increase in homozygosity for deleterious alleles and a decrease in heterozygosity at heterotic loci. Such a fitness decline with increasing levels of inbreeding is known as an inbreeding depression. The inbreeding depression can be quantified by the number of lethal equivalents. The number of lethal equivalents are calculated by assuming that all the fitness depression associated with inbreeding is due to homozygosity for lethal recessive alleles at independently segregating loci. This number turns out to be twice the regression coefficient of the natural logarithm of viability on the inbreeding coefficient. Let this regression slope be designated by $B$. (For a more detailed discussion of how an inbreeding depression is measured, see Morton, Crow, and Muller 1956). Of course the actual fitness decline can be due to a mixture of heterotic and recessive deleterious effects, and the recessive deleterious effects can be due to loci other than those that are absolutely lethal. Nevertheless, the number of lethal equivalents provides a convenient index to the severity of an inbreeding depression. Moreover, by estimating the intercept at $f = 0$ of the regression line of the natural logarithm of viability against inbreeding coefficient (designated hereafter by $A$), it is possible to theoretically discriminate between the relative contributions of heterotic and recessive deleterious effects to the overall inbreeding depression. This theory (Morton, Crow, and Muller 1956) shows that recessive deleterious alleles contribute very little to mortality under outbreeding (measured by $A$), but make a major contribution under inbreeding (measured by $B$). On the other hand, heterotic loci should contribute significantly to both the $A$ and $B$ terms. As a consequence, the $B/A$ ratio can serve as an indicator of the relative contribution of heterotic versus deleterious alleles to the inbreeding depression. If most of the inbreeding depression is due to heterotic alleles, the $B/A$ value is expected to be small, typically around 2 to 3. Large values of $B/A$ indicate that most of the depression is due to recessive, deleterious alleles. Unfortunately, due to some theoretical difficulties and the impact of environmental mortality, Crow (1963) concluded that a large $B/A$ ratio can still legitimately be used to infer the primary importance of deleterious alleles as the cause of an inbreeding depression, but that small $B/A$ are uninterpretable.

If a population has a history of inbreeding, the theory discussed above indicates that little or no inbreeding depression should be encountered because recessive deleterious alleles are much rarer, as are heterotic loci. This basic theoretical prediction has been confirmed in a number of studies. For example, most human populations from Europe and their North American derivatives have a long history of wide outbreeding. When inbreeding does occur in these normally outbred human populations, an inbreeding depression is generally encountered, with the number of lethal equivalents being 3 to 5 for early mortality (Morton, Crow, and Muller 1956), and 5 to 8 for mortality over the entire life span (Stine 1977). However, not all human populations are outbreeding; in India regularly 30% to 50% of Tamil marriages take place between relatives, primarily cousins, and this pattern has apparently existed for many generations. In contrast to the studies on European populations, no significant inbreeding depression is detectable in the Tamils (Rao and Inbaraj 1980). A total lack of inbreeding depression also characterizes another tribe in South India in which all matings are consanguineous (Ghosh and Majumder 1979).

Another example is provided by captive populations of the European bison or wisent (*Bison bonasus*). This species almost became extinct early in the twentieth century, experiencing a bottleneck with an effective size of no more than 12 (Slatis 1960). Extensive inbreeding occurred, and Slatis's analysis indicates that the herd initially had about 6 lethal equivalents per individual with respect to juvenile death. After a few generations of inbreeding, this inbreeding depression vanished, the number of lethal equivalents being no longer significantly different from zero. These examples provide empirical confirmation of the theoretical prediction that populations that regularly outcross should display inbreeding depressions, while normally inbred populations should display no such depression.

Although inbred populations should be free of inbreeding depression, they are more likely than outbred populations to evolve coadapted gene complexes that can yield an outbreeding depression (Shields 1982). A coadapted gene complex is one that yields phenotypes having high fitness due to strong epistatic interactions between loci. Hence, a coadapted complex depends upon the persistence of associations between genes at different loci. Outbreeding disrupts such multilocus associations, whereas inbreeding preserves them (Shields 1982). However, different inbred lines can evolve different coadapted gene complexes. When such lines are hybridized, genetic recombination disrupts the parental gene complexes yielding recombinants with low fitness in the $F_2$ or backcrosses. This phenomenon is known as an outbreeding depression.

These theoretical predictions are supported by experimental evidence. For example, Malmberg (1977) selected for proflavine resistance in T4 bacteriophage raised under conditions that either promoted extensive recombination (outbreeding) or made recombination extremely unlikely. The bacteriophage evolved resistance under both rearing conditions, but the resistance in the outbred phage was based upon an additive genetic system, whereas the resistance in the phage subjected to little or no recombination was based upon a coadapted gene complex characterized by strong epistasis. Hence, the types of genetic systems that evolve are strongly influenced by the amount of outbreeding present in the evolving population.

More extensive work on the evolution of coadapted gene complexes under inbreeding has been performed with the fruit fly, *Drosophila mercatorum*. This fruit fly normally reproduces sexually, and natural populations are characterized by levels of isozyme polymorphism and heterozygosity that are typical for the genus and indicative of a basically outbreeding population structure (Clark, Templeton, and Sing 1981). Like most other *Drosophila* (Templeton 1983), virgin females isolated directly from the natural, sexual populations can reproduce parthenogenetically in the laboratory (Templeton, Sing, and Brokaw 1976). Moreover, the mode of parthenogenetic reproduction is such that it results in total homozygosity instantaneously; that is, each parthenogenetic strain has an inbreeding coefficient of one. As expected from theory, these various parthenogenetic strains have strongly coadapted gene complexes characterized by extensive fitness epistasis (Templeton, Sing, and Brokaw 1976; Templeton 1979). These coadapted gene complexes evolve very rapidly under inbreeding, and incompatible coadapted gene complexes can evolve in different inbred strains derived from the same natural population sampled at the same time (Templeton 1979).

When incompatible strains are hybridized, an outbreeding depression occurs in the $F_2$ and backcrosses. Dominance effects generally prevent it from being expressed in the $F_1$. This outbreeding depression can be quantified in a manner analogous to that of inbreeding depression. Recall that an outbreeding depression is caused by genetic recombination between incompatible gene complexes during meiosis in the parents of the animals that actually suffer the fitness decline. Let $h$ be the hybridity coefficient that measures the proportion of the parents' genotype available for recombination between genes derived from different ancestral stocks (Templeton and Read 1984). The hybridity coefficient can be easily manipulated in the *D. mercatorum* studies through use of visible genetic markers on the various chromosomes of that species (Templeton, Sing, and Brokaw 1976). These markers were used to breed flies with levels of hybridity ranging from 1.0

to 0.4, with the original parental strains having a hybridity level of 0. Table 17.1 lists some of the data from Templeton, Sing, and Brokaw (1976) that show a strong decline in viability with increasing levels of hybridity. This decline is a clear indication of an outbreeding depression.

The theoretical and experimental results outlined earlier yield two fundamental predictions about captive populations. First, if inbreeding occurs in a captive population derived from a normally outbreeding species, inbreeding depression should be encountered but no outbreeding depression. On the other hand, if the captive population is derived from a normally inbred species, no inbreeding depression should be experienced, but if the original import animals used to establish the captive population came from different local populations with incompatible coadapted gene complexes, then outbreeding depression can be expected. Because the status of the import animals is often unknown, the absence of an outbreeding depression cannot be used to infer outbreeding in nature. These fundamental predictions can now be used to make inferences about the population structure in nature of various mammalian species based upon the response to inbreeding of recently established captive populations.

## OUTBREEDING VERSUS INBREEDING DEPRESSION IN MAMMALS

Over the last twenty years, many zoos have begun captive breeding programs for a large number of mammalian species. Motivation stems in part from the great expense of importing wild-caught animals, the desire not to collect from dwindling natural populations, and the intention to use zoos as "arks" to save species that are extinct or nearly extinct in nature. In the vast majority of these captive breeding programs, the number of original import animals was very small, so inbreeding soon occurred. Hence, there is a potentially large data base for using the types of predictions outlined in the previous section to infer the natural population structure of a large number of mammalian species.

The initial analyses of this data base indicated that mammals in general are highly outbred in nature. Inbreeding depressions were detected in

TABLE 17.1 Adult Parthenogenetic Offspring Produced per Female in Parthenogenetic Stocks of *Drosophila mercatorum,* in Hybrids between the Stocks, and in Partial Hybrids

| HYBRIDITY COEFFICIENT | OFFSPRING PER FEMALE ($N$) |
|---|---|
| 0.0 | 14.54 |
| 0.4 | 10.25 |
| 0.6 | 5.36 |
| 1.0 | 1.63 |

*Note:* The extent of hybridity is measured by $h$, the hybridity coefficient, which gives the proportion of the parental genotype derived from two different parthenogenetic strains.

almost all of the species with sufficient sample sizes, and the trends in the smaller samples were likewise indicative of inbreeding depression (Ballou and Ralls 1982; Bouman 1977; Flesness 1977; May 1980; Ralls and Ballou 1982; Ralls, Brugger, and Ballou 1979; Ralls, Brugger, and Glick 1980; Templeton and Read 1983; Treus and Lobanov 1971).

Shields (1983, pers. comm.) has recently questioned whether these studies have actually documented an inbreeding depression. He points out that detailed information concerning the import animals is generally lacking. Hence, import animals may have come from different local populations. If the species is highly inbred in nature, these different local populations could have incompatible coadapted gene complexes. Shields next points out that because many of these breeding programs are recent, most of the inbred progeny are from backcrosses to import animals or from the $F_2$ generation. If the different import animals really do have different coadapted gene complexes, then an outbreeding depression would be expected to occur in the $F_2$ and backcrosses, precisely the same generations primarily used to infer inbreeding depression. Hence, Shields argues that there is a confoundment between inbreeding and outbreeding depression in many of the studies, making the inference of inbreeding depression unwarranted at this time.

There are two basic approaches to dealing with Shields's argument. The first strategy is to obtain genetic information on the import animals. Unfortunately, the best information usually obtainable is whether or not the import animals were caught in the same general locality. Ballou and Ralls (1982) have documented "inbreeding depressions" in zoo populations established from import animals caught in the same area at the same time. In such cases, the founders most likely came from the same local breeding population, and the outbreeding depression explanation would not be applicable. However, even knowing that animals were captured in the same general area is not enough to exclude Shields's hypothesis because genetic subdivision can exist on a very fine geographical scale (Endler 1977; Levin 1984; Price and Waser 1979).

It is therefore necessary to go to a second strategy in which the effects of inbreeding and hybridity are separated statistically. This strategy was first used by Templeton and Read (1984) on a captive population of Speke's gazelle (*Gazella spekei*). All zoo populations of Speke's gazelle in North America trace their ancestry to just four import animals (one male and three females) brought into the St. Louis Zoological Garden between 1969 and 1972 (Read and Frueh 1980). As a result of this extreme founder effect, intense inbreeding was soon encountered in the captive populations and viability began to plummet. (Viability will hereafter refer to survivorship to 30 days after birth. As shown by Templeton and Read 1983 and 1984, almost all genetic deaths occurred in this time period, with later mortality having a much larger environmental component.) The drop in viability was initially interpreted by Templeton and Read (1983) as an inbreeding depression, with each import animal bearing about 6 lethal equivalents for juvenile mortality. Moreover, the $B/A$ ratio is 19.8 (from data in table 2 of Templeton and Read 1984), indicating that the inbreeding depression is primarily due to recessive deleterious alleles. However, the inbreeding coefficient and hybridity coefficient were highly correlated in these initial generations ($r^2 = 0.66$), so this initial conclusion required further examination in light of Shields's objection. Fortunately, there was sufficient complexity in the breeding patterns even in these initial generations to separate out the effects of hybridity and inbreeding upon viability. This complexity is shown in figure 17.1. As shown in that figure, a cross between two full sibs would result in an inbreeding coefficient of 1/4 and a hybridity coefficient of 1. On the other hand, a backcross to an imported parent would also result in an inbreeding coefficient of 1/4, but in this case the hybridity coefficient would be only 1/2. Similarly, a backcross to an unrelated import animal would still have a hybridity coefficient of 1/2, but now the inbreeding coefficient would be 0.

Utilizing these and other crosses, it is possible to fix the level of inbreeding and examine the impact of varying levels of hybridity, and vice versa. The results of this type of analysis are shown in table 17.2 and 17.3 (adapted from Templeton and Read 1984). As seen from table 17.2, hybridity had no effect upon viability in noninbred offspring, a result inconsistent with the outbreeding depression hypothesis. Hybridity did have a significant effect (at the 5% level) upon viability of inbred offspring, but the effect was to *increase* viability, not decrease it! As detailed by Templeton and Read (1984), this beneficial effect of hybridity in the inbred animals is exactly what is expected to occur when an inbred line is being selected from outbred ancestors. On the other hand, table 17.3 clearly shows that inbreeding had

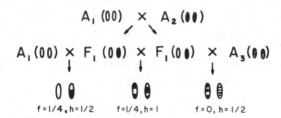

FIGURE 17.1. Some hypothetical crosses illustrating the distinction between the inbreeding coefficient, $f$, and the hybridity coefficient, $h$. $A_1$, $A_2$, and $A_3$ represent import animals, and a representative chromosome is indicated in parentheses, with open areas indicating genetic material derived from import animal $A_1$, shaded areas indicating genetic material derived from $A_2$, and striped areas indicating genetic material derived from $A_3$. The animals are bred to form various $F_1$ individuals, which in turn are crossed with one another, backcrossed to a parent, or crossed to an unrelated import animal. All of these represent common breeding situations in zoo populations. The bottom of the figure then illustrates a representative chromosome pair emerging from these various crosses, with the $f$ and $h$ coefficients given underneath.

TABLE 17.2 Effect of the Level of Hybridity ($h$) upon 30-day Survivorship in the Noninbred ($f = 0$) and Inbred ($f > 0$) Offspring of Noninbred Parents of Speke's Gazelle as Measured by the Probabilities of the Fisher's Exact Test (FET)

| $f$ | $h$ | ALIVE ($N$) | DEAD ($N$) | FET |
|---|---|---|---|---|
| $f = 0$ | 0 | 15 | 6 | .18 |
| | .5 | 6 | 0 | |
| $f > 0$ | .5 | 5 | 11 | .03 |
| | 1 | 15 | 7 | |

a significant deleterious effect upon viability, regardless of the hybridity level. Taken together, these results strongly indicate that an inbreeding depression was encountered in the Speke's gazelle rather than an outbreeding depression.

This conclusion is strengthened by analyzing the later generations in which inbred animals themselves are now parents. This study is possible in the Speke's gazelle because females become reproductively mature in about a year and males in a year and a half. Consequently, this captive population has had sufficient time to go well beyond simple backcrosses and $F_2$ generations. The correlation between hybridity and inbreeding is rapidly broken down in these later generations, particularly for the Speke's gazelle population. When inference is confined to those offspring having at least one inbred parent, the correlation between the inbreeding and hybridity coefficients is reduced to $-0.099$, which is not significantly different from zero (Templeton and Read 1984). Hence, for this group of gazelles, the regression of viability against hybridity is not confounded by the regression against inbreeding. The regression against the inbreeding coefficient is significant at the 5% level and implies a load of 3.18 lethal equivalents per individual. Note that this lethal equivalent load is about half the load detected in the original import animals because all animals

TABLE 17.3 Effect of the Level of Inbreeding ($f$) upon 30-day Survivorship in the Offspring of Noninbred Parents of Speke's Gazelle within a Fixed Level of Hybridity ($h$) as Measured by the Probabilities of the Fisher's Exact Test (FET)

| $h$ | $f$ | ALIVE ($N$) | DEAD ($N$) | FET |
|---|---|---|---|---|
| .5 | $f = 0$ | 6 | 0 | .01 |
| | $f > 0$ | 5 | 11 | |
| 1 | $f < .25$ | 11 | 1 | .02 |
| | $f = .25$ | 4 | 6 | |

in this group have at least one inbred parent. These parents, being inbred, have already been selected under inbreeding, and hence the lethal equivalent load is expected to decline. However, the important point is that a significant inbreeding depression still exists.

In contrast, the regression against hybridity coefficient yields a slope of 0.41 (Templeton and Read 1984). Although not significantly different from 0, this slope is positive—just the opposite of what is expected under an outbreeding depression. The Speke's gazelle data therefore very clearly support an inbreeding depression and reject an outbreeding depression. This finding in turn implies a natural population structure characterized by outbreeding rather than inbreeding. In particular, the following population structures for the Speke's gazelle can be eliminated. First, these gazelles obviously do not have a system of mating characterized by regular inbreeding or sufficient population subdivision to result in an average inbreeding level of even .02 to .03, since the studies on human populations in India show that such average levels of inbreeding are sufficient to eliminate lethal equivalents. Second, the Speke's gazelle has not undergone any recent founder or bottleneck events, or any other kind of temporary intense inbreeding, because the studies on the European bison and the follow-up studies on the Speke's gazelle indicate that short but intense episodes of inbreeding also eliminate lethal equivalents.

It is also possible to define, at least to some extent, what is meant by a "recent" episode of intense inbreeding. The studies on the European bison (Slatis 1960) and Speke's gazelle (Templeton and Read 1984) clearly demonstrate that only a few generations of inbreeding are needed to eliminate virtually all lethal equivalents from a population, but population genetic theory indicates that lethal equivalents are only slowly reaccumulated through mutations if outbreeding is reestablished. This theoretical expectation has been confirmed by experiments with *Drosophila melanogaster*. Wallace (1956) made a strain isogenic for chromosome 2, thereby eliminating all lethal genes from that chromosome. He then monitored the reaccumulation of lethal and semilethal genes through mutation; it took about 50 generations for the percentage of lethal and semilethal genes to plateau. Moreover, this percentage was a third or less of its plateau value up to about generation 25. If mutation rates per generation are comparable in mammals and *Drosophila* and if percentage of lethal or semilethal mutations are also comparable, Wallace's work implies that the Speke's gazelle has had no major episodes of inbreeding and has been outbreeding with an average $f$ certainly smaller than .02 for at least 25 generations prior to being inbred in captivity.

## DISCUSSION

The genetic analysis performed upon the captive population of Speke's gazelles implies a population structure characterized by extensive outcrossing and little or no inbreeding for many generations prior to breeding in captivity. An important question is Can this result be generalized to other mammals or is the Speke's gazelle somehow an atypical mammal that has features biasing it toward outcrossing?

First, nothing is known about Speke's gazelle to suspect any peculiar bias toward outcrossing; indeed, just the opposite is true. This gazelle has a harem-type system of mating (Read and Frueh 1980)—a system that tends to reduce effective size. Moreover, this species has one of the most limited geographical distributions of African gazelles, being confined to the border area between Ethiopia and Somalia (Effron et al. 1976). In addition, the subfamily that includes Speke's gazelle has experienced extensive chromosome evolution (Effron et al. 1976), a phenomenon believed by Wilson et al. (1975) to imply a highly inbred population structure. Hence, in spite of harem formation, restricted geographical range, and extensive chromosome evolution, Speke's gazelle has a highly outcrossed population structure that results in a lethal equivalent load comparable to that of highly outcrossed human populations.

The Speke's gazelle population is also not an isolated example among captive populations of mammals to show an inbreeding depression. Ralls and Ballou (pers. comm.) applied the design and statistical methods developed by Templeton and Read (1984) to distinguish between inbreeding versus outcrossing depression in a variety of mammalian species that includes ungulates, primates, and carnivores. In some captive populations, there is a total confoundment between inbreeding and hybridity coefficients, just as Shields had warned, that precludes any further analysis at this time. However, for many other species, the analysis could be performed, although the sample sizes are generally much smaller than those associated with the Speke's gazelle. In every case yielding statistical significance, the inference is one of inbreeding depression and not outcrossing depression. In the nonsignificant cases (all associated with small sample sizes), the trend is indicative of inbreeding depression and opposed to outcrossing depression. This trend over several species is significant using a sign test.

These results do not mean that no inbred mammalian species exist; indeed, the work of O'Brien et al. (1983) indicates that the cheetah is a good candidate for a highly inbred mammalian species. Rather, these studies on inbreeding depression in captive populations indicate that many mammals, perhaps most, have a population structure that promotes outcrossing and that extensive inbreeding occurs much less frequently than about once in 25 generations.

In contrast, Chepko-Sade and Shields et al. (chap. 19, this vol.) argue that most mammals have small effective sizes with at least mildly inbred demes. However, as indicated earlier in this chapter, mild but regular levels of inbreeding are sufficient to eliminate inbreeding depressions, so the data on captive populations are incompatible with this conclusion. One possible explanation for this discrepancy is that the inbreeding studies are not measuring current population structure, but rather are sensitive to the average population structure that existed over the past several generations prior to captivity. In contrast, much of the data presented by others in this volume relate to current patterns of dispersal, demography, and biogeography. Human activity has had a tremendous impact on the biological world, particularly during this century. Many species, and particularly mammals, have experienced severe reductions in population size, habitat fragmentation, and disruption of normal dispersal routes. The prairie dogs studied by Halpin (chap. 7, this vol.) or the wolves studied by Mech (chap. 4, this vol.) are but two examples of species whose numbers and geographical distribution have been severely altered by human activities over the last century. Consequently, many mammals may indeed have a current population structure that indicates subdivision and mild inbreeding, yet the mammals may have come from an evolutionary lineage that until very recently was characterized by extensive outcrossing. Hence, a true contradiction may not exist between the conclusions of this chapter and those of Chepko-Sade and Shields et al. (chap. 19, this vol.).

However, the case made by Chepko-Sade and Shields et al. for small effective sizes with at least mild inbreeding is not compelling at present. First, the theoretical models employed are inadequate to accurately estimate effective or neighborhood size. These models refer to an idealized population and fail to incorporate many deviations that are commonly observed. For example, Mech (chap. 4, this vol.) says that the observed dispersal in his wolf populations has a sex-dependent directional bias. Also noted by Mech, a nonrandom directional dispersal by the two sexes would promote outcrossing, yet such nonrandom sex-dependent directional dispersal is not incorporated into any of the standard models for effective or neighborhood size used by Chepko-Sade and Shields et al. (chap. 19, this vol.).

The data also are often inadequate to support strong conclusions about effective size. For example, rare long-distance dispersers can have a tremendous impact in promoting an outcrossing population structure, but this type of dispersal is the most difficult type of all to quantify or even detect. For example, Smith (chap. 9, this vol.) primarily studied intrapatch dispersal in pikas and noted that we have very little direct information about the dynamics of interpatch dispersal, except for isolated accounts. However, Smith pointed out that populations on talus patches are probably subject to frequent extinction, yet pikas are found at close to saturation densities on even extremely isolated talus patches throughout most of their geographic range. Smith therefore concludes that pikas can and do disperse frequently between patches of talus. Smith also observed four unmarked immigrants that became successfully established in his study site. Unfortunately, no information is available about how far these individuals traveled, whether or not they came from the same source patch, and so forth. Without this information, estimates of effective size are premature.

The frequent local extinction and recolonization noted by Smith (chap. 9, this vol.) for pikas and by Bowen and Koford (chap. 12, this vol.) and Lidicker and Patton (chap. 10, this vol.) for other rodent species raises another important theoretical issue that can only be settled by long-term studies. Slatkin (1977) has investigated the impact of extinction-recolonization cycles on population structure. He discovered that this impact depends strongly upon how recolonization takes place. If most of the colonists are drawn from a single-source deme, cycles of extinction and recolonization promote increased population subdivision and inbreeding. On the other hand, if colonists are drawn from several different source populations, recolonization is a powerful factor promoting outcrossing. The genetic data and computer simulations of Bowen and Koford (chap. 12, this vol.) indicate that for at least *Microtus californicus,* extreme outcrossing occurs as a result of recolonization.

In summary, the common occurrence of inbreeding depression in captive mammal populations indicates that many mammals have had an outcrossing population structure in the recent past. This outcrossing structure may exist in present-day populations as well, although some mammals currently appear to be more subdivided and inbred due to human activities.

## ACKNOWLEDGMENTS

I wish to thank Peter Smouse, William M. Shields, and Peter Waser for their valuable comments upon an earlier draft of this chapter. This work was supported by the National Institutes of Health (RO1-GM31571).

## REFERENCES

Baker, A. E. M. 1981. Gene flow in housemice: Behavior in a population cage. *Behavioral Ecology and Sociobiology* 8:83–90.

Ballou, J., and K. Ralls. 1982. Inbreeding and juvenile mortality in small populations of ungulates: A detailed analysis. *Biological Conservation* 24:239–72.

Bouman, J. 1977. The future of Przewalski horses in captivity. *International Zoo Yearbook* 17:62–68.

Buettner-Janusch, J., T. J. Olivier, C. L. Ober, and B. D. Chepko-Sade. 1983. Models for lineal effects in rhesus group fissions. *American Journal of Physical Anthropology* 61:347–53.

Cheney, D. L., and R. M. Seyfarth. 1983. Nonrandom dispersal in free-ranging vervet monkeys: Social and genetic consequences. *American Naturalist* 122:392–412.

Clark, R. L., A. R. Templeton, and C. S. Sing. 1981 Studies of enzyme polymorphisms in the Kamuela population of *D. mercatorum.* I. Estimation of the level of polymorphism. *Genetics* 98:597–611.

Crow, J. F. 1963. The concept of genetic load: A reply. *American Journal Human Genetics* 15:310–15.

Daly, J. C. 1981. Effects of social organization and environmental diversity on determining the genetic structure of a population of the wild rabbit, *Oryctolagus cuniculus*. *Evolution* 35:689–706.

Dracopoli, N. C., F. L. Brett, T. R. Turner, and C. J. Jolly. 1983. Patterns of genetic variability in the serum proteins of the Kenyan vervet monkey (*Cercopithecus aethiops*). *American Journal of Physical Anthropology* 61:39–49.

Effron, M., M. H. Bogart, A. T. Kumamoto, and K. Bernischke. 1976. Chromosome studies in the mammalian subfamily Antilopinae. *Genetics* 46:419–44.

Endler, J. A. 1977. *Geographic variation, speciation, and clines.* Princeton: Princeton University Press.

Fisher, R. A. 1958. *The genetical theory of natural selection.* 2d ed. New York: Dover.

Fix, A. G. 1978. The role of kin-structured migration in genetic microdifferentiation. *Annals of Human Genetics* 41:329–39.

Flesness, N. R. 1977. Gene pool conservation and computer analysis. *International Zoo Yearbook* 17:77–81.

Foltz, D. W. 1981. Genetic measures of inbreeding in Peromyscus. *Journal of Mammalogy* 62:470–76.

Foltz, D. W., and J. L. Hoogland. 1981. Analysis of the mating system in the black-tailed prairie dog (*Cynomys ladovicianus*) by likelihood of paternity. *Journal of Mammalogy* 62:706–12.

Ghosh, A. K., and P. P. Majumder. 1979. Genetic load in an isolated tribal population of South India. *Human Genetics* 51:203–8.

Hoogland, J. L. 1982. Prairie dogs avoid extreme inbreeding. *Science* 215:1639–41.

Hoogland, J. L., and D. W. Foltz. 1982. Variance in male and female reproductive success in a harem-polygynous mammal, the black-tailed prairie dog (Sciuridae: *Cynomys ludovicianus*). *Behavioral Ecology and Sociobiology* 11:155–63.

Levin, D. A. 1984. Inbreeding depression and proximity-dependent crossing success in *Phlox drummondii*. *Evolution* 38:116–27.

Lewin, R. 1983. Brotherly alliances help avoid inbreeding. *Science* 222: 148–51.

McCracken, G. F., and J. W. Bradbury. 1977. Paternity and genetic heterogeneity in the polygynous bat, *Phyllostomus hastatus*. *Science* 198:303–6.

Malmberg, R. L. 1977. The evolution of epistasis and the advantage of recombination in populations of bacteriophage T4. *Genetics* 86:607–21.

May, R. M. 1980. Inbreeding among zoo animals. *Nature* 283:430–31.

Melnick, D. J., C. J. Jolly, and K. K. Kidd. 1984. The genetics of a wild population of Rhesus monkeys (*Macaca mulatta*), pt. 1: Genetic variability within and between social groups. *American Journal of Physical Anthropology* 63:341–60.

Moll, R. H., and C. W. Stuber. 1971. Comparisons of response to alternative selection procedures initiated with two populations of maize (*Zea mays* L.). *Crop Science* 11:706–11.

Moran, P. A. P. 1962. *The statistical processes of evolutionary theory.* Oxford: Clarendon Press.

Morton, N. E., J. F. Crow, and J. J. Muller, 1956. An estimate of the mutational damage in man from data on consanguineous marriages. *Proceedings of the National Academy of Sciences* (USA) 42:855–63.

Murray, R. D., and E. O. Smith. 1983. The role of dominance and intrafamilial bonding in the avoidance of close inbreeding. *Journal of Human Evolution* 12:481–86.

O'Brien, S. J., D. E. Wildt, D. Goodman, C. R. Merril, and M. Bush. 1983. The cheetah is depauperate in genetic variation. *Science* 221:459–62.

Olivier, T. J., C. Ober, J. Buettner-Janusch, and D. S. Sade. 1981. Genetic differentiation among matrilines in social groups of rhesus monkeys. *Behavioral Ecology and Sociobiology* 8:279–85.

Packer, C. 1979. Inter-group transfer and inbreeding avoidance in *Papio enubis*. *Animal Behaviour* 27:1–36.

Patton, J. L., and J. H. Feder. 1981. Microspatial genetic heterogeneity in pocket gophers: Non-random breeding and drift. *Evolution* 35:912–20.

Price, M. V., and N. M. Waser. 1979. Pollen dispersal and optimal outcrossing in *Delphinium nelsoni*. *Nature* 277:294–97.

Prout, T. 1981. A note on the island model with sex dependent migration. *Theoretical and Applied Genetics* 59:327–32.

Ralls, K., and J. Ballou. 1982. Effects of inbreeding on infant mortality in captive primates. *International Journal of Primatology* 3:491–505.

Ralls, K., K. Brugger, and J. Ballou. 1979. Inbreeding and juvenile mortality in small populations of ungulates. *Science* 206:1101–3.

Ralls, K., K. Brugger, and A. Glick. 1980. Deleterious effects of inbreeding in a herd of captive Dorcas gazelle. *International Zoo Yearbook* 20:137–46.

Rao, P. S. S., and S. G. Inbaraj. 1980. Inbreeding effects on fetal growth and development. *Journal of Medical Genetics* 17:27–33.

Read, B., and R. J. Frueh. 1980. Management and breeding of Speke's gazelle (*Gazella spekei*) at the St. Louis Zoo with a note on artificial insemination. *International Zoo Yearbook* 20:99–104.

Rothman, E. D., C. F. Sing, and A. R. Templeton. 1974. A model for analysis of population structure. *Genetics* 78:943–60.

Schwartz, O. A., and K. B. Armitage. 1980. Genetic variation in social mammals: The marmot model. *Science* 207:665–67.

Shields, W. M. 1982. *Philopatry, inbreeding, and the evolution of sex*. Albany: State University of New York Press.

———. 1983. Genetic considerations in the management of the wolf and other large vertebrates: An alternative view. In *Wolves in Canada and Alaska: Their status, biology, and management*, ed. L. Carbyn, 90–92. Canadian Wildlife Service Publication No. 45. Edmunton, Alberta.

Singleton, G. R., and D. A. Hay. 1983. The effect of social organization on reproductive success and gene flow in colonies of wild house mice, *Mus musculus*. *Behavioral Ecology and Sociobiology* 12:49–56.

Slatis, H. M. 1960. An analysis of inbreeding in the European bison. *Genetics* 45:275–88.

Slatkin, M. 1977. Gene flow and genetic drift in a species subject to frequent local extinctions. *Theoretical Population Biology* 12:253–69.

Stine, G. J. 1977. *Biosocial genetics*. New York: Macmillan.

Templeton, A. R. 1979. The unit of selection in *Drosophila mercatorum*, pt. 2: Genetic revolutions and the origin of coadapted genomes in parthenogenetic strains. *Genetics* 92:1283–93.

————. 1983. Natural and experimental parthenogenesis. In *The genetics and biology of Drosophila,* vol. 3, ed. M. Ashburner, H. L. Carson, and J. N. Thompson, 343–98. London: Academic Press.

Templeton, A. R., and B. Read. 1983. The elimination of inbreeding depression in a captive herd of Speke's gazelle. In *Genetics and conservation: A reference for managing wild animal and plant populations,* ed. C. M. Schonewald-Cox, S. M. Chambers, B. MacBryde, and L. Thomas, 241–61. Reading, Mass.: Addison-Wesley.

————. 1984. Factors eliminating inbreeding depression in a captive herd of Speke's gazelle (*Gazella spekei*). *Zoo Biology* 3:177–99.

Templeton, A. R., C. F. Sing, and B. Brokaw. 1976. The unit of selection in *Drosophila mercatorum,* pt. 1: The interaction of selection and meiosis in parthenogenetic strains. *Genetics* 82:349–76.

Treus, V. D., and N. V. Lobanov. 1971. Acclimatization and domestication of the Eland at Askenya-Nova Zoo. *International Zoo Yearbook* 11:147–56.

Wallace, B. 1956. Studies on irradiated populations of *Drosophila melanogaster. Journal of Genetics* 54:280–93.

Waser, P. M., and W. T. Jones. 1983. Natal philopatry among solitary mammals. *Quarterly Review of Biology* 58:355–90.

Wilson, A. C., G. L. Bush, S. M. Case, and M. C. King. 1975. Social structuring of mammalian populations and the rate of chromosomal evolution. *Proceedings of the National Academy of Sciences* (USA) 72:5061–65.

Wright, S. 1932. The roles of mutation, inbreeding, crossbreeding, and selection in evolution. *Proceedings of the Sixth International Congress of Genetics* 1:356–66.

————. 1941. On the probability of fixation of reciprocal translocation. *American Naturalist* 75:232–40.

————. 1965. The interpretation of population structure by F-statistics with special regard to systems of mating. *Evolution* 19:395–420.

# 18. Kin Selection in Complex Groups: Mating Structure, Migration Structure, and the Evolution of Social Behaviors

*Michael J. Wade and Felix J. Breden*

When the genotypic composition of local populations exerts an important influence on individual fitness, evolution in subdivided populations can be very different from evolution in large, homogeneously distributed populations. If local groups differ significantly from one another in genotypic composition, then the genotype with the highest fitness in one locality will not necessarily be the genotype with the highest fitness in some other locality. That is, the fitness of an individual in one group may not be a good predictor of its fitness in some other group. In addition, maximum relative fitness and maximum absolute fitness need not coincide when individual fitnesses are influenced by social interactions. Thus, selection may operate between groups in a direction or with an intensity different from that of selection within groups. The importance of considerations such as these for our understanding of evolution depends upon the extent of genetic differentiation between local groups, which, in turn, depends upon the factors influencing the distribution of individuals among groups.

Many of the most interesting examples of social behavior occur in organisms that spend their lives in small, more or less closely related groups. These groups may range from single families to multiple or extended families. The interaction of an organism's breeding biology, demography, and local ecology determines not only how individuals are distributed within and between such kin groups but also how they exert effects on the fitnesses of themselves and other conspecifics. By *breeding biology* we mean the distribution of numbers of mates per female, the apportionment of paternity among progeny of males mated to the same female, the distribution of the numbers of females reproducing in common nests or other groupings, and the apportionment of total reproduction within a group among the founding females. In addition, there is the evident possibility of genetic correlations existing among mates or among the migrating founders of new groups. In this chapter we illustrate the influence of such genetic correlations on the evolution of social behaviors.

For certain population structures, in the absence of selection, we know that subdivided populations will become genetically differentiated owing to random genetic drift, the sampling variance in the transmission of genes from one generation to the next. The important population parameters necessary for describing the rate and the extent of this differentiation are the effective size of breeding groups, $N_e$, and the amount of migration among local breeding groups per generation, $m$. In general, groups that exchange migrants are more similar to one another genetically than groups that do not, and Wright's widely recognized formula for the expected degree of genetic differentiation among groups at equilibrium, $F_{ST} = 1/(4Nm + 1)$, is a decreasing function of the migration rate, $m$ (Wright 1931). It would be desirable to have a similar description of the effects of the different components of an organism's breeding biology on its capacity for social evolution.

In this chapter we present the results of a theoretical investigation of the effects of variations in the breeding biology and population structure on social evolution. Using a population genetic model similar to that of Hamilton (1964a, 1964b), we examine the rate of change in gene frequency of an altruistic social behavior. Specifically, we ask: (1) How is the evolution of a social behavior affected by variations in breeding biology, such as variation in the numbers of mates of founding females or variation in the apportionment of paternity among males mated to the same female? and (2) How is the evolution of a social behavior affected by genetic correlations between mates (inbreeding) or between the migrants founding new groups (e.g., the fissioning of groups along matrilines)? In answering these questions, we take the breeding biology and the mating system as fixed; given a specific breeding biology or a mating system with a fixed level of inbreeding, we ask what are its effects on social evolution? In these studies the population structure itself does not evolve and we do not address questions of Why should the mating system or the breeding biology be this way?

In a later section of this chapter, we directly address the problem of the evolution of population structure itself. We do so by introducing genetic variation for the mating system in the form of a second locus that controls the tendency toward sib mating (Breden, n.d.). In these models *there is no direct selection on the mating system*. That is, there is no selective advantage to inbreeding or outbreeding in the form of inbreeding depression or adaptive philopatry. Any changes in gene frequency that occur at the mating locus occur as a result of indirect selection owing to a genetic correlation between the altruistic social behavior locus and the mating locus. This genetic correlation or linkage disequilibrium is initially zero but arises as a result of selection on the social behavior in structured populations. Once it exists, a fraction of the *direct selection* on the locus controlling the altruistic social behavior becomes *indirect selection* on the mating locus owing to the genetic correlation. Thus, the first finding of these models is that, in the absence of any direct selection, the population structure can be expected to evolve as a *correlated response* to kin selection in subdivided populations. A second important finding is that there is positive feedback between evolution at the social behavior locus and evolution at the mating locus. The higher the frequency of the allele for sib mating (inbreeding at the mating locus), the faster is the evolution of the altruistic social behavior. And the faster the evolution of the altruistic social behavior, the higher the genetic correlation between this social behavior and the mating system, and, consequently, the stronger the indirect selection on the mating locus. The result is a "run away" evolutionary process similar to that between male characters and female mating preferences in sexual selection (Lande 1980; Kirkpatrick 1982; Heisler 1984). We find a qualitatively similar effect when we introduce a second locus that influences the degree of kin recognition (Wade, unpub.).

## THE GENERAL MODEL

In order to accommodate the complexity of the breeding biology described earlier, it is necessary to adopt an exceedingly simplified genetic model for the social behavior. This is a fairly typical modeling strategy, and it clearly represents an oversimplification of the genetic basis of social behavior. Nevertheless, the model of a single locus affecting the tendency toward altruistic social behavior is similar in all important respects to the model adopted in the classic work of Hamilton (1964a, 1964b) and in most theoretical research since (see review in Michod 1982). The details of this genetic model have been derived in several different papers (Wade 1980, 1985a, 1985b), and only the general features of the model are presented here.

We let allele $a$ in frequency $q$ be the "altruistic" allele and $A$ in frequency $(1 - q) = p$ be the "nonaltruistic" allele. Define the trait $M_i$ as the frequency of $a$ within individuals. Clearly, for our single locus model, $M$ can take only three values, 0. 0.5, or 1, depending upon whether the individual in question is homozygous $AA$, heterozygous $Aa$, or homozygous $aa$, respectively. We let the population be subdivided into groups as a consequence of the reproductive behavior of the females. That is, we imagine females mate and produce progeny in such close proximity to one another that a group consists of the progeny of a number of females. We let $F_j$ be the frequency of groups with genotypic composition given by $G_{ij}$, which is the frequency of genotype $i$ in groups of type $j$. By specifying how individuals mate and aggregate to reproduce, we specify the frequency distribution of groups of different genotypic composition.

For any breeding biology, it can be shown that the rate of gene frequency change is given by the sum of two components—the net effects of selection within groups and the effects of selection between groups (Wade 1980, 1985a, 1985b). Each of these components of gene frequency change is equivalent to a covariance between relative fitness and gene frequency. The expression for the change in the gene frequency is given by

$$\Delta q = \text{Cov}(w_{ij}, M_{ij}) \tag{18.1a}$$

$$= \text{Cov}_{\cdot}(\bar{w}_{ij}, M_{ij}) + \text{Cov}(\bar{w}_{\cdot j}, M_{\cdot j}), \tag{18.1b}$$

where $w_{ij}$ is the global relative fitness of an individual of genotype $i$ in groups of type $j$, $\bar{w}_{ij}$ is the local relative fitness, and $\bar{w}_{\cdot j}$ is the group mean relative fitness. The first covariance term on the right-hand side of equation (18.1b) gives the effect of individuals within groups averaged over all groups, denoted by $\text{Cov}_{\cdot}$, and the second covariance term represents the effects of selection between groups. Because we are considering an altruistic allele, the first covariance term is necessarily negative, that is, individual selection within groups opposes the evolution of altruistic traits. However, the second covariance term is positive, and the fitness of local groups is an increasing function of the frequency of the altruistic allele within the groups.

At this point we adopt a second simplifying assumption regarding how the altruistic social behavior affects the fitnesses of the individuals performing the behavior and the fitnesses of the other group members. We let the absolute fitness of an individual of genotype $i$ in group $j$ be given by

$$W_{ij} = 1 + cM_{ij} + bM_{\cdot j}, \tag{18.2}$$

where $b > 0 > c$ and $M_{\cdot j}$ is the frequency of the $a$ allele in the group. This is the linear additive model of fitness that characterizes most theoretical studies of kin selection (cf. Michod 1982 for a recent review). The components of covariance under this model of fitness become simple functions of the within-group and between-group genic variances:

$$\text{Cov}_j (W_{ij}, M_{ij}) = c\text{Var}_j(M_{ij}); \tag{18.3a}$$

$$\text{Cov}(W_{\cdot j}, M_{\cdot j}) = (c + b)\text{Var}(M_{\cdot j}). \tag{18.3b}$$

Because $c < 0$, the average covariance within groups is negative, and when $b > |c|$ the between-group component of covariance is positive. In order for selection between groups to override the opposing effects of selection within groups, $br > c$, where $r$ is the intraclass correlation coefficient or the ratio of $\text{Var}(M_{\cdot j})$, the genetic variance between groups, to the total genetic variance. The simplified genetic and fitness models permit us to consider any kind of complexity in the breeding biology or formation of groups through its effects on $r$. We now apply this model to the three different questions raised in the introduction: (1) How do variations in the breeding biology of organisms affect the potential for social evolution in the absence of genetic correlations between mates and/or foundresses? (2) What are the effects of genetic correlations between mates or foundresses on the potential for social evolution? (3) How does the evolution of social behaviors influence the evolution of population structure (e.g., migration among social groups)?

## BREEDING BIOLOGY AND SOCIAL EVOLUTION

Using this model, Wade (1985a) investigated a variety of different breeding biologies in terms of their effects on social evolution. Specifically, he examined the effects of four different distributions separately and in combination: (1) the distribution of mates of females; (2) the apportionment of paternity within broods of females; (3) the distribution of numbers of foundresses per group; and (4) the apportionment of reproduction among foundresses of new groups. The variations in the breeding biology introduced by each distribution were assumed to be random with respect to the entire population so there were no genetic correlations among mates or foundresses and no genetic correlations between the allele for social behavior and the apportionment of paternity or reproduction. The introduction of even weak correlations can have very important consequences for the evolution of social behaviors, as we describe later. The purpose of the model was to incorporate into the theory of social evolution some of the diversity of breeding systems manifest by social organisms.

The first finding of this work (Wade 1985a) was that the *harmonic mean* number of mates per female and the *harmonic mean* number of foundresses are the most important summary statistics of the breeding biology. The harmonic mean is always smaller than the arithmetic mean whenever there is variation for the parameter in question. In this way, the theoretical study is useful as a guide to the *presentation* of data. In most instances, only the arithmetic mean number of mates or foundresses and its standard error are presented and it is not possible to reconstruct the harmonic mean.

A second important finding was that the harmonic mean number of foundresses was a more important determinant of the potential for social evolution than the harmonic mean number of mates per female. That is, a given amount of variance in the numbers of foundresses results in a greater variety of rates of social evolution than does the same amount of variation in the numbers of mates. In this way, the theoretical study is a useful guide to the investment of time or research energies at least in the crucial initial stages of research. Consider the case where we suspect that an organism has both variation in the numbers of foundresses of new colonies and variation among foundresses in the numbers of mates, but we have only limited research resources. The theory indicates that we should invest our research efforts in determining the foundress distribution because it has a greater potential for explaining variations in sociality than does the distribution of the numbers of mates.

The third important result of this investigation (Wade 1985a) was that *different* breeding biologies can be *equivalent* in terms of their influence on the evolution of social interactions such as altruistic behavior. Variation in breeding biology does not necessarily result in variation in the potential for social evolution. It is not a necessary result because the equation determining the conditions for and rate of evolution of altruistic social behaviors is a complex function of the means and variances of the four distributions described in the introduction to this section. Several different combinations of these parameters result in the same criteria for social evolution. Theoretical investigations of this sort can provide a guide to comparative studies across species or other taxa. Some variations in the breeding biology should, on theoretical grounds, result in differences in the potential for social evolution; other breeding biologies on these same grounds should not. More "fine-scale" theoretical predictions should improve the power of comparative studies among taxa for identifying important or unusual circumstances in the evolution of social interactions.

## GENETIC CORRELATIONS BETWEEN MATES OR FOUNDRESSES

In an experimental study of kin selection using flour beetles (genus *Tribolium*), Wade (1980) found that the observed evolutionary response of cannibalism rate, an important social behavior (Hausfater and Hrdy 1984), coincided with the predictions of kinship theory *only in those treatments in which there was a genetic correlation between mates*. This empirical finding was explained by the theoretical investigations of Breden and Wade (1981) and Wade and Breden (1981), who found that relatively weak genetic correlations between mates (as low as 0.1) had a very strong accelerating effect on the rate of evolution of altruistic social behaviors in single-family groups. We have extended our theoretical investigations to multiple family groups and explored the effects of genetic correlations between mates and between foundresses separately and in combination.

The population structure shown in figure 18.1 is used to explore these issues. Pairs of females mate and rear their progeny in sufficient proximity

FIGURE 18.1. Schematic diagram for the population structure modeled in the text. Each circle represents a family. Pairs of females mate and rear their progeny in sufficient proximity to one another that the progeny of these two females can legitimately be considered a group.

to one another that the progeny of these two females can legitimately be considered a group. More complicated social structures are an obvious extension of this scheme. This population structure can be considered to consist of three levels of organization and consequently three levels of selection: (1) between individuals within families; (2) between families within groups of families; and (3) between groups of families. In this way the modeled population structure is similar to that studied by Wade (1982). The novel feature added in this model is the degree of genetic correlation both among mates and foundresses.

We introduced the genetic correlations by permitting the individuals after selection and before mating to choose their mates in one of the three following ways: (1) at random within families; (2) at random within groups; or (3) at random over the entire population. By varying the proportion of the population that mated in each manner we can vary arbitrarily the genetic correlation between mates. Similarly, after mating, we permitted foundresses to establish new two-family groups by pairing in one of three ways: (1) at random within families; (2) at random within groups of families; or (3) at random over the entire population. By varying the proportion of the population that establishes new groups in any one of these three ways we introduced variation in the genetic correlation among foundresses. The model was formulated in such a way that we could introduce variation in the mating structure independent of variation in the migration structure. For example, we can have individuals mate at random over the entire population but establish new groups with full sibs. This type of population structure might be representative of those cases in which males disperse more or less at random before maturity but new groups are established by fissioning along matrilines. In describing the results, we refer to variations in the genetic correlation among mates as the *mating structure* and to variations in the genetic correlation among foundresses as the *migration structure*.

We explored a variety of different strengths of selection by varying the fitness costs and benefits and a variety of different degrees of dominance or recessiveness (Breden and Wade, in prep.). Here we report a summary of those findings in the form of a table contrasting the effects of extreme variations in the three genetic correlations on the rate of evolution of social behaviors for weak selection and intermediate dominance (table 18.1). We initiated each population with a gene frequency of 0.05 for the altruistic allele, in Hardy-Weinberg proportions with exactly the same costs and benefits associated with the behavior. We then recorded the number of generations required for the social allele to increase to a frequency in excess of 0.999. Thus, we are concerned with the effects of the population's mating and migration structure on the rate of evolution of social behaviors.

The results of a representative subset of our study are presented in table 18.1. In this table the extreme cases of complete within-family, within-group, or random mating are reported in all possible combinations, with

TABLE 18.1 Number of Generations to Reach Fixation of Altruistic Allele for Three Levels of Mating and Migration Structure

| MIGRATION STRUCTURE | MATING STRUCTURE | | | |
|---|---|---|---|---|
| | Random within Family | Random within Group | Random within Population | Row Average |
| Random within family | 30 | 35 | 300 | 120.3 |
| Random within group | 30 | 35 | 370 | 145.0 |
| Random within population | 70 | 225 | 520 | 271.6 |
| Column average | 43.3 | 98.3 | 396.6 | |

complete within-family assortment (no migration among families), within-group assortment (no migration among groups), or random assortment (free migration among all families and groups in the population). We see from the column marginal averages that varying the mating structure from complete full-sib mating to random mating changes the rate of evolution from 43.3 generations to 396.6 generations, almost an order of magnitude. In contrast, changing the migration structure (row marginal averages) from full-sib assortment to random assortment changes the rate of evolution by only a factor of two (full-sib assortment 120.3 generations; random assortment 271.6 generations). Clearly, genetic correlations are an important evolutionary parameter in both components of the population structure, but the mating structure exerts a greater effect on the rate of evolution than does the migration structure. Therefore, the potential for social evolution is more sensitive to variations in the genetic correlations among mates than it is to genetic correlations among migrants.

Another feature of table 18.1 is noteworthy. There is a clearly nonadditive interaction between these two components of the population structure. The combined effects of both sorts of genetic correlations are greater than our expectation from either component alone. For example, changing both assortment and mating from random within the population to random within groups decreases the number of generations to fixation from 520 to 35. This difference is much larger than would be expected from simply adding the effect of changing from random mating within the population to within groups, 520 to 225 generations or a 50% decrease, and changing from random assorting within the population to within groups, 520 to 370 or less than a 50% decrease. This finding will be further documented in Breden and Wade (in prep.).

## THE EVOLUTION OF POPULATION STRUCTURE AND SOCIAL BEHAVIOR

In the studies reported earlier, the population structure was taken to be a nonevolving parameter in the model. We have explored the effects of the evolution of social behaviors on population structure by introducing genetic variation at a second locus that affects the tendency of individuals to mate with their sibs (Breden, n.d.) or the tendency of individuals to behave differently toward related and unrelated conspecifics (Wade, in prep.). These loci might be considered as reflecting genetic variations in the tendency to mate prior to as opposed to after dispersal from a common nesting site or in the tendency toward kin recognition. In these two-locus models, there is *no direct selection* on the population structure locus. All selection occurring at that locus is a result of indirect selection through association with the social allele (linkage disequilibrium). The striking conclusion of these investigations is that population structure can be expected to evolve even in the absence of direct costs and benefits to inbreeding, migrating, or kin recognition, that is, with no relative fitness differences between inbreeding and outbreeding individuals, or between migrating and nonmigrating individuals. Our findings are similar in many respects to those investigations of the evolution of recombination rates (Turner 1967). Furthermore, the rate of change of the population structure allele is very rapid as a result of the positive feedback between it and the social allele. A small change in the frequency of the population structure allele results in a faster rate of evolution of the social behavior, which, in turn, increases the strength of indirect selection on the population structure allele. The net result for both of the cases we have studied is a "runaway process" in which the altruistic allele evolves at several times the rate observed in the same model, with no genetic variation at the population structure locus, and the degree of population subdivision is always increased. One interpretation of these findings is that there may be few intermediately social organisms because, in those cases where the appropriate genetic variation for social behavior and inbreeding is present, social evolution is expected to be very rapid.

## DISCUSSION

The first two of these simplified population genetic models indicate the degree of variation in the potential for social evolution that can result from variations in the breeding biology and population structure of organisms. Thus, these theoretical discussions show that it is not necessarily obvious how the mating and migration structures being measured in natural populations will affect population structure and social evolution. As an example, the comparison of multiple inseminations and multiple foundress associations showed that the number of foundresses, a component of the migration structure, affects the potential for social evolution more strongly than the number of inseminations per foundress, a component of the mating structure. However, when comparing genetic correlations among foundresses (migration structure) versus between foundresses and their mates (mating structure), it was shown that the mating structure was more important than the migration structure. Thus, although much of the data on the population structure of natural populations is gathered under the intuitive understanding that increased levels of genetic correlations between associating and mating individuals will increase the potential for social evolution, even qualitatively correct predictions about the relative strengths of these effects are only possible through examination of these types of family-structured and group-structured models of the evolution of social behavior.

The results of these models can be explained in terms of how variations in the migration structure and mating structure change the relative amounts of within-group and between-group genetic variances. Thus, referring to equations (18.1) through (18.3), the linear effects of the costs and benefits of the behaviors determine the strength of selection on the social behaviors. However, the response to that selection, that is, how quickly or under what conditions gene frequencies will change, depends not only on the strength of selection determined by costs and benefits but also on the genetic variances within and between groups. Consider as an example the question of how multiple inseminations affect the potential for the evolution of social behaviors. The theoretical predictions can be explained in terms of how much an extra insemination per female increases the average variation within families but decreases the variance among family groups. The direct result of multiple inseminations is a decrease in the component of gene frequency change owing to selection among groups and a corresponding increase in gene frequency change owing to selective differences among individuals within families. Perhaps the most important result of this approach to the question of the effect of population structure on the evolution of social behavior is that it stresses that to understand these processes in natural populations, both the fitness differences among behavioral phenotypes and the amount of genetic variation at each organizational level in the species must be known.

The last type of model discussed in this chapter indicates that genetic variations in the social behavior of organisms can create selective forces that will operate to change population structure. Furthermore, the evolution of social behavior and the evolution of population structure are not independent processes. They can interact in a demonstrably nonadditive manner we might call "runaway" social evolution exhibiting many of the same features of "runaway" sexual selection (e.g., Lande 1980). Our results depend on modeling complex social interactions in terms of cost to the individual and average benefit to group members, and on the assumption of no direct selective differences among different inbreeding phenotypes. However, given these assumptions, one important result from this model in terms of our understanding of the distribution of social structures among species is that this self-accelerating system predicts that the evolution of social behavior would be a very fast process and would concurrently tend to a high level of sociality and a viscous population structure.

## REFERENCES

Breden, F. n.d. The effect of genetic variation for inbreeding on the evolution of altruistic social behavior. Typescript.

Breden, F., and M. J. Wade. 1981. Inbreeding and evolution by kin selection. *Ethology and Sociobiology* 2:3–16.

Hamilton, W. D. 1964a. The genetical evolution of social behaviour, pt. 1. *Journal of Theoretical Biology* 7:1–16.

———. 1964b. The genetical evolution of social behaviour, pt. 2. *Journal of Theoretical Biology* 7:17–52.

Hausfater, G., and S. B. Hrdy. 1984. *Infanticide.* New York: Aldine.

Heisler, I. L. 1984. A quantitative genetic model for the origin of mating preferences. *Evolution* 38:1283–95.

Kirkpatrick, M. 1982. Sexual selection and the evolution of female choice. *Evolution* 36:1–12.

Lande, R. 1980. Sexual dimorphism, sexual selection, and adaptation in polygenic characters. *Evolution* 34:292–305.

Michod, R. E. 1982. The theory of kin selection. *Annual Review of Ecology and Systematics* 13:23–55.

Turner, J. R. G. 1967. Why does the genotype not congeal? *Evolution* 21:645–46.

Wade, M. J. 1980. An experimental study of kin selection. *Evolution* 34:844–55.

———. 1982. The evolution of interference competition by individual, family, and group selection. *Proceedings of the National Academy of Sciences* (USA) 79:3575–78.

———. 1985a. The influence of multiple inseminations and multiple foundresses on social evolution. *Journal of Theoretical Biology* 112:109–21.

———. 1985b. Soft selection, hard selection, kin selection, and group selection. *American Naturalist* 125:61–73.

Wade, M. J., and F. Breden. 1981. Effect of inbreeding on the evolution of altruistic behavior by kin selection. *Evolution* 35:844–58.

Wright, S. 1931. Evolution in Mendelian populations. *Genetics* 16:97–159.

# VI. CONCLUSIONS

# 19. The Effects of Dispersal and Social Structure on Effective Population Size

*B. Diane Chepko-Sade and William M. Shields with Joel Berger,*
*Zuleyma Tang Halpin, W. Thomas Jones, Lynn L. Rogers, Jon P.*
*Rood, and Andrew T. Smith*

In the preceding chapters of this book, dispersal patterns in populations of several species of large and small carnivores, ungulates, large and small rodents, pikas, humans, and nonhuman primates have been examined. We have examined some of the demographic and genetic correlates of population structures in a few species (several *Microtus* species, other small-bodied mammals, rhesus monkeys, and two human groups) in parts 3 and 4. In part 5, we have presented mathematical models that attempt to predict dispersal distance distributions (Waser), opportunities for kin selection in complex groups (Wade and Breden), and inferences one might make about population structure from genetic studies (Templeton).

Nevertheless, a large jump seems to occur between the descriptive empirical studies in part 2 and the more theoretical treatments in later sections of the book. How can we relate what animals do on the ground to the evolutionary theories and models discussed in the latter part of the book? All of the empirical studies in part 2 attempt to describe and quantify the population structure of the species studied, and yet each study, because of the peculiarities of the species' social structure and the particular interests of the investigators, seems to focus on different aspects of population structure. We are left in the end with a wealth of specific data on mating patterns and/or dispersal patterns, but little basis for comparing one species with another in terms of the evolutionary implications of the various social structures described. In this chapter we attempt to calculate for each population in part 2 a measure of population size that will take into account the various effects of social factors such as mating structure, rate and distance of dispersal, and degree of philopatry or vagility. This measure is effective population size, and it can be used to compare the evolutionary potential of one population to that of another.

## EFFECTIVE POPULATION SIZE

Sewall Wright (1931, 1956, 1965, reviewed in 1969, 1978a) developed his effective population size ($N_e$, defined in Shields, chap. 1, this vol.) as a construct to allow biologists to compare the population structure of otherwise disparate species or populations of single species. Both the theory and methods for estimating and interpreting $N_e$ have been elaborated and discussed extensively (e.g., Crow and Kimura 1970; Kimura and Ohta 1971; Wright 1978a; Shields 1982). Theoretically, $N_e$ is the size of an "ideal" population, that is, a population of $N$ breeding individuals, randomly mating, which produces $N$ new individuals each generation by random union of $N$ male and $N$ female gametes regarded as random samples from the population of the parent generation (Crow and Kimura 1970, 34). Since most real populations do not fit this description, a number of formulae have been developed to estimate what size ideal population would produce the same genetic correlations (expected genotype frequencies) as those in the population under study. For example, if our population has 200 adults, but half of them are effectively prevented from breeding because of insufficient nest sites, then $N_e$ for that population would be 1/2 $N$ or 100.

Any factor that puts constraints on the random sampling of male and female gametes from the parental population or on their random union with one another will affect $N_e$. Such factors include (1) sexual (as opposed to asexual) reproduction; (2) uneven sex ratio; (3) unequal contribution of gametes to the descendant generation; and (4) changes in the size of the population from one generation to another.

Social and demographic factors that result in differences between $N_e$ and the census number of a population include (1) some proportion of a population that does not mate (e.g., due to immaturity, senescence, or repression of reproduction as a result of social dominance or shortage of some limiting factor, such as nest sites); (2) philopatric dispersal patterns (such that individuals within a given area show a degree of relatedness greater than would be expected had they been drawn at random from a larger population); (3) social factors in which intraspecific behavioral interactions produce a large variance in the proportion of offspring produced by each member of the parental generation (e.g., polygynous mating structures or frequency of mating determined by a dominance hierarchy); and (4) expanding, decreasing, or fluctuating population size.

One of the sources of confusion between population biologists and population geneticists is the use of some of the same words but with entirely different meanings. *Migration* is commonly defined by vertebrate population biologists to mean "the relatively long-distance movements made by large numbers of individuals in approximately the same direction at approximately the same time, and is usually followed by a regular return migration" (Endler 1977, 20). This type of migration, as long as the starting points and end points are the same (as they are in many species), results in a great deal of movement but little or no gene flow between groups. However, population geneticists define migration specifically as the movement of alleles between semiisolated subpopulations, a process that by definition involves gene flow between subpopulations.

*Dispersal* has been defined variously as "the roughly random and nondirectional small-scale movements made by individuals rather than by groups, continuously rather than periodically, as a result of their daily activities" (Endler 1977, 21), but has also been defined as "the movement of an organism or propagule from its site or group of origin to its first or subsequent breeding site or group" (Shields, chap. 1, this vol.). The latter definition is certainly not migration in the sense used by population biologists, but may well result in genetic migration. In this chapter we are interested in the effects of dispersal patterns on the genetic structure of populations. Therefore we will use the latter definition of dispersal by Shields. Dispersal can be short range (within a randomly mating subpopulation) or long range (between populations or subpopulations), depending on the distance moved and the average distance between organisms. Since the genetic effects are quite different for long-range as opposed to short-range dispersal, we will refer to long-range dispersal that involves gene flow between populations or subpopulations as *genetic migration* in order to distinguish it from both the periodic mass movements of animals and short-range dispersal.

The effective population size has important implications for evolutionary processes. In essence the smaller the $N_e$, the greater the opportunity for genetic drift and the greater the level of inbreeding characterizing a species. A single isolated and randomly mating population with a small $N_e$, which remains small and isolated over long periods of evolutionary time (e.g., $2N$ generations), will eventually lose all its genetic variability (except what is reintroduced by mutation, which may be negligible in small populations) and hence its potential to evolve to meet changing environmental challenges. Such a population faces the strong probability of extinction when environmental changes occur. A single randomly mating population with a large $N_e$, in contrast, favors the very slow but predictable mass selection processes discussed and elaborated by Fisher ([1930] 1958).

A more complex possibility, the existence of a number of partially isolated local subpopulations within a larger population, each with small $N_e$ but with occasional migration of genes between subpopulations, favors the occurrence of Wright's shifting balance process in evolution (Wright 1931, 1932, 1980). Here, genetic variability is maintained, even if the population as a whole is no larger than the first population described. While each subpopulation tends to become homozygous for neutral and near-neutral alleles, each subpopulation will tend to lose different alleles because the loss of these alleles is random. Then when genetic migration occurs infrequently between subpopulations, old alleles that were lost in a subpopulation are reintroduced and new gene combinations arise. Thus different gene combinations are constantly being formed, and the loss of alleles from the population as a whole is considerably slowed down. When a gene combination occurs that is adaptive (rather than selectively neutral), it can become established by selection in a small subpopulation far more quickly than in a larger population by mass selection. This subpopulation will then experience higher fitness than the other subpopulations in the area and will tend to grow more quickly than other subpopulations, producing a disproportionately large number of genetic migrants. These migrants will introduce the favorable new genetic combination to other subpopulations until the new combination becomes widespread in the entire population and the population as a whole evolves to a new and higher level of fitness, or an "adaptive peak" (Wright 1982; also discussed in Wright 1932, 1956, 1978b, 1980).

Laboratory experiments and livestock breeding programs have found the latter evolutionary process to be considerably faster and more effective than the conceptually simpler method of mass selection (Wright 1978b; Wade 1976; Slatkin and Wade 1978). For example, artificial selection for traits in a population subdivided into a number of relatively inbred groups with occasional crosses between groups was found to be far more effective than the development of distinct purebred lines in terms of maintaining adequate levels of fecundity and viability long enough for the desired traits to become fixed (Wright 1978b). Furthermore, the shifting balance process provides an effective mechanism for the evolution of traits produced by combinations of genes (polygenic traits, or pliotropy) that include most of the adaptive traits of which we know.

While Wright (1931, 1978b) was convinced that the model for evolution he had developed was a far more efficient one than the more widely accepted mass selection scheme, little field data were available until recently to assess whether natural populations were organized in effectively subdivided populations or whether large panmictic populations were more the rule. Although information has been available on geographical distributions of natural populations, we must now examine how mating and dispersal patterns and other aspects of social structure translate into terms useful to a population geneticist (whether mating is random with respect to genotype within large or small groups, and how much genetically effective migration takes place). One of the tasks of the field biologist interested in the evolutionary process, then, is to estimate a species' population structure by estimating $N_e$. A second important task is to estimate genetic migration rates between populations. In this chapter we attempt to estimate $N_e$ for the species described in part 2, but in most cases we can only make educated guesses about genetic migration rates between populations (however, see Sade et al., chap. 15, this vol., for a method by which migration rates can be calculated even for very complex social systems).

Crow and Kimura (1970) distinguish between the *inbreeding effective population number,* which is related to the number of unique parental genotypes contributing to the descendant generation, and the *variance effective population number,* which is related to random drift in gene frequencies because of sampling variance. They note that "whereas the inbreeding effective number is naturally related to the number in the parent (or with separate sexes, the grandparent) generation, the variance effective number is related to the number in the progeny generation. This is to be expected, since the probability of identity by descent depends on the number of ancestors whereas the sampling variance depends on the size of the sample, i.e., the number of offspring" (p. 357). Except where otherwise indicated, in this chapter we are concerned with the inbreeding effective population size.

## ESTIMATING EFFECTIVE POPULATION SIZE

The inbreeding effective population size is really a measure of the diversity of ancestors of, or ancestral alleles likely to be carried by, a particular group. As such it is influenced by many factors, including the actual size of the mating group in space (neighborhood size) and time (influenced by dispersal rates among neighborhoods and fluctuations in population size), the sex ratio of breeders, and the variance in successful progeny production characterizing those breeders. Each factor can influence $N_e$ independently and should be accounted for in a best estimate of effective size. Currently there is no commonly accepted method for combining these different factors into a single estimate. Perhaps the easiest might be to calculate a final $N_e$ by using the formula associated with each factor sequentially, with the $N_e$ derived from the first factor being used as the initial estimate of $N$ in the next formula and so on until the contributions of all factors have been included.

While this approach makes sense intuitively, a theoretician has yet to demonstrate formally its correctness; there may be inconsistencies, depending on the order in which different factors are taken into account. Nevertheless, it seems that this final estimate of static $N_e$ would be more accurate than those accounting for just one or two of the important influences. Of course, it would still not have accounted for the contributions of successful long-distance dispersers (genetic migration between populations) and the gene flow they engender over time. $N_e$ alone cannot distinguish between a small isolated population and a small deme that exchanges infrequent genetic migrants with other semi-isolated demes. This distinction would require additional models, which are extensively reviewed in Kimura and Ohta (1971) and are discussed in this volume by Sade et al. (chap. 15), Smouse and Wood (chap. 14), Lidicker and Patton (chap. 10), and O'Brien (chap. 13). Estimates of $F_{ST}$ (defined by Crow and Kimura [1970, 108] as the probability that two homologous genes, chosen at random from the subpopulation, are both descended from a single ancestral gene in the subpopulation) can be used in conjunction with $N_e$ to provide a more accurate picture of population structure than is provided by $N_e$ alone. A population with a small $N_e$ based on population structure but a low $F_{ST}$

based on genetic measures (indicating low levels of homozygosity due to inbreeding) either may be recently formed from a larger population or may be a deme receiving occasional genetic migrants from other demes within a larger population (i.e., a population that fits Wright's shifting balance model). In contrast, an isolated deme with small $N_e$ based on population structure and high $F_{ST}$ values based on genetic measures (indicating strong differentiation from the larger population) is a population that has been genetically isolated long enough to be drifting toward fixation of neutral alleles and the loss of genetic variability.

A series of demes with small $N_e$ based on population structure but low $F_{ST}$ based on genetic structure could recently have formed from a larger population or may have enough genetic migration between demes to limit differentiation. In contrast, a series of demes with small $N_e$ based on population structure but high $F_{ST}$ based on genetic measures may have been isolated from one another long enough to be drifting toward fixation of neutral alleles within the separate demes. This differentiation would occur without the loss of genetic variability from the population as a whole since each deme would become fixed for different alleles. Therefore, a more accurate characterization of the evolutionary potential of a population would require the simultaneous examination of $N_e$ and $F_{ST}$ together for a series of demes (or subpopulations) rather than either alone for a single deme (see Lidicker and Patton, chap. 10, this vol.). However, at the present time genetic studies are lacking for many populations in which social and demographic factors have been studied, so the static $N_e$ for a single deme is frequently the best estimate of evolutionary potential we can presently calculate.

The primary determinant of effective population size is the size of the randomly mating group, or deme (equivalent to the term *subpopulation* used in much of the genetic literature and used frequently in this chapter). Deme size, in turn, is controlled by the dispersal and dispersion characterizing a population. Dobzhansky and Wright (1943) and later Wright (1943, 1969, 1978a) suggested that a good estimate of neighborhood area for species distributed in two dimensions was the root mean square of the dispersal distance. $N_e$ based on neighborhood size is then estimated as

$$N_e = 4 \pi\sigma^2 d, \tag{19.1}$$

where $d$ is the density of breeders in the area (i.e., $n/4 \pi\sigma^2$) and $\sigma^2$ is the one-way variance (= 1/2 the standard deviation squared) of the distance from the birthplaces of the separate parents relative to that of the offspring.

As Wright (1969) stated, his method theoretically gives the number of breeders in a circle of radius $2\sigma$ that would include 86.5% of the parents of central individuals, if dispersal distances were distributed normally. As Wright and many others noted, however, dispersal distributions are rarely normal. Since the goal is to estimate the size of the breeding group that mates approximately at random, a simple correction allowing for any dispersal distribution is possible. Neighborhood extent could be defined as the 85th or 86th percentile dispersal distance and estimated from an empirically derived distribution. This distance would be estimated in the same way as a median (the 50th percentile distance) and would be equivalent to the radius of a circular neighborhood containing 85% of the parents regardless of the shape (skew or kurtosis) of the dispersal distribution. This distance could then be converted to an area estimate to be multiplied by local densities (or divided by home range size or normal adult interindividual distances) to estimate an $N_e$ (neighborhood size) unbiased by assumption of normality.

Using the two methods on the same distributions can generate either similar or disimilar estimates (e.g., table 19.1). The difference depends on the actual shape of the distribution (how closely does it approximate normal?) and on the frequency of overly long-distance propagules (which can inflate variances disproportionately). In the examples in table 19.1 and figure 19.1, the traditional method compares well with the unbiased distribution for the song sparrow, but gives considerably larger estimates of $N_e$ for the less normal distributions characterizing the deer mouse and the rusty lizard. Since dispersal distributions are rarely normal, perhaps the percentile method warrants further study.

Areal $N_e$ is an appropriate measure of population size when individuals or groups of individuals are fairly continuously distributed over a large area (i.e., larger than the greatest dispersal distances). Some of the species described in part 2 of this volume are distributed in this way, such as Jones's kangaroo rats, Mech's wolves, and Rogers's bears. However, other species live in distinct groups, separate from others of their species, with little or no genetic migration between groups, such as Halpin's prairie dogs and Berger's horses. In these cases, an areal $N_e$ is meaningless and a census of breeding-age adults is a much more meaningful measure of $N_e$. Therefore, in cases in which there are obvious large discontinuities in the distribution

TABLE 19.1 Effective Population Sizes of Three Vertebrates

| SPECIES AND SEX | WRIGHT'S $N_e$ | 85TH PERCENTILE $N_e$ |
|---|---|---|
| P. maniculatus (deer mouse) | 308 | 109 |
| M. melodia (song sparrow) | 244 | 226 |
| S. olivaceus (total) (rusty lizard) | 472 | 187 |
| Male | 140 | 86 |
| Female | 332 | 101 |

Note: Calculated via Wright's root mean square method and via the 85th percentile distance of the original dispersal distribution.

Sources: P. Maniculatus, Dice and Howard 1951; M. melodia, Nice 1964; S. olivaceus, Blair 1960.

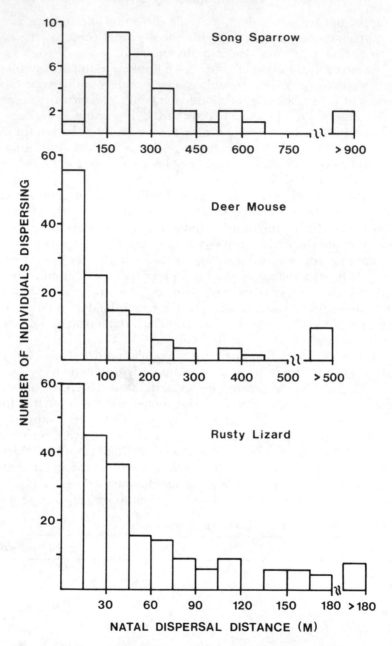

FIGURE 19.1. A sampling of natal dispersal distance distributions that indicate how the shape of the distribution can affect the use of the variance of a distribution in estimating effective population size (table 19.1). The song sparrow illustrates a platykurtic, the deer mouse a strong leptokurtic, and the rusty lizard a mild leptokurtic distribution.

of animals, we will use the population census number of breeders as our starting $N_e$ in place of an areal $N_e$.

Once an areal $N_e$ (or census $N_e$ where appropriate) has been generated, it can be used as a base to account for other factors. Species with biased breeding sex ratios (e.g., polygynous mammals such as Berger's horses and Rogers's bears described in this volume or the rusty lizard of table 19.1) may even generate different neighborhood sizes for each sex because of sexual differences in average home range sizes (table 19.1). The reduction in the numbers of parents of one sex that results from such a mating system tends to reduce effective population size such that

$$N_e = \frac{4N_m N_f}{(N_m + N_f)} \tag{19.2}$$

where $N_m$ is the number of breeding males and $N_f$ the number of breeding females in a neighborhood. While sex ratio can have a profound effect on $N_e$, it does so only at fairly extreme sex biases (fig. 19.2).

Once area and sex ratio have been accounted for, temporal variations in population size can be included by using the harmonic mean of population size over a number of generations:

$$\frac{I}{N_e} = \frac{1}{t} \Sigma_i \frac{1}{N_i} \tag{19.3}$$

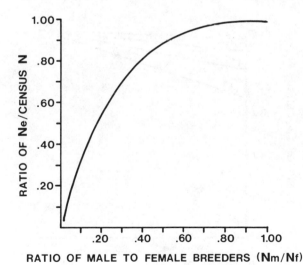

FIGURE 19.2. The relationship between effective population size and census population size as a function of the sex ratio of actual breeders. As polygamy increases, effective population size decreases, and the rate of decrease accelerates the most at extreme sex biases.

where $N_i$ is the effective population size based on dispersal and sex ratio in each of $t$ generations. If a population goes through an extreme bottleneck occasionally, or regularly cycles through different population sizes (as do the *Microtus* species described by Lidicker and Patton, chap. 10, Bowen and Koford, chap. 12, and Gaines and Johnson, chap. 11, in this vol.), then the smaller-sized generations (bottlenecks) will decrease the number of ancestors and can reduce $N_e$ substantially (fig. 19.3).

The final major factor to be accounted for is variation in progeny production. If just a few of the parents in any generation produce all of the progeny, then the mean relatedness of the population must increase. The stronger the differences or selection among parents, the greater the variation in progeny production and the greater the reduction in effective population size. This effect can be produced by intrinsic variation in adaptation or quality, by random processes, or by social structures that limit breeding to one or a few of the mature individuals in social groups (e.g., in wolves—Mech chap. 4, this vol.; mongooses—Rood, chap. 6, this vol.; and many primates—Moore and Ali 1984; Sade et al., chap. 15, this vol.). The actual reductions can be estimated using

$$N_e = \frac{(4N - 2)}{(V_k + 2)} \tag{19.4}$$

where $V_k$, the parental variance in progeny production, is a measure of selective and chance effects on reproductive success. Since variability in

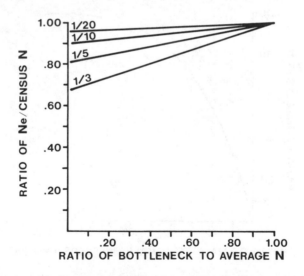

FIGURE 19.3.  The relationship between effective population size and census size as a function of how census numbers vary over time. Bottlenecks (size minima) can have disproportionate effects but only if they are very small or occur frequently. The lines indicate bottlenecks that occur from once in 3 to once in 20 generations.

reproductive success is likely to be the rule in nature, and since this variation can have a very strong effect (fig. 19.4), this factor is likely to have a greater effect on $N_e$ than anything except the basic dispersal pattern characterizing a particular species. This version of the effect of progeny variation assumes a stable or stably oscillating population. To determine the effect in either growing or declining populations it is necessary to estimate the variance (versus the inbreeding) effective population size (Crow and Kimura 1970; Kimura and Ohta 1971; Wright 1978a).

Finally, minor factors, such as behavioral avoidance of extreme inbreeding (incest avoidance), affect static effective population size. Wright (1969, 212) has explored the effects of avoidance of consanguinous mating on $N_e$. He showed that eliminating selfing in an otherwise randomly mating monoecious species resulted in $N_e$ increasing to $N + 0.5$, and excluding sib mating in an otherwise randomly mating dioecious group results in $N_e = N + 1.5$. As he put it, "the deviations of $N_e$ from $N$ in all of those cases are trivial unless $N$ is very small." This suggests that incest avoidance is greatly overemphasized as evolutionarily important in the literature. In fact, sex linkage has a greater effect on $N_e$ than consanguinous matings for most population sizes. Perhaps the size of a population controls the evolutionary process in a more fundamental fashion than the pattern of easily observable pedigree consanguinity? It is certainly the case that avoiding

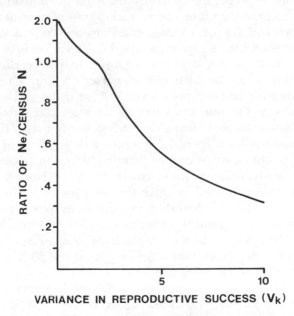

FIGURE 19.4. The relationship between effective and census population size as a function of variance in reproductive success of all potential breeders.

close pedigree inbreeding in otherwise moderate-sized demes may have little effect on the background level of relatedness and inbreeding.

## MAMMALIAN EFFECTIVE POPULATION SIZES

While few studies of any organisms, including mammals, have estimated $N_e$ using all of the contributing factors (but see Begon 1976, 1978; and Daly 1981), studies that attempt some estimate are appearing with greater regularity (e.g., see Lidicker and Patton, chap. 10, and Bowen and Koford, chap. 12, this vol.; others reviewed in Wright 1978a and Shields 1982). In the following section we have used the formulae reviewed earlier and data provided by the junior authors of this chapter whose long-term studies of some mammals are reported in part 2 of this volume to estimate directly effective population sizes for their species (table 19.2). In many cases the data available are inadequate for calculating a reliable estimate of $N_e$ for the population in question. The purpose of this section is not to make reliable estimates of $N_e$ for all of the populations presented here. Rather, it is to demonstrate how data can be applied to the formulae presented earlier and to point out some of the problems in calculation and interpretation that may be encountered when doing so.

### Calculating $N_e$ for Nelson and Mech's Deer

Nelson and Mech (chap. 2, this vol.) report that deer density for their study area varied from 0.2 to 0.4 deer/km². Accurate individual dispersal distances for yearlings (based on radio tracking) were provided by Nelson and Mech (pers. comm.) for 13 male yearlings and 20 female yearlings. Dispersal distances for male yearlings varied from 4.0 km to 38.4 km from their birth site, with a mean of 11.9 km. Only 1 female yearling out of 20 studied dispersed at all, but that one dispersed 22 km. Thus the mean dispersal distance for females was 1.1 km and for males and females combined was 5.35 km. The one-way variance for male and female dispersal distances combined was 39.1 km². Using the earlier figures for density and the one-way variance for dispersal distances of all yearlings of both sexes, we calculated Wright's areal $N_e$ using equation (19.1). The one-way variance (which equals one-half the square of the standard deviation), rather than the two-way variance, is used because the distribution is assumed to be centered on zero, with equivalent dispersal distances considered as a series of concentric rings moving out from the origin (Wright 1969). We obtained an $N_e$ of 98.4 when deer density is 0.2 deer/km² and an $N_e$ of 196.7 when deer density is 0.4 deer/km². When we calculated the 85th percentile $N_e$, which is the number of breeding age adults within a circle whose radius is the greatest distance traveled by the closest 85% of dispersers, we obtained an area of 289.53 km² ($r = 9.6$ km). Using Nelson and Mech's density figures we obtained an $N_e$ of 57.9 (for density = 0.2 deer/km²) and 115.8 (for density = 0.4 deer/km²).

TABLE 19.2 Effective Population Size Estimates of Populations Presented in Part 2

| Species (Author) | $N$ | Wright's Areal $N_e$ | 85th $N_e$ Percentile | Sex Ratio $N_e$ | Temporal Variance $N_e$ | Census of Subpop. $N_e$ |
|---|---|---|---|---|---|---|
| Deer (Nelson and Mech) | 1,290[a] | 98.4–196.7 | 57.9–115.8 | n.d. | n.d. | 45–800[a] |
| Horses (Berger) | 171[a] | n.a. | n.a. | 50.2 | n.d. | 112 |
| Wolves (Mech) | 18–20[a] | 804 | 1,660.7 | n.c. | n.d. | n.a. |
| Bears (Rogers) | 115 | 843.8 (428.0) | 810.3 | 583.6 (291.7) | n.d. | n.a. |
| Mongooses (Rood) | 38–137 | 31.4 | 20.7 | 11.0 | n.d. | 1.4[b] |
|  |  | 248.5 | 163.8 | 87.0 |  | 11.9[b] |
| Prairie dogs (Halpin) | 21–34 | n.a. | n.a. | 17.1–27.7 | 23.1 | 21–34 |
| Kangaroo rats (Jones) | 142 | 15.9 | 7.3 | n.c. | n.d. | n.a. |
| Pikas (a) (Smith) | 41 | 15.4–7.2 | undef. | n.c. | n.c. | 12–27 |
| (j) | 10 | 3.6–1.7 | 3.3–1.5 |  |  |  |
| (j + mig) | 14 | 58.8–27.4 | 60.0–28.3 |  |  |  |

*Notes:* $N_e$ was calculated via Wright's root mean square method and via the 85th percentile distance of the original dispersal distribution. Sex ratio $N_e$ and Temporal variance $N_e$ were calculated where possible, using the Areal $N_e$ as a starting figure before correcting for sex ratio or temporal variance. $N_e$ based on a census of an isolated subpopulation is also given where appropriate. See text for reservations regarding the accuracy of estimates for each case.

$N$ = number of breeding-age adults in sample upon which calculations are based unless otherwise indicated; n.c. = no change; n.d. = no data; n.a. = not appropriate; undef. = undefined; a = adult; j = juvenile; mig = migrant.

a. Individuals of all ages. Number of breeding age adults not reported.

b. Reproductive variance $N_r$.

The figures for the 85th percentile $N_e$ (adjusting for non-normal distribution of dispersal distances) are substantially smaller than those found using the traditional method, implying that the earlier estimate is biased by a few long-distance dispersers that inflate the dispersal variance. This is true: only 3 of 33 individuals traveled farther than 22 km, while the rest remained within 14.5 km of their birthplaces. Because dispersal distances are non-normally distributed, the 85th percentile $N_e$ is probably the more accurate estimate.

In all of these calculations, deer density is based on counts of deer of *all* ages, not just those of reproductive age. Therefore, all of the previous figures are overestimates of $N_e$. If reproductive-age individuals make up only half of the population, then $N_e$ would be half of the figures reported earlier. On the other hand, dispersal variance is based only on yearling dispersal distances. If older individuals disperse, and if adding their dispersal information would increase the variance of the dispersal distances, then $N_e$ would be increased. The variance in this equation is meant to be the one-way variance of the distances from the birthplaces of the separate parents, relative to that of the offspring (Wright 1978a). Therefore, if an individual disperses a certain distance as a yearling, and then another distance as an adult, but before breeding, the dispersal distance used should be the straight-line distance between the place of birth and the place where that individual finally breeds.

We were unable to calculate the sex ratio $N_e$ because the ratio of breeding males to breeding females was not available. If fewer males than females breed in a given breeding season (which is generally the case with deer), $N_e$ would be further reduced according to equation (19.2). If such information could be collected in follow-up studies, a more accurate approximation of $N_e$ would be possible.

All of the above calculations are based on the assumption that deer are uniformly distributed and randomly mating over the 2,500 km² study area. We know, however (from Nelson and Mech, chap. 2, this vol.), that they are not, but are subdivided into four semi-isolated breeding groups of different sizes. Even when dispersal distances are great enough to carry dispersers into another subpopulation (or deer yard), the direction of dispersal was such that dispersers usually remained within their own subpopulation or yard. These subpopulations ranged in census size from 45 to 800. If most mating occurs on summer ranges and involves no additional movements other than those already described, we could use 800 deer as our starting point and apply the equation for areal $N_e$ to it, using deer density on the summer range for the density part of the equation. The latter figure was not reported, however. If most deer breed after moving to their winter deer yards, where density is much higher, the census number itself should be used as a first approximation of $N_e$. If most breeding takes place on the summer grounds, the subdivision of the breeding population would further reduce $N_e$ below the figures in table 19.2. If a substantial amount of breeding takes place on the wintering grounds, $N_e$ should range from 45 to 800 before correction for breeding age and sex ratio.

## Calculating $N_e$ for Berger's Horses

The population of horses studied by Berger increased in size from 49 to 171 over the 6 years he observed them, with virtually no emigration and no immigration during this time (Berger, chap. 3, this vol.). Using the raw dispersal distances for 32 females and 31 males provided by Berger (pers. comm.) we were able to calculate Wright's areal $N_e$. Dispersal distances from birth area to area at the time of last observation for males range from 3 km to 24 km, while those for females range from 2 km to 8 km. The one-way variance in dispersal distances for both sexes combined is 16.42 km². While the population more than tripled in size over the study period, more area was being colonized by the migrants during this time too, so winter density of horses 1 year old and older on the range varied between 2.09 horses/km² and 3.26 horses/km² over the 6 years of the study (Berger, pers. comm.).

Using the one-way variance of dispersal distances and the highest and lowest densities provided by Berger, we calculated Wright's areal $N_e = 431.16$ for a density of 2.09 horses/km² and 672.67 for a density of 3.26 horses/km². These values assume that horses are uniformly distributed at these densities throughout a large area. However, Berger's entire population was composed of only 49 to 171 horses, which were isolated from other populations by geographical barriers as well as long distances. Therefore, the assumption of uniform density is clearly violated, and the direct census count is a more appropriate first approximation of $N_e$ for this population.

The census $N_e$ includes individuals of all ages, while the effective population size should include only breeding-age individuals. Berger (chap. 3, this vol.) indicates that the mean age of puberty for females is 2.1 years; while females frequently breed at this age, age at first reproduction for males was much older. (Only 2 of 16 males had sired offspring by the time they had been away from natal bands for 3 years. Since males leave their natal groups at about 2.3 years of age, less than 15% of males had bred by the time they were 5 years old.) Berger (pers. comm.) reports that in 1984 there were 58 breeding-age females (i.e., females 3 years old and older) and approximately 54 adult males. If we count all of the adult females and males, we obtain an $N_e$ of 112 breeding-age animals of the 171 animals in the population.

Horses live in harems consisting of one adult male and a number of reproductive-age females. Reproductively mature males that have not acquired harems are usually prevented from breeding to a great degree by this social structure. Thus the sex ratio of breeding animals is biased toward females. Of the approximately 54 adult males in the population in 1984, only 16 were stallions in possession of harems of one or more females (there were at least 38 nonbreeding bachelor males). Of the females, some had foals the previous year and were therefore less likely to breed in 1984. However, if we calculate the sex ratio $N_e$ using equation (19.2), with 16 males and 58 females, we obtain an $N_e$ of 50.2. This estimate would probably be further reduced if we were to subtract the number of females with foals from the number of females ($N_f$) before calculating $N_e$.

Berger (1986) indicates that there was a great deal of variance in progeny production among the Granite Range horses, but that the variance was much greater for males than for females. For females, the variance in progeny production was largely a function of longevity, but for males, the variance in progeny production was a function of tenure as a harem owner and size of a male's harem. While the mean age of death was similar for males and females (7.86 for females and 7.23 for males), the variance was much greater for males, and the median for longevity for males was less than for females (4.5 years for males, 6 years for females). Based on age-specific fecundity measures, Berger calculates that at the mean age of death an average female would leave behind 3.91 foals, while an average male would only leave 1.60. However, males showed a much greater variation in reproductive success than females, with the most successful male siring over 20 foals by the time he was 9 years old, but with many males dying before reproducing at all. Because few animals were followed throughout their lives, complete breeding histories could not be compiled for very many individuals. Berger did calculate an estimation of variation in maximal reproductive success based on age-specific fecundity for males and females (Berger 1986, 220) and found that the variation in male reproductive success was approximately five times that of females. The differences were not reported in the form of variances and so could not be used in our equation (19.4) to calculate the effect of variation in progeny production on $N_e$. However, it is clear from the calculations made on Rood's mongoose data (discussed later) that such differences would substantially reduce our estimation of $N_e$.

## Calculating $N_e$ for Rogers's Bears

$N_e$ for Rogers's bears was calculated using raw dispersal distances provided by Rogers (pers. comm.) and density estimates from his chapter. Since the density estimate for bears in the study area (1 bear/4.5 km²) presented in Rogers's chapter (chap. 5, this vol.) was based on individuals of all ages, we calculated the ratio of young to adults from a table of numbers of bears of each age and sex category from Rogers (1987). This figure (1.75 young: 1 adult) was then used to calculate the density of adult bears in the study area (1 adult bear/12.37 km²). Of the 18 male bears for whom dispersal information was available, dispersal distances ranged from 13 km to 219 km. These distances were leptokurtically distributed, however, with all but

1 male dispersing a distance of 105 km or less. Females dispersed a maximum of 11 km, but 28 of 31 females did not disperse at all. The 1 long-distance male disperser affects the dispersal variance so much, however, that Wright's areal $N_e$ (including the long-distance disperser) equals 843.8, while the same measure, excluding the one long-distance disperser, produces an $N_e$ of only 428.0.

Rogers indicates that his study was specifically designed to favor the collection of dispersal data within the study area. That is, individuals that did not disperse or dispersed to new locations within the study area (mainly females) could be easily located using signals from their radio collars. Information on longer-range dispersers (most males) was obtained only when their ear tags were found (usually by hunters) and returned. Although chance of ear-tag recovery in Canada was lower, as human population density was lower there, the chance of ear-tag recovery outside the study area was fairly uniform for shorter- or longer-distance dispersers throughout northern Minnesota and northern Wisconsin, as bear habitat was nearly continuous there, with scattered human population centers. Therefore it is not unreasonable to consider the longest dispersal distance (more than twice the next greatest dispersal distance) as genetic migration between neighborhoods rather than dispersal within a neighborhood. In this case the $N_e$ of 428 is not an unreasonable estimate for bears.

When the 85th percentile $N_e$ was calculated, using the greatest distance traveled by the 85% closest dispersers (56.5 km) as the radius of a circle to determine the area of the neighborhood, and then calculating the number of breeding-age adults within this area, we obtained an area of 10,023.67 km². Using the density figure calculated earlier (1 adult/12.37 km²), we calculated a neighborhood size of 810.3 breeding adults, which is more similar to the larger figure calculated for the areal $N_e$, uncorrected for bias in dispersal distances and including the long-distance disperser. For this population the 85th percentile $N_e$ is almost as great as the areal $N_e$ (and greater than the areal $N_e$ corrected for very-long-distance dispersal, i.e., migration) because bear dispersal is bimodally distributed, with short-distance dispersers (primarily females) forming one peak and long-range dispersers (primarily males) forming a second peak.

In his description of bear dispersal, Rogers indicated that males tend to disperse out of their mothers' territory while daughters tend to be philopatric, taking over a part of the mother's territory or establishing a territory nearby. This sex-biased pattern of dispersal results in a lineal effect such that females within a geographical area tend to be closely related to one another (though males do not). Therefore, while males tend not to mate with their own relatives, the females they mate with tend to be related to one another. This lineal distribution of females over a geographical area is not taken into account in our calculation of $N_e$, and in fact the method for including it has not yet been worked out. Nevertheless, it is an important characteristic of many mammalian populations including bears, deer, human and nonhuman primates, lions, and many other species. The effect of such a sex-biased pattern would be to lower the variance between the combined distances between birth sites of parents and birth sites of offspring, since each measure would include a near-zero number for the mother plus a larger number for the father. Thus sex-biased dispersal patterns (in which one sex is consistently more philopatric than the other) should further reduce $N_e$ beyond the measures indicated here.

Rogers, in censusing the population, found a sex ratio of 1 male to 3.5 females for adults (individuals 4 years old and older for both sexes). The Areal $N_e$ of 843.8 was divided into males and females according to this ratio, and the sex ratio $N_e$ was calculated according to equation (19.2). This brought about a reduction of $N_e$ from 843.8 to 583.6. If the areal $N_e$ calculated without the one longest-distance disperser is used, we see a reduction in $N_e$ from 428.0 to 291.7. If the 85th percentile $N_e$ is used as the initial figure before calculating the sex ratio $N_e$, we see a reduction of $N_e$ from 810.3 to 552.2.

We had initially calculated population size and sex ratio using mean "territory" sizes for male and female bears reported by Rogers. Rogers informed us, however, that because male "territories" are not exclusive, but overlap extensively, these calculations were inaccurate. It is important to be aware of the assumptions involved in interpolating from one type of information to another. If mean territory size is used to calculate population density, it is necessary that the territories be exclusive and nonoverlapping, as they are in prairie dogs and pikas, rather than overlapping as are home ranges of male bears.

## Calculating $N_e$ for Rood's Mongooses

Although Rood's study (chap. 6, this vol.) includes information on 97 mongoose pack–years (one pack year = one pack monitored for one year), this information was collected over a period of 10 years and is summed in the tables in chapter 6. The summed information for the 10-year period includes data on 32 different mongoose packs at the onset of 115 birth seasons, collected at 3 different study sites (2 of which were combined in the latter part of the study when the area monitored was expanded). The calculation of areal $N_e$ requires a measurement of population density and variance of dispersal distances traveled by reproductive-age animals in a single population at a given point in time. The additional information needed for this analysis was supplied by Rood (pers. comm.).

Between 1975 and 1979, population density of the original study site of 2.2 km² in October, just prior to the onset of the birth season, varied from 17.3 to 30.9 mongooses of all ages per square kilometer (Rood 1983). In the enlarged Sangere River study area, which contained sections of habitat unsuitable for mongooses, population density varied from 3.9 to 4.6 mongooses per km² (Rood, chap. 6, this vol.). Dispersal distances supplied by Rood (pers. comm.) were generally natal dispersal distances and were biased toward short-range dispersal, since most dispersal observed was between neighboring packs, and long-range dispersers that left the study area would not usually be found. Dispersal distances for 37 males ranged from 0.0 km to 4.0 km, with a one-way variance of 0.22 km². Dispersal distances for 56 females ranged from 0.0 km to 5.5 km, with a one-way variance of 0.92 km². The data suggest that females disperse farther than males, but Rood doubts this and believes that larger sample sizes may show that males disperse at least as far as females (Rood, pers. comm.). Furthermore, many more females than males were found to breed in their natal groups (29 females versus 3 males over the 10-year period). The one-way variance of combined male and female dispersal distances ($N$ = 93) is 0.64 km².

Using the one-way variance of the combined male and female dispersal distances and the four density figures indicated earlier, we calculated the following areal $N_e$'s using equation (19.1). In the less densely populated larger Sangere River study area, for a population density of 3.9/km², $N_e$ = 31.4; for a population density of 4.6/km², $N_e$ = 37.0. In the more densely populated original study site, for a population density of 17.3/km², $N_e$ = 139.1, and for a population density of 30.9/km², $N_e$ = 248.5. The larger figures would only apply to a population of mongooses that was uniformly distributed at these higher densities over a large enough area for adult population numbers to reach these levels. This was not the case in the area studied by Rood, and so the lower figures are more realistic for this population.

To calculate the 85th percentile $N_e$, we calculated the population density of a circle whose radius equals the distance traveled by the longest disperser of the closest 85% of all dispersers. This distance was 1.3 km, giving a circle whose area is 5.3 km². For the less densely populated larger Sangere River area, $N_e$ = 20.7 for a population density of 3.9/km², and 24.4 for a population density of 4.6/km². For the more densely populated part of the study area, $N_e$ = 91.7 for a population density of 17.3/km², and 163.8 for a population density of 30.9/km². These figures are somewhat lower than those calculated using Wright's areal $N_e$ equation (equation 19.1), reflecting the leptokurtic distribution of dispersal distances, with more mongooses dispersing shorter distances (or not at all) than longer distances.

In all of these calculations, the total number of mongooses in each pack was used to calculate population density. However, Rood points out that the social structure of dwarf mongooses is such that only the alpha male and the alpha female in each pack actually reproduce. Therefore, $N_e$ must be corrected for the ratio of breeding adults to others in the pack. The total number of adults at the start of the mating season in the area studied over a 10-year period varied from 23 to 98, but the number of packs during this period varied from 4 to 17. Since only 1 male and 1 female in each pack reproduce, the number of breeding adults was 2 times the number of packs, with a range of 8 to 34 over the 10-year period. We calculated the percentage of breeders to nonbreeders for each year of the 10-year period and found a mean percentage of adults that breed of 35%, with a range of 29% to 40%. Taking 35% of the areal $N_e$'s generated above, $N_e$ was reduced from 31.4 to 11.0 for a density of 3.9 mongooses/km², and from 37.0 to 13.0 for a density of 4.6 mongooses/km². $N_e$ was reduced from 139.1 to 48.7 for a density of 17.3 mongooses/km² and from 248.5 to 87.0 for a density of 30.9 mongooses/km².

Rood indicates that the population density at the start of the birth season at the enlarged Sangere River study area varied between 1980 and 1984 from 3.9 to 4.6 mongooses/km². These temporal variations in population size could cause further reductions in $N_e$.

Rood's is the only data set that includes lifetime reproductive success for a number of both females and males, and thus the only population for which we can calculate the effect of variation in progeny production on $N_e$, using equation (19.4). Rood was able to supply us with the complete breeding histories of 13 females and 5 males of known ages, as well as that of 10 females and 15 males whose ages were estimated by tooth wear. The total number of litters produced by females ranged from 1 to 13, with a range of 0 to 18 offspring that survived to the start of the subsequent birth season. The total number of litters produced by males ranged from 1 to 24, with a range of 0 to 23 offspring that survived to 1 year of age. Using the criterion of number of offspring that survived to the start of the next birth season, the mean number of yearlings raised by the 23 females was 3.43, with a reproductive variance of 17.47. The mean number of yearlings raised by the 20 males was 5.30, with a reproductive variance of 36.4. Although the reproductive variance of males was more than twice that of females for this sample, Rood believes that larger sample sizes would show that the variance for females is very similar to that found for males. Four females, for whom only partial breeding histories were known, raised far more yearlings than any of those for whom complete histories were known. Therefore we have combined the data for males and females, finding a mean of 4.30 yearlings raised by the 43 individuals, with a variance of 27.14. Using 27.14 for $V_k$ in equation (19.4), and the $N_e$'s calculated above based on the number of breeding adults for our beginning $N$'s, we found a reduction in $N_e$ from 11.0 to 1.4 for a population density of 3.9 mongooses/ km², a reduction from 13.0 to 1.7 for a population density of 4.6 mongooses/ km², a reduction from 48.7 to 6.6 for a population density of 17.3 mongooses/km², and a reduction from 87.0 to 11.9 for a population density of 30.9/km². Thus reproductive variance of the magnitude documented in this species substantially reduces effective population size.

Of all the factors affecting $N_e$, reproductive variance appears to have the greatest effect, particularly in a species that suffers a fairly high mortality rate among juveniles. (Rood 1983 reports that only 46% of 148 juveniles counted when less than 2 months old were still present at 1 year of age.)

### Calculating $N_e$ for Mech's Wolves

Mech (pers. comm.) was able to supply us with dispersal information on 10 individuals: 5 females and 5 males. These distances are the straight-line distances between where an animal was born and where it eventually either bred or died (or both). The only 2 animals definitely known to have bred did so 11 km from their home territory. Two of the long-distance dispersers were trapped as solitary wolves and probably did not breed. The third was with a pack and may have bred. The range of dispersal distances is similar for males (0–306 km) and females (10–290 km), but the distribution of dispersal distances in this small sample is extremely bimodal, with 7 wolves dispersing no farther than 80 km from their birthplaces while 3 traveled more than 200 km from where they were born. The sample size is too small to determine whether this bimodal pattern is typical of wolf dispersal or whether it is an artifact of small sample size and/or availability of suitable habitat.

Wolves live in groups of 2 to 17 individuals in territories averaging approximately 230 km² in size. Although there may be more than 2 adults in a territory, wolf social structure is such that only 2 adults in each territory breed, so the density of breeding individuals is 2/230 km² or .0087 wolves/km². The one-way variance of wolf dispersal distances for the sample is 7,355.7 km². Using these figures for density and dispersal variance in equation (19.1), we obtain an $N_e$ of 804 wolves.

To calculate the 85th percentile $N_e$ we defined a circle whose radius was midway between the dispersal distances of the eighth closest (203 km) and the ninth closest (290 km) dispersal distances. Thus the radius of the circle was 246.5 km and the area was 190,890.23 km². Multiplying this figure by a density of .0087 wolves/km² we obtained an $N_e$ of 1660.7, which is more than twice that found with the traditional method. Here the sizable proportion of long-distance dispersers (30%) suggests that the variance method underestimates $N_e$.

Note that all of these values assume that wolf habitat is uniformly suitable and that wolf densities are the same over a very large area. We know that this is not true where Lake Superior abuts the study area and dispersing wolves were seen to change course upon reaching the lake (Mech, chap. 4, this vol.). We also know that the wolves avoided heavily populated areas such as Duluth. However, Mech (pers. comm.) indicates that wolf habitat and wolf population are fairly uniformly distributed over much of northern Minnesota and southern Canada and that wolf populations in the two geographical areas are contiguous. There are an estimated 1,200 wolves in Minnesota and an estimated 50,000 wolves in Canada. Since wolf distribution is fairly uniform with a large proportion (30%) of long-distance dispersers, it is not unlikely that $N_e$ is really as large as these figures indicate. However, dispersal variance here is based on a very small sample size (only 10 dispersers) from a single pack whose population varied from 18 to 20 members over the course of the study. Because of the small sample size these figures should be considered only a first approximation of $N_e$ for wolves that should be verified against a larger body of information when it becomes available.

Furthermore, all of these estimates for $N_e$ assume that all dispersal distances are effective dispersal distances (that is, that the dispersers succeed in breeding in their new locations). However, of the wolves, only the two dispersers that moved only 11 km were actually known to have bred. If more long-distance dispersers fail to breed than short-distance dispersers, as was found by Endler (1979) for *Drosophila* and is implied here for the wolves, then the actual effective population size would be much lower than these estimates.

Since wolves are monogamous, the sex ratio of breeding males to females would be 1:1, so there would be no change in $N_e$ due to sex ratio.

## Calculating $N_e$ for Halpin's Prairie Dogs

$N_e$ for Halpin's prairie dogs was calculated using raw data on individual dispersal distances provided by Halpin (pers. comm.). Halpin points out that her prairie dog town may not be typical for a number of reasons. The prairie dog town was recently founded and was smaller than those in a number of other studies. Also, a shooting incident in 1981 significantly decreased the population and probably disrupted social structure to some extent (Halpin, pers. comm.). King (1955) mentions that in black-tailed prairie dogs there is some adult dispersal as well as juvenile dispersal, but only juvenile dispersal was monitored in Halpin's study.

Over the course of the 4-year study, the number of adults of breeding age varied from a high of 34 in 1979 to a low of 21 in 1980. Animals lived in four to nine coteries, usually with 1 male and 1 to 6 females in each coterie. There was an average of 9.7 (+ or − 4.95) individuals in each coterie, including adults, yearlings, and young confirmed to be surviving at the end of the summer.

Of the 81 females trapped and marked as juveniles over the 4 years, 51 remained in their natal coteries and 30 disappeared and were presumed to have died. There was no evidence for female dispersal, either between coteries or between prairie dog towns. Of the 75 males born in the study area and marked as juveniles, only 10 joined coteries within the prairie dog town and bred there. The other 65 disappeared, and it is not clear how many successfully dispersed out of the prairie dog town and how many died. A number of males were last seen on the outskirts of the town and so were presumed to have been attempting to disperse. As Halpin indicates in her study (chap. 7, this vol.), she surveyed the surrounding area for evidence of prairie dog activity and found two other small prairie dog towns in the vicinity, one located 3 km and the other 6.5 km from the study area. Although she was unable to ascertain whether any prairie dogs from her study area ever successfully dispersed out of the town, over the course of 4 breeding seasons one male from outside did set up a territory in the town and did breed there. It was supposed that he came from one of the nearby prairie dog towns.

Male juvenile dispersal appears to be relatively random among coteries within this prairie dog town (though mating is limited to coterie members); however, little genetic migration appeared to occur between prairie dog towns. Therefore, we consider the prairie dog town a deme (or subpopulation) with low levels of migration between demes, and so we use the census size (21–34) as our first approximation of $N_e$.

Female prairie dogs may breed at 1 year of age, while males usually do not breed until they are 2 years old. Therefore the adult (breeding age) sex ratio is skewed—2.5 females to 1 male. Using this sex ratio to divide the figures obtained for census $N_e$ into males and females, and equation (19.2), we calculated the further reduction in $N_e$ to be expected as a result of the uneven sex ratio. If 34 is used as the starting figure, the sex ratio $N_e$ is 27.74. If 21 is used as the starting figure, the sex ratio $N_e$ is 17.14.

In order to calculate the effect of temporal variation on $N_e$ we would need to calculate $N_e$ by the above methods for each of a series of generations in order to use equation (19.3) to calculate the effect of changes in population size on $N_e$. In most mammals, including prairie dogs, generations are overlapping, and so more complex equations are more appropriate to the task. Nevertheless, for purposes of demonstration, we will use the prairie dog data to calculate $N_e$ via equation (19.3), considering each year of the study as a separate generation.

We have used numbers of adults present and sex ratios for each of the 4 years of the study provided by Halpin (pers. comm.). From these figures we have calculated first the sex ratio $N_e$ (equation 19.2) for each year, and then $N_e$ for the population based on the temporal variations over the 4 years. The data used are shown in table 19.3. Entering these data into equation (19.3) we obtain

$$\frac{1}{N_e} = \frac{1}{t} \, \Sigma_i \frac{1}{N_i}$$

$$= \frac{1}{4}\left(\frac{1}{25.8} + \frac{1}{29.2} + \frac{1}{18.0} + \frac{1}{22.4}\right)$$

$$= \frac{1}{4}\left(.173\right)$$

$$= .0433.$$

$$N_e = 23.1.$$

Table 19.3 Breeding-Age Adult Prairie Dogs in Halpin's Sample for the Years 1978–1981

| Year | No. Adults | Male:Female Ratio | Sex Ratio $N_e$ |
|---|---|---|---|
| 1978 | 30 | 8:22 | 25.8 |
| 1979 | 34 | 11:23 | 29.2 |
| 1980 | 21 | 6:15 | 18.0 |
| 1981 | 26 | 7:19 | 22.4 |

*Note:* Adults include all males and females 2 years old or older, plus those yearling females breeding in a given year.

*Source:* Halpin, pers. comm.

It has been argued that a single genetic migrant per generation will raise estimations of $N_e$ substantially. Since one long-distance disperser did successfully immigrate into the study area during one of these years, let us try to estimate what effect his migration could have on $N_e$. Assuming that one year is the approximate generation time for prairie dogs, let us suppose that that one genetic migrant raises the $N_e$ as high as 1328.8 (the value obtained when the areal $N_e$ is calculated according to equation (19.1), making the [incorrect] assumption of uniform distribution of prairie dogs and using dispersal distances reported by Halpin plus an additional dispersal distance of 3 km, assuming that the immigrant came from the closest nearby colony). We consider this very much an overestimate, since prairie dogs are not uniformly distributed, as noted earlier. If we substitute the value 1/1328.8 for 1/22.4 in the previous equation, we obtain

$$\frac{1}{N_e} = \frac{1}{4}\left(\frac{1}{25.8} + \frac{1}{29.2} + \frac{1}{18.0} + \frac{1}{1328.8}\right)$$
$$= \frac{1}{4}(.1293)$$
$$= .032.$$
$$N_e = 30.9.$$

Because the harmonic mean is used to calculate temporal variation in $N_e$, the one year when $N_e$ is very large is swamped by the years when $N_e$ is small, so the $N_e$ over several generations is much smaller than if a long-distance migrant were received in each generation.

### Calculating $N_e$ for Jones's Kangaroo Rats

For Jones's kangaroo rats, the areal $N_e$ was an appropriate estimation of $N_e$ since kangaroo rats are relatively uniformly distributed. The dispersal variance was calculated using only the dispersal distances of the 79 juveniles whose natal burrows were known. Raw data were generously supplied by Jones (pers. comm.), and dispersal distances varied from 0 m to 388 m but were strongly leptokurtic (Jones, chap. 8, this vol.) with a median of 17 m for males and 30 m for females. Mean dispersal distance for male and female juveniles combined was 41.9 m, with a one-way variance of 2,478.78 m$^2$. Kangaroo rat density was calculated by using the mean distance between burrows (50 m) as home range diameter, and, assuming that territories approximate circles. We calculated the area of a circle of radius equals 25 m, and obtained an average territory size of 1,963.50 m$^2$. Since there is only one adult per territory, density equals 1/1,963.50 m$^2$. Using these figures for dispersal variance and density of breeding-age adults in equation (19.1), we obtained an $N_e$ of 15.9.

When the 85th percentile $N_e$ is calculated using the longest dispersal distance of the 85% of dispersers that moved the shortest distances as the radius of a circle encompassing the neighborhood, we obtain a radius of 68 m and an area of 14,526 m$^2$. Multiplying this figure by a density of 1 kangaroo rat/1,963.5 m$^2$ we obtained an $N_e$ of 7.3.

Since the sex ratio of kangaroo rats does not differ substantially from 1:1, the sex ratio $N_e$ would be the same as the areal $N_e$. Since kangaroo rat burrows appear to be relatively continuously distributed over the study area, rather than in patches like the prairie dogs' and the pikas', and since a concerted effort was made to find long-distance dispersers (see Jones, chap. 8, this vol.), it seems that these figures, though surprisingly low, are probably fairly accurate.

### Calculating $N_e$ for Smith's Pikas

Smith and Ivins (1983) reported densities of adult pikas on their study site as 8.6 pikas/ha in 1979, 8.2/ha in 1980, and 4.0/ha in 1981 (after a particularly bad winter). We have used the highest and the lowest of these density figures in calculating our estimates of $N_e$ in order to show the effect of density on $N_e$ within the density levels observed. Smith (pers. comm.) provided us with raw data on individual dispersal distances for 41 adult pikas (18 males and 23 females) and 10 juveniles (4 males and 6 females). It is important to remember that pikas are found only on talus, which is distributed in patches of varying sizes separated from each other by forested areas or meadows that extend for varying distances and that act as greater or lesser barriers to dispersal, depending on the distance between talus slopes and altitude (see Smith, chap. 9, this vol.). Of the adults, only 4 males and 2 females dispersed, while 35 remained where they were. Thus the average dispersal distance for adults was 16.83 with a one-way variance of 1,428.45 m$^2$. Using this figure for dispersal variance in equation (19.1) we obtained an $N_e$ of 15.4 when density equals 8.6 pikas/ha and an $N_e$ of 7.18 when density equals 4.0 pikas/ha.

Among the 10 juvenile pikas that were born on the study area and eventually settled there, only 6 dispersed and only 1 of these dispersed more than 50 m from where she was born. Dispersal distances ranged from 0 m to 80 m with a mean of 22.0 m, and a one-way variance of 331.02 m$^2$.

Using this figure in equation (19.1) we obtained an $N_e$ of 3.58 pikas when density equals 8.6 pikas/ha and an $N_e$ of only 1.66 when $N_e$ equals 4.0 pikas/ha.

Smith did observe 4 unmarked immigrants that settled on the study site and successfully set up territories there over the course of 3 years. When asked to make an educated guess as to how far these immigrants may have migrated, Smith was reluctant to do so because he lacked evidence to support such a guess. He did give us the distance to the center of the closest "pool" of high-population density up the slope from the study area. We have used these figures as a rough approximation of possible dispersal distances, though one should be aware that these are only approximations and may either underestimate or overestimate actual distances traveled. The estimated dispersal distances varied from 100 m to 300 m. When these immigrants' dispersal distances are added to those of the juveniles, we obtained a mean dispersal distance of 76.43 m and a one-way variance of 5,443.4 m². Using this figure for dispersal variance in equation (19.1) we obtained an $N_e$ of 58.8 for a density of 8.6 pikas/ha and an $N_e$ of 27.4 pikas for a density of 4.0 pikas/ha.

When we attempted to calculate the 85th percentile $N_e$ for adults, $N_e$ was undefined because all dispersal distances were zero. In calculating the 85th percentile $N_e$ using dispersal distances for juveniles alone we obtained a circle whose radius is 35 m and whose area equals 3,848.4 m². The $N_e$ obtained using this method equals 3.3 for a density of 8.6 pikas/ha and 1.5 for a density of 4.0 pikas/ha. These figures agree very well with those obtained using equation (19.1). When the 85th percentile $N_e$ is calculated using the juveniles plus the 4 long-distance dispersers, a radius of 150 m is obtained, giving an area of 70,685.8 m² and an $N_e$ of 60 when density equals 8.6 pikas/ha and an $N_e$ of 28.3 when density equals 4.0 pikas/ha.

Both sets of figures for the 85th percentile $N_e$ compare well with those for Wright's areal $N_e$ even though pikas, like prairie dogs, are not uniformly distributed but occur in patches of acceptable habitat widely separated by uninhabitable terrain. Because the uniform density criterion is violated for pikas, it is likely that an $N_e$ based on census numbers is a more accurate approximation of effective population size for this species. Smith and Ivins (1983) report that the number of pikas (adults plus immatures) in their White Rock study area at the end of summer varied from 29 to 52 over a 3-year period, but that only 12 to 27 of these were adults of breeding age. Based on censuses of this one subpopulation we would then estimate an $N_e$ of 12 to 27. Additional censuses of other subpopulations may yield different figures, but these numbers seem reasonable considering the very low dispersal rates and distances described for pikas.

Pikas are monogamous, with a sex ratio of 1:1, so there is no change in $N_e$ to be expected due to sex ratio. Temporal variance and reproductive variance are also expected to have little to no effect on $N_e$. Population density of pikas is considered to be one of the most constant over time of any mammal yet investigated (Smith 1978; Southwick et al. 1986). Similarly, Smith (chap. 9, this vol.) mentions that there is little to no reproductive variance in pikas, and indicates that what little variance may occur is presumably nullified because mothers are long-lived and chance effects of one year are unlikely to affect them in subsequent years. Therefore, neither of these parameters is believed likely to further lower $N_e$ in most populations most of the time (despite the threat of an occasional avalanche).

Nevertheless, even the highest estimates for $N_e$ of subpopulations of pikas are extremely low with very low levels of genetic migration between subpopulations. Consistent with these observations is the fact that of the 6 individuals born on the study site and for whom both pedigree information and mating information was available, 3 mated with close relatives (2 mother-son matings and 1 father-daughter mating). The low effective population sizes and high levels of inbreeding indicate that this species is almost certainly evolving in a manner similar to Wright's shifting balance process rather than according to a mass selection scheme.

## SUMMARY

We do not intend for the estimates of $N_e$ presented in table 19.2 to be taken as the final word on effective population sizes for these species. They are merely estimates based on certain measures (often made on inadequate sample sizes) and a number of simplifying assumptions (see Lidicker and Patton [chap. 10, this vol.], Templeton [chap. 17, this vol.]). Since these measures are only based on data for specific populations, they may not be representative of the species as a whole. We hope we have made clear what the assumptions are and how reasonable they may or may not be in each case. Further, as mentioned earlier, all $N_e$'s calculated here are static $N_e$'s—that is, the effective population size for a given slice of time, which may or may not persist over many generations and which may or may not accurately represent the effect of genetic migration between subpopulations. Future work may indicate that long-distance dispersers in some species are able to set up territories and breed successfully much more frequently than the present data indicate. If this is the case, our estimates of $N_e$ for those species are much too low. On the other hand, if lifetime reproductive success is highly variable (which was true in the only case in which we had sufficient data to assess), $N_e$'s would be substantially smaller than those reported here.

The overall picture that emerges, however, is that in almost all cases examined here, $N_e$ is quite low. How low is low enough for Wright's shifting balance model of evolution to apply? Wright (1978a, 56–57) says, "In terms of effective neighborhood sizes, in a species with fairly uniform density and amounts of diffusion throughout its range, this implies a very effective process [of interdeme selection by excess diffusion from the superior centers] if $N_e$ averages about 50, a somewhat effective process if about 100 or even 200 but probably not if of the order of 500." We would therefore expect, based on the estimates of $N_e$ obtained here, that the shifting balance process is likely to occur very effectively in Berger's horses, Rood's mongooses, Halpin's prairie dogs, Jones's kangaroo rats, and Smith's pikas. A less effective process appears to be operating in Nelson and Mech's deer. However, Rogers's bears and Mech's wolves appear more likely to be evolving under a mass selection scheme (though sample size is a problem here).

### Tendency of Many Mammals to Have a Subdivided Population Structure

The general conclusion one may draw from these direct and indirect estimates of $N_e$ (table 19.2) and other measures of population structure reported in this volume as well as in similar studies reported elsewhere (reviewed in Wright 1978a; Shields 1982) is that many mammals live in small, partially isolated, and *therefore* at least mildly inbred demes. The *inevitable* increase in inbreeding in smaller populations is sometimes overlooked. For example, Cheney and Seyfarth (1983) were able to document nonrandom dispersal in their vervet monkeys that would reduce effective population size much below what would have occurred with more random dispersal. Despite the increased inbreeding that would result, they discussed the pattern as an inbreeding avoidance mechanism (apparently referring to extreme pedigree inbreeding, or incest, rather than to the milder random inbreeding occurring as a result of finite population size). Cheney & Seyfarth (1983, 393) dismissed the inbreeding effects by citing Wright (1921): "This effect [reduced heterozygosity and inbreeding depression] appears to be limited to close kin, however, since even repeated mating between individuals less closely related than full first cousins produces no significant increase in homozygosity." Wright's reference, however, was to consanguinous matings in an otherwise *infinite* population.

In any finite population, like their vervets, every potential mate will come to trace common ancestry via multiple paths, and the degree of inbreeding can be very high even when mates are not immediate close pedigree kin. The only field studies that can provide data on actual levels of inbreeding must be truly long term and document a number of generations where pedigrees are known with some surety (for example, the Hutterite population studied by O'Brien, chap. 13, this vol.).

As more of these studies are done, it seems likely that the initial conclusion that many mammalian species are subdivided into more or less isolated demes will continue to be confirmed. Such a subdivided population structure with small $N_e$ and some genetic migration among demes has been documented for enough mammals that brief consideration of some of its expected consequences might be profitable.

### Genetic Consequences of Subdivided Population Structure

The subdivided population structure observed in many mammals is expected to have numerous genetic consequences. Small to moderate effective population sizes ($N_e < 500$) and low rates of interdeme migration will increase the level of inbreeding and the opportunity for genetic drift relative to the wider dispersal and larger $N_e$'s characterizing other taxa (e.g., marine invertebrates, reviewed in Shields 1982). If much of the allelic variability characterizing mammals were neutral, or nearly so, with respect to phenotypic effects on fitness, then individual heterozygosity would be expected to decline or remain at relatively low levels. Empirical estimates of structural allele heterozygosity are consistent with this view, as mammals often show lower levels of average heterozygosity and mean population polymorphism than many other taxa (e.g., Selander and Johnson 1973; Powell 1974; Nevo 1978).

The pattern of mammalian chromosomal evolution is also consistent with a view that mammals live in relatively small demes in subdivided metapopulations in that many mammalian taxa have relatively high rates of chromosomal evolution (for reviews, see Wilson et al. 1975; Bush et al. 1977; and Imai, Maruyama, and Crozier 1983). In fact, using reasonable estimates of the rates of chromosome mutation and fixation, Lande (1979) concluded that mammals breed in demes with long-term effective sizes of 30 to 200 and that the social structure and mating systems characterizing mammals probably contributed to the limited deme sizes he estimated. Thus, many lines of evidence suggest that many, if not all, mammals breed in small, semi-isolated, and therefore relatively inbred demes (but see Templeton, chap. 17, this vol.).

It is not true, however, that a deficiency of heterozygotes is always going to be in evidence given a demic population structure. As many have noted, that expectation *requires* that allelic and genotypic variability be effectively neutral. If any form of diversifying selection is operating, then heterozygosity may be maintained at relatively high levels despite significant subdivision and inbreeding. Indeed, relatively mild selection pressures can maintain Hardy-Weinberg genotype frequencies even in the face of significant inbreeding (reviewed in Crow and Kimura 1970; Wright 1978a; and Shields 1982). Given that the vast majority of progeny born are going to die without issue anyway (or populations would grow continuously), any form of "soft" or truncation selection (Wallace 1968) can maintain much variability, especially in the face of mild or moderate inbreeding.

If there is strong genetic inertia, such that some set of alleles is useful regardless of geographic or temporal environmental flux, then each population might carry the same allele complexes at similar frequencies despite being more or less isolated from each other. Finally, even if the alleles and genotypes in question are effectively neutral, cycles of isolation followed by periods of effective gene flow can periodically reintroduce heterozygosity and restore the genetic continuity of a group of subpopulations even when such demes are sufficiently small and isolated to allow considerable drift over one or a few generations (e.g., Lidicker and Patton, chap. 10, and Bowen and Koford, chap. 12, this vol.).

Thus, we cannot go out into the field, measure allele or genotype frequencies, calculate $F_{ST}$ or genetic distances, and draw *unequivocal* conclusions about whether a species is inbreeding or outbreeding or is structured or not. Recently, however, Slatkin (1980, 1981) has provided a novel genetic technique that used different assumptions to estimate the levels of gene flow in nature. His method suggests that the distribution of conditional average frequencies of alleles in subpopulations is, based on simulation, relatively insensitive to mutation and selection and quite responsive to relative levels of gene flow. In this sense, then, it offers a reasonable index for exploring gene migration or gene dispersal in the absence of data on propagule dispersal or mating systems (for an analysis of this and older methods, see Larson, Wake, and Yanev 1984). Despite such advances, however, the only unequivocal method for determining population structure is to obtain long-term pedigrees and knowledge of genetically effective migration among local breeding groups. Demographic and behavioral information must be combined with genetic surveys of allele frequencies and patterns of differentiation to get a realistic view of the effects of dispersal and social structure on population genetics.

## GUIDELINES FOR FUTURE FIELD STUDIES

In calculating the estimates of $N_e$ we ran into a number of difficulties with the data that were supplied for the purpose. While each of the previously discussed field studies is one of the most complete and detailed of its kind, certain types of data often either were not collected or were not reported in a way that was appropriate for the calculations necessary to estimate $N_e$. In many cases the appropriate data had been collected, but it was necessary for the contributors to recollate their data to produce the measures required for our calculations. In other cases field-workers could easily have collected the information needed but did not do so because they were unaware of its usefulness. We would therefore like to list the types of information that should be collected and reported on in future field studies.

### Calculating Areal $N_e$

This method is to be used only for animals that are uniformly (or nearly uniformly) distributed over a given area.

1. The equation (19.1) requires a measure of population density of animals of breeding age, only for an area over which the population is fairly uniformly distributed. Such a measure should be based either on accurate censuses over a large area, average home range size and overlap, or interindividual distances between breeding adults. If one of the latter two measures is used it should be checked against census figures for a large area to be sure that it is consistent over the range of the study area. If the population borders on geographical barriers such as lakes, cities, or other uninhabitable areas, these areas should be subtracted from the areal measure before density is computed.

2. The equation also requires the one-way variance of dispersal distances traveled by individuals between time of birth and time of reproduction. The more commonly reported mean and range of dispersal distances are certainly of interest, but it is the variance that is used in the equation for areal $N_e$ and this is often not reported. If dispersal distance varies for males and females, the variance should be reported separately for the sexes as well as combined. If individuals migrate between episodes of reproduction, more complex methods are appropriate (see Sade et al., chap. 15, this vol.).

### Estimating $N_e$ of a Subpopulation

This method should be used for animals whose distribution is discontinuous or patchy, or where an isolated population is studied.

The first approximation of $N_e$ is simply the total number of reproductively mature individuals. Size of subpopulation including young is also of interest, but $N_e$ includes only reproductive individuals.

### Calculating Sex-Ratio $N_e$

Sex ratio $N_e$ is calculated using the number of breeding males and the number of breeding females for a single breeding season in equation 19.2. These numbers should be presented separately for each breeding season if more than one breeding season was observed. When combining values, the totals for separate mating seasons should be averaged, not summed. If individuals are reproductively mature but do not mate they should not be counted in the sex-ratio $N_e$.

### Finding Temporal Variations in $N_e$

The number of breeding males and females *per generation* is needed to calculate this estimate of $N_e$. Temporal variation is particularly important in populations that experience large variances in population size, such as the boom and bust years of *Microtus* sp. (Lidicker and Patton, chap. 10, and Bowen and Koford, chap. 12, this vol.). However, for populations in which generations are overlapping, one should refer to the life table equations discussed by Charlesworth (1980), which are more appropriate than the simple equation presented here.

### Finding Effects of Variation in Number of Progeny on $N_e$

Here the measure needed is the variance in progeny production over the *lifetime* of individuals. The collection of such data involves following each of a substantial number of individuals over the course of their lifetimes and collecting data on how many offspring each produces. In long-lived individuals this can be an ambitious undertaking. Alternatively one can compute such a value from a cross section of a population by compiling a life table for the population (see Keyfitz 1968). Indeed the life table values are the most appropriate measures to use in populations of animals with overlapping generations (which includes most mammals). Once this has been done using a reasonable sample size for the population, the variance in number of progeny produced over the course of an individual's life must then be calculated, for it is the variance that is used in calculating this estimate of $N_e$. Variance in progeny production may differ for males and females, and so the sexes should be calculated both separately and combined.

### Effective Migration Rate

In addition to this information necessary to the calculation of $N_e$, an attempt should be made to make an accurate assessment of the per generation genetically effective migration rate for a population. This assessment requires first an estimation of generation time, which for mammals with overlapping generations is best calculated from a life table (Keyfitz 1968). Second, genetic migration must be separated from mortality, a truly difficult task in mammals that range over long distances (e.g., Halpin, chap. 7, Smith, chap. 9, Rogers, chap. 5, and Mech, chap. 4, all in this vol.). Finally, an individual that moves from one location to another cannot be considered a genetically effective migrant until he has mated and produced offspring in his new location (see Gaines and Johnson, chap. 11, and Smouse and Wood, chap. 14, this vol.). The collection of this information is difficult, but an accurate measure of genetic migration rates is impossible without it.

### Genetic Polymorphisms

Finally, all of the population and life-history information indicated earlier should ideally be accompanied by genetic information on a large sample of the population studied in the form of genetically inherited traits which show some variability in the population, such as blood group polymorphisms, isozymes, or chromosome banding studies. Since evolution involves changes in the genetic structure of populations, genetic sampling, like censusing, should be an ongoing project so that changes in gene frequencies can be monitored over time. These changes can then be measured and related to stochastic events and/or selection pressures.

## CONCLUSIONS

While it is true that effective population size calculated in this chapter is only one of many population genetical parameters and is certainly not the main objective of population genetics, it is the one most appropriate measure for the assessment of the genetic effects of different social structures and dispersal. It is also a measure that allows us to compare very different species and social organizations with each other from the standpoint of the expected rate and mode of evolution.

The present volume is not intended to be the final word on the effects of mammalian dispersal patterns on the genetic structure of populations. Rather it is an indication of the state of the art and deals with the interface between two fields that have rarely been successfully brought together before. It is meant to stimulate interest and to generate new, more comprehensive research that will take into account some of the problems, methods, and opportunities for further study indicated in these chapters.

## ACKNOWLEDGMENTS

This chapter has profited from comments from all of the participants in the symposium volume, most particularly, Bonnie Bowen, Jim Cheverud, Rolf Koford, Donald Sade, Peter Smouse, Alan Templeton, Michael Wade, and Peter Waser, whose direct comments on the manuscript and stimulating general discussions resulted in significant revisions. We must also thank those who were not participants in the symposium but who provided manuscripts, ideas, and critical comments in abundance, including J. Crook, L. Wolf, and E. Waltz. Kerry Knox has been very helpful in supplying copies of journal articles to the senior author at remote locations. The North Country Institute for Natural Philosophy and Northwestern University have provided institutional support.

## REFERENCES

Begon, M. 1976. Dispersal, density and microdistribution in *Drosophila subobscura* Collin. *Journal of Animal Ecology* 45:441–56.

———. 1978. Population densities in *Drosophila obscura* Fallen and *D. subobscura* Collin. *Ecological Entomology* 3:1–12.

Berger, J. 1986. *Wild horses of the Great Basin: Social competition and population size.* Wildlife Behavior and Ecology Series. Chicago: University of Chicago Press.

Blair, W. F. 1960. *The rusty lizard: A population study.* Austin: University of Texas Press.

Bush, G. L., S. M. Case, A. C. Wilson and J. L. Patton. 1977. Rapid speciation and chromosomal evolution in mammals. *Proceedings of the National Academy of Sciences* (USA) 74:3942–46.

Charlesworth, B. 1980. *Evolution in age-structured populations.* Cambridge: Cambridge University Press.

Cheney, D. L., and R. M. Seyfarth. 1983. Nonrandom dispersal in free-ranging vervet monkeys: Social and genetic consequences. *American Naturalist* 122:392–412.

Crow, J. F., and M. Kimura. 1970. *An introduction to population genetics theory.* Minneapolis, Minn.: Burgess Publishing.

Daly, J. C. 1981. Effects of social organization and environmental diversity on determining the genetic structure of a population of the wild rabbit, *Oryctolagus cuniculus. Evolution* 35:689–706.

Dice, L. R., and W. E. Howard. 1951. Distance of dispersal by prairie deermice from birthplace to breeding sites. *University of Michigan Laboratory of Vertebrate Biology Contributions* 50:1–15.

Dobzhansky, T., and S. Wright. 1943. Genetics of natural populations. X. Dispersion rates in *Drosophila pseudoobscura. Genetics* 28:304–40.

Endler, J. A. 1977. *Geographic variation, speciation, and clines.* Princeton: Princeton University Press.

———. 1979. Gene flow and life history patterns. *Genetics* 93:263–84.

Fisher, R. A. [1930] 1958. *The genetical theory of natural selection.* Rev. ed. New York: Dover.

Imai, H. T., T. Maruyama, and R. H. Crozier. 1983. Rates of mammalian karyotype evolution by the karyograph method. *American Naturalist* 121:477–88.

Keyfitz, N. 1968. *Introduction to the mathematics of population.* Reading, Mass.: Addison-Wesley.

Kimura, M., and T. Ohta. 1971. *Theoretical aspects of population genetics.* Princeton: Princeton University Press.

King, J. A. 1955. Social behavior, social organization, and population dynamics in a black-tailed prairie dog town in the Black Hills of South Dakota. *University of Michigan Laboratory of Vertebrate Biology Contributions* 67:1–123.

Lande, R. 1979. Effective deme sizes during long-term evolution estimated from rates of chromosomal rearrangement. *Evolution* 33:234–51.

Larson, A., D. B. Wake, and K. P. Yanev. 1984. Measuring gene flow among populations having high levels of genetic fragmentation. *Genetics* 106:293–308.

Moore, J., and R. Ali. 1984. Are dispersal and inbreeding avoidance related? *Animal Behaviour* 32:94–112.

Nevo, E. 1978. Genetic variation in natural populations: Patterns and theory. *Theoretical Population Biology* 13:121–77.

Nice, M. M. 1964. *Studies in the life history of the song sparrow,* vol. 1: *A population study of the song sparrow and other passerines.* New York: Dover.

Powell, J. R. 1974. Protein variation in natural populations of animals. *Evolutionary Biology* 8:79–119.

Rogers, L. L. 1987. Effects of food supply and kinship on social behavior, movements, and population dynamics of black bears in northeastern Minnesota. *Wildlife Monographs* 97.

Rood, J. P. 1983. The social system of the dwarf mongoose. In *Recent advances in the study of mammalian behavior,* ed. J. F. Eisenberg and D. G. Kleiman. Special Publication No. 7, American Society of Mammalogists, Shippensburg, Pa.

Selander, R. K., and W. E. Johnson. 1973. Genetic variation among vertebrate species. *Annual Review of Ecology and Systematics* 4:75–91.

Shields, W. M. 1982. *Philopatry, inbreeding, and the evolution of sex.* Albany: State University of New York Press.

Slatkin, M. 1980. The distribution of mutant alleles in a subdivided population. *Genetics* 95:503–23.

———. 1981. Estimating levels of gene flow in natural populations. *Genetics* 99:323–35.

Slatkin, M., and M. J. Wade. 1978. Group selection on a quantitative character. *Proceedings of the National Academy of Sciences* 75:3531–34.

Smith, A. T. 1978. Comparative demography of pikas (*Ochotona*): Effect of spacial and temporal age-specific mortality. *Ecology* 59:133–39.

Smith, A. T., and B. L. Ivins. 1983. Colonization in a pika population: Dispersal vs philopatry. *Behavioral Ecology and Sociobiology* 13:37–47.

Southwick, C. H., S. C. Golian, M. R. Whitworth, J. C. Halfpenny, and R. Brown. 1986. Population density and fluctuations of pikas (*Ochotona princeps*) in Colorado. *Journal of Mammalogy* 67:149–53.

Wade, M. J. 1976. Group selection among laboratory populations of *Tribolium*. *Proceedings of the National Academy of Sciences* (USA) 73:4604–7.

Wallace, B. 1968. *Topics in population genetics*. New York: W. W. Norton.

Wilson, A. C., G. L. Bush, S. M. Case, and M. C. King. 1975. Social structuring of mammalian populations and rate of chromosome evolution. *Proceedings of the National Academy of Sciences* (USA) 72:5061–65.

Wright, S. 1921. Systems of mating, pts. 1–4. *Genetics* 6:111–78.

———. 1931. Evolution in Mendelian populations. *Genetics* 16:97–159.

———. 1932. The roles of mutation, inbreeding, cross-breeding and selection in evolution. *Proceedings of the Sixth International Congress of Genetics*, 356–66.

———. 1943. Isolation by distance. *Genetics* 28:114–38.

———. 1956. Modes of selection. *American Naturalist* 90:7–24.

———. 1965. The interpretation of population structure by f-statistics with special regard to the system of mating. *Evolution* 19:395–420.

———. 1969. *Evolution and the genetics of populations*, vol. 2: *Theory of gene frequencies*. Chicago: University of Chicago Press.

———. 1978a. *Evolution and the genetics of populations*, vol. 4: *Variability within and among natural populations*. Chicago: University of Chicago Press.

———. 1978b. The relation of livestock breeding to theories of evolution. *Journal of Animal Science* 46:1192–1200.

———. 1980. Genic and organismic selection. *Evolution* 34:825–43.

———. 1982. Character change, speciation, and the higher taxa. *Evolution* 36:427–43.

# Contributors

Joel Berger
Department of Range, Wildlife
  and Forestry
University of Nevada
Reno, Nevada 89512
  *and*
Conservation and Research Center
Smithsonian Institution
Front Royal, Virginia 22630

Bonnie S. Bowen
Department of Biological Sciences
State University of New York at
  Albany
Albany, New York 12222

Felix J. Breden
Department of Biology
University of Missouri-Columbia
Columbia, Missouri 65211

B. Diane Chepko-Sade
Department of Anthropology,
Northwestern University,
Evanston, Illinois 60201
  *and*
The North Country Institute for
  Natural Philosophy, Inc.
R. D. #3
Mexico, New York 13114

James Cheverud
Departments of Anthropology,
Cell Biology and Anatomy, and
Ecology and Evolutionary Biology
Northwestern University
Evanston, Illinois 60201

Malcolm Dow
Department of Anthropology
Northwestern University
Evanston, Illinois 60201

Michael S. Gaines
Department of Systematics and
  Ecology
University of Kansas
Lawrence, Kansas 66045

Zuleyma Tang Halpin
Department of Biology
University of Missiouri-St. Louis
8001 Natural Bridge Rd.
St. Louis, Missouri 63121

Michael L. Johnson
Department of Systematics and
  Ecology
University of Kansas
Lawrence, Kansas 66045

W. Thomas Jones
Department of Biological Sciences
Purdue University
West Lafayette, Indiana 47907

Rolf R. Koford
Department of Biological Sciences
State University of New York at
    Albany
Albany, New York 12222

William Z. Lidicker, Jr.
Museum of Vertebrate Zoology
University of California
Berkeley, California 94720

L. David Mech
U.S. Fish and Wildlife Service
Patuxent Wildlife Research Center
Laurel, Maryland 20708;
    *Mailing address:*
North Central Forest Experiment
    Station
1992 Folwell Ave.
St. Paul, Minnesota 55108

Michael E. Nelson
U.S. Fish and Wildlife Service
Patuxent Wildlife Research Center
Laurel, Maryland 20708;
    *Mailing address:*
Box 7200
Star Route #1
Ely, Minnesota 55731

Elizabeth O'Brien
Department of Human Genetics,
University of Utah
Salt Lake City, Utah 84112

James L. Patton
Museum of Vertebrate Zoology,
University of California
Berkeley, California 94720

Lynn L. Rogers
USDA—Forest Service
North Central Forest Experiment
    Station
1992 Folwell Avenue
St. Paul, Minnesota 55108

Jon P. Rood
Conservation and Research Center
National Zoological Park
Smithsonian Institution
Front Royal, Virginia 22630

Donald Stone Sade
Department of Anthroplogy
Northwestern University
Evanston, Illinois 60201
    *and*
The North Country Institute for
    Natural Philosophy, Inc.
R.D. #3
Mexico, New York 13114

William M. Shields
College of Environmental Science
    and Forestry
SUNY
Syracuse, New York 13210

Andrew T. Smith
Department of Zoology
Arizona State University
Tempe, Arizona 85287

Peter E. Smouse
Department of Human Genetics
University of Michigan
Ann Arbor, Michigan 48109

Alan R. Templeton
Department of Biology
Washington University
St. Louis, Missouri 63130

Michael J. Wade
Department of Biology
University of Chicago
1103 E. 57th Street
Chicago, Illinois 60637

Peter M. Waser
Department of Biological Sciences
Purdue University
West Lafayette, Indiana 47907

James W. Wood
Population Studies Center and
    Reproductive Endocrinology
    Program
University of Michigan
Ann Arbor, Michigan 48109

# Index